DātaMyte

DātaMyte
HANDBOOK
Fourth Edition

A practical guide to
computerized data collection for
Statistical Process Control

by DataMyte Corporation

DataMyte Corporation
14960 Industrial Road
Minnetonka, Minnesota 55345
612-935-7704
FAX 612-935-0018

DataMyte Corporation

DataMyte Corporation
14960 Industrial Road
Minnetonka, Minnesota 55345
612-935-7704
FAX 612-935-0018

For the location of your nearest DataMyte
Sales Representative, call 612-935-7704.

International Sales Offices

**EUROPE/MIDDLE EAST/AFRICA
HEADQUARTERS**
Amsterdamseweg 15
1422 AC Uithoorn
The Netherlands
Tel: (31) 2975-60611
FAX: (31) 2975-60222

ASIA/PACIFIC HEADQUARTERS
Allen-Bradley (Hong Kong) Limited
2901 Great Eagle Center
23 Harbour Road
G.P.O. Box 9797
Wanchai, Hong Kong
Tel: (852)5-739391
Telex: 780 64347
FAX: (852)5-745326

CANADA HEADQUARTERS
Allen-Bradley Canada Limited
135 Dundas Street
Cambridge, Ontario N1R 5X1
Canada
Tel: (519)623-1810
Telex: 069 59317
FAX: (519)623-8930

LATIN AMERICA HEADQUARTERS
DataMyte Corporation
14960 Industrial Road
Minnetonka, Minnesota 55345
U.S.A.
FAX: (612)935-0018

TABLE OF CONTENTS

Part II Applications

Part III Products

FOREWORD TO THE FOURTH EDITION

Over two years have passed since the previous edition was published. More than 160,000 copies of the DataMyte Handbook are now in print.

DataMyte has grown even more dominant in the Quality business. Our revenues in the past twelve months exceeded that of our next five competitors combined. Our direct sales force and our customer service staff have more than doubled. The focus of everything we do is Total Customer Satisfaction.

A recently completed independent national survey of our customers indicate that our efforts are noticed and appreciated. We now serve over 4,000 customer locations worldwide and have well over 20,000 systems installed.

We have developed several new hardware, software, and systems products and they are in this book. The application stories have been revised, updated, and expanded. The theory section was reviewed, completely revised, and expanded.

The authors of this handbook made every effort to continue to provide the best reference book in this rapidly changing field. Listed below are the major revisions to the fourth edition of the DataMyte Handbook:

- A new chapter, Chapter 4, on short-run charting techniques.

- A section on geometric tolerancing in Chapter 6.

- Sections on Mil-Std 2000 for solderability and inspection theory added to Chapter 7.

- The DataMyte OVERVIEW™ quality management system described in detail in Chapter 18.
- More information about training and customer support in Chapter 22.
- A revised Chapter 23 on Allen-Bradley quality management products, including Pyramid Integrator.
- A new Chapter 24 on BBN Software Products Corporation.
- A new Chapter 25 on Digital Equipment Corporation products for the quality market.

DataMyte Corporation
October, 1989

FOREWORD TO THE THIRD EDITION

The DataMyte Story — Update

In the three year period since the first DataMyte Handbook was published, the company has quadrupled in size. With over 10,000 systems installed in more than 2000 factories worldwide, DataMyte clearly has emerged as the leader in factory data collection for quality and productivity. Much of the growth has been fueled by a constant stream of new products, new applications, and an ever increasing number of gage interfaces. The company built a strong direct sales force, added selected distributors, created a knowledgeable technical support and customer service group and just recently launched a training division, all designed to ensure total customer success and satisfaction.

What started out as a predominantly automotive industry dominated business spread to all types of discrete and hybrid manufacturing operations as attended by the diversity of the one hundred applications in this Handbook. Initially, DataMyte made only handheld data collectors that could collect only variable (measurement) data and only from a limited number of gages. The current offerings described in the product section of this Handbook include both handheld and fixed station data collectors for attribute and variable data interfacing to hundreds of different analog and digital gages. Not only has DataMyte's software capability grown tremendously, but through cooperation with many third party software vendors providing bridges to their software, DataMyte has become the de facto standard. DataMyte products can be connected by the various FAN®, Factory Area Networks, thus elevating our products to a total systems solution.

In December 1986, DataMyte became part of the Allen-Bradley Company, Inc. (a Rockwell International Company). Allen-Bradley has long been a leader in providing automation products to the factory floor. The increased resources of our new parent organization (DataMyte will function as a wholly owned subsidiary) should accelerate both our growth and our ability to serve our customers better worldwide.

The authors of this Handbook made every effort to continue to provide the best reference book in this rapidly growing field. Here are the revisions to the Handbook Third Edition:

- 624 pages compared to 560 pages in the Second Edition.
- 128 pages of a two-color products section compared to 96.
- Theory section expanded to include more on regression analysis (chi square test, kurtosis).
- Applications section has 100 applications compared to 65 in the Second Edition.
- A major change was made to arrange the applications section into eight industry-specific chapters:

Chapter 7 — Aerospace and defense
Chapter 8 — Automotive suppliers
Chapter 9 — Electronics and computers
Chapter 10 — Food, cosmetics and health care
Chapter 11 — Furniture and appliance
Chapter 12 — Glass, plastics, paper and chemical
Chapter 13 — Metalworking and machinery
Chapter 14 — Transportation

- Almost twice as many gage interfaces as the Second Edition.
- Chapter 21 on Allen-Bradley quality management products, represents the largest offering of such products in the world.
- Chapter 22 on factory automation for quality includes product offerings from Apollo, AT&T, DEC and HP.

DataMyte Corporation
May 1987

FOREWORD TO THE SECOND EDITION

Since June of 1984, when the First Edition of this Handbook was published, we have distributed over 20,000 copies. The overwhelming majority of the comments which we received have been favorable and complimentary. We have also received some constructive criticism and requests for expanding the treatment on certain subjects. In preparing this expanded Second Edition, we have taken into account all the requests which we could accommodate. The changes and additions enlarge this book by over a third, to 576 pages.

DataMyte itself has also experienced significant growth. Many new products were developed and put to ever increasing new applications. In addition to our original handheld data acquisition systems, the fixed station DataMyte 750 family of products was developed and the new FAN™ (Factory Area Network) systems were introduced. The latest development, the DataMyte 2000 system, brings totally new capabilities to the concept of roving inspection. All these products plus new software developments and sensor/gage interfaces are described in the product section of this Handbook and are featured in the greatly expanded applications section. Listed below are the major changes and the four totally new additions:

Chapter 4, on measurement: an addition on gage capability studies.

Chapter 5, on acceptance sampling: an addition on Dr. Deming's statements about acceptance sampling.

Chapter 6, (new) on Just-In-Time (JIT) manufacturing and Total Quality Control (TQC).

Chapter 7, (new) on plant-wide quality control, including 16 new FAN applications. (Of the 40 applications in the First Edition, seven were eliminated and 32 new ones added, bringing the total in this book to 65 applications.)

Chapter 13, (new) on computerized data collection and the FAN (Factory Area Network) systems.

Chapter 17, (new) on SPC software available from other sources. In keeping with our open software architecture, many interfaces were developed between DataMyte data collectors and third party software vendors.

Chapter 18, (new) on SPC training. In addition to DataMyte's training programs on our own equipment, two cooperative programs are described. They are SMIP, from Control Data Corporation, a computer delivered SPC training course, and a video delivered training course from Technicomp Corporation.

It is the continuing interest of the authors and editors of this Handbook to make it a most useful, relevant and informative source of knowledge regarding all aspects of industrial data collection for the improvement of quality and productivity. To this end we again wish to solicit the constructive comments of our readers. These comments, together with the results of our continually expanding search for valuable, pertinent technical information, will be incorporated into future editions of this book.

DataMyte Corporation
January 1986

FOREWORD TO THE FIRST EDITION

The DataMyte Story

DataMyte Corporation is a rapidly growing private company in Minnetonka (a suburb of Minneapolis), Minnesota. Started in 1967, the Company has specialized in the design and development of handheld data acquisition systems for the past ten years.

Because DataMyte products are rugged, they are uniquely suited for the factory data collection environment. Because they are a complete user friendly system, comprised of automatic input sensors, a handheld computer and specialized software, they can provide both the data collection and data processing functions required for Statistical Quality Control (SQC).

Interestingly, SQC, which was developed several decades ago as a theory in the United States, found its first wide acceptance among the manufacturing companies of Japan. Superior quality and the resulting competitive advantage explain why.

Well managed companies in this country have adopted quality as a corporate strategy also, and made the necessary top management commitment "to do it right the first time." Hewlett-Packard estimated that as much as 25% of their total manufacturing cost was spent on rework. Others found equally dramatic savings opportunities. The experience of many companies shows that Statistical Quality Control reduces the cost of Quality. DataMyte Corporation has become successful because our systems simplify the implementation of an SQC program and, in turn, pay for themselves in a short time.

Our first volume users were the major automotive companies. Stung by Japanese competition and changing customer preferences for higher quality automobiles, they increasingly turned to SQC and began using DataMytes. We are pleased to have played a part in their overall quality improvement effort.

SQC, SPC — What's the Difference?

Statistical Quality Control (SQC) and Statistical Process Control (SPC) sometimes are used interchangeably, yet to some they mean different things. Without entering the semantic debate, it can be said that SQC is the broader "umbrella" term.

Ultimately, manufacturers hope to have perfect parts, perfect processes and perfect designs, and "just in time delivery." To aim for this perfection, companies are beginning to treat manufacturing as a yield driven *process*, i.e., they set specific defect rates at each step. Key elements of such a program are: performance measures, a test philosophy and Statistical Quality Control.

Why This Book

Data collection in Quality Control is labor intensive. As the use of SQC spreads in American industry, managers as well as practitioners, become more interested in data collection for testing rather than for screening out rejects. To change a process for the desired effect, data must be collected rapidly, accurately, automatically, and should be processed "real time," "on-line" rather than later. Here is where DataMyte fills a need.

This handbook is a convenient reference, and a practical guide to the underlying theory, real life applications, and the DataMyte product offerings for the implementation of an SQC program. The presentation is aimed at both the managers and the quality professionals with emphasis on practical problem solutions. Where there are differing theories or methods, this book takes no sides but presents the different views and provides full reference literature where applicable, thus, hopefully, facilitating proper communications. Another aim of this book is to make the implementation and use of an SQC program easier by attempting to share the accumulated technical knowledge of our company, which has specialized in this field for the past decade.

How To Use The Book

This book was prepared as a compendium of articles on many related subjects. The editors made no attempt to homogenize the styles of the different authors. An effort was made however, to make the chapters self-contained and to organize the subject matter in a meaningful way. The definitions in Part I should help readers decide how much background they need to cover.

Some of the information presented is standard reference material, and some probably will not be found elsewhere. Some chapters are quite theoretical, while others focus on the more practical "how to" aspects. Generally the book is organized in three major parts: Theory, Applications, and Products.

This is a handbook, meant to be consulted as a reference as problems rise, rather than being read cover to cover as a student's textbook.

About the Authors

Twenty-three people contributed to the preparation of this book. They comprise the technical staff of DataMyte Corporation, with backgrounds in hardware and software engineering, application and sales support, R & D and training and — most of all — practical problem solving. The authors collectively have over 300 man-years of experience in SQC and data collection -data processing.

In addition to their regular daily work, the authors gave their time generously to make this book possible.

Chapter 15 and 16 require special mention. DataMyte is pleased to have had the cooperation and assistance of two other companies, Hewlett Packard (Chapter 15) and Allied Corporation (Chapter 16), in presenting a systems approach to quality control. These companies are thanked for providing most of the material contained in these two chapters.

DataMyte Corporation
June 1984

PART I Theory

DEFINITIONS

a (alpha) Risk — The maximum probability of saying a
process or lot is unacceptable when, in fact, it is acceptable.
See also Producer's Risk. See p. 7-4.

Acceptable Quality Level (AQL) — When used in acceptance sampling: maximum percentage or proportion of nonconforming units which can be considered acceptable as a
process average. See p. 7-11.

Acceptance Sampling — Sampling inspection in which decisions concerning acceptance or rejection of materials or
services are made. Also includes procedures used in determining the acceptability of the items in question based on
the results of inspection. See p. 7-2.

Accuracy (of measurement) — Difference between the average result of a measurement with a particular instrument,
and the true value of the quantity being measured. See p. 6-7.

AOQ (Average Outgoing Quality) — The expected average
outgoing quality following the use of an acceptance sampling plan for a given level of quality. Any lots not meeting
this level of quality must be inspected and nonconforming
units removed and replaced. These lots are then included
with the acceptable lots.

AOQL (Average Outgoing Quality Limit) — For a given set
of sampling criteria (sampling plan), the maximum AOQ
over all possible levels of incoming quality. See p. 7-11.

Assignable Causes (of variation) — Significant, identifiable
changes in the relationships of materials, methods, machines and people. See p. 1-9.

Attributes Data (quality) — Data coming basically from GO/
NO-GO, pass/fail determinations of whether units conform
to standards. Also includes noting presence or absence of a
quality characteristic. May or may not include weighting by
seriousness of defects, etc. See p. 2-2.

See p. 1-12. *Average* — See Mean.

See p. 7-4. *β (beta) Risk* — The maximum probability of saying a process or lot is acceptable when, in fact, it should be rejected. See also Consumer's Risk.

$$P(B_n | A_m) = \frac{P(A_m | B_n)P(B_n)}{\sum_j P(A_m | B_j)P(B_j)}$$

Bayes Theorem Formula

See p. 7-19.

Bayes Theorem — A theorem of statistics relating conditional probabilities: Suppose that $P(A_i | B_j)$ is the probability that $A = A_i$ given that $B = B_j$, and suppose $P(B_n)$ is the probability that $B = B_n$, where B_n is any outcome other than B_j. Then if we know that $A = A_m$, then $P(B_n)$ is given by the formula. The theorem is used for determining the probability $P(B_n | A_m)$ when we have knowledge of the conditional probability $P(A | B)$.

Bell-shaped Curve — A curve or distribution showing a central peak and tapering off smoothly and symmetrically to "tails" on either side. A normal (Gaussian) curve is an example.

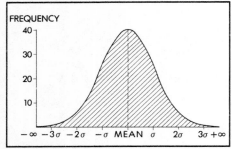

Bell-Shaped Curve See p. 1-4.

Bias (in measurement) — Systematic error which leads to a difference between the average result of a population of measurements and the true, accepted value of the quantity being measured.

See p. 5-14. *Bimodal Distribution* — A frequency distribution which has two peaks. Usually an indication of samples from two processes incorrectly analyzed as a single process.

$$P(r,n) = \frac{n!}{r! \, (n - r)!} \, p^r q^{n-r}$$

Binomial Distribution Formula

See p. 7-6.

Binomial Distribution (probability distribution) — Given that a trial can have only two possible outcomes (yes/no, pass/fail, heads/tails), of which one outcome has probability p and the other probability q ($p + q = 1$), the probability that the outcome represented by p occurs r times in n trials is given by the binomial distribution.

See p. 3-35. *c-Chart* — For attributes data: A control chart of the number of defects found in a subgroup of fixed size. The c-chart is used where each unit typically has a number of defects.

See p. 6-5. *Calibration (of instrument)* — Adjusting an instrument using a reference standard to reduce the difference between the average reading of the instrument and the "true" value of

D-2

Definitions

the standard being measured, i.e., to reduce measurement bias.

Camp-Meidell Conditions — For frequency distribution and histograms: A distribution is said to meet Camp-Meidell conditions if its mean and mode are equal and the frequency declines continuously on either side of the mode. See p. 5-19.

Capability (of process) — The uniformity of product which a process is capable of producing. Can be expressed numerically using CP, CR, CpK, and Zmax/3 when the data is normally distributed. See p. 5-6.

Cause and Effect Diagram — A pictorial diagram showing all the cause (process inputs) and effect (resulting problem being investigated) relationships among the factors which affect the process. See p. 2-5.

Cell (of frequency distribution and/or histogram) — For a sample based on a continuous variable, a cell is an interval into which individual data points are grouped. The full range of the variable is usually broken into intervals of equal size and the number of points in each cell totalled. These intervals (cells) make up a frequency distribution or histogram. This greatly reduces the amount of information which must be dealt with, as opposed to treating each element (data point) individually. See p. 2-13.

Centerline — For control charts: the horizontal line marking the center of the chart, usually indicating the grand average of the quantity being charted. See p. 3-11.

Central Limit Theorem — If samples of a population with size n are drawn, and the values of $\bar{x}$ are calculated for each sample group, and the distribution of $\bar{x}$ is found, the distribution's shape is found to approach a normal distribution for sufficiently large n. This theorem allows one to use the assumption of a normal distribution when dealing with $\bar{x}$. "Sufficiently large" depends on the population's distribution and what range of $\bar{x}$ is being considered; for practical purposes, the easiest approach may be to take a number of samples of a desired size and see if their means are normally distributed. If not, the sample size should be increased. See p. 1-16.

See p. 1-12. *Central Tendency* — A measure of the point about which a group of values is clustered; some measures of central tendency are mean, mode, and median.

See p. 2-19. *Characteristic* — A dimension or parameter of a part which can be measured and monitored for control and capability.

Check Sheet — A sheet for the recording of data on a process or its product. The check sheet is designed to remind the user to record each piece of information required for a particular study, and to reduce the likelihood of errors in recording data. The data from the check sheet can be typed into a computer for analysis when the data collection is complete.

See p. 5-13. *Chi-square* (χ^2) — As used for goodness-of-fit: a measure of how well a set of data fits a proposed distribution, such as the normal distribution. The data is placed into classes and the observed frequency (O) is compared to the expected frequency (E) for each class of the proposed distribution. The result for each class is added to obtain a chi-square value. This is compared to a critical chi-square value from a standard table for a given α (alpha) risk and degrees of freedom. If the calculated value is smaller than the critical value, we can conclude that the data follows the proposed distribution at the chosen level of significance.

$$\chi^2 = \Sigma \frac{(O - E)^2}{E}$$

Chi-Square Formula

See p. 3-21. *Chronic Condition* — Long-standing adverse condition which requires resolution by changing the status quo.

See p. 1-9. *Cluster* — For control charts: a group of points with similar properties. Usually an indication of short duration, assignable causes.

Common Causes — Those sources of variability in a process which are truly random, i.e., inherent in the process itself.

Concerns — Number of defects (nonconformities) found on a group of samples in question.

Confidence Interval — Range within which a parameter of a population (e.g. mean, standard deviation, etc.) may be expected to fall, on the basis of measurement, with some specified confidence level.

Confidence Level — The probability set at the beginning of a hypothesis test that the variable will fall within the confidence interval. A confidence level of 0.95 is commonly used.

Confidence Limits — The upper and lower boundaries of a confidence interval.

Conformance (of product) — Adherence to some standard of the product's properties. The term is often used in attributes studies of product quality, i.e., a given unit of the product is either in conformance to the standard or it is not.

Constant-Cause System — A system or process in which the variations are random and are constant in time.   See p. 1-9.

Consumer's Risk — The maximum probability of saying a process or lot is acceptable when, in fact, it should be rejected. See also β Risk.   See p. 7-4.

Continuous Data — Data for a continuous variable (variable data). The resolution of the value is only dependent on the measurement system used.   See p. 2-2.

Continuous Variable — A variable which can assume any of a range of values; an example would be the measured size of a part.

Control (of process) — A process is said to be in a state of statistical control if the process exhibits only random variations (as opposed to systematic variations and/or variations with known sources). When monitoring control with control charts, a state of control is exhibited when all points remain between set control limits.   See p. 1-17.

Control Chart — A plot of some parameter of process performance, usually determined by regular sampling of the product, as a function (usually) of time or unit number or other chronological variable. The control limits are also plotted for comparison. The parameter plotted may be the mean value of a particular measurement for a product sample of specified size ($\bar{x}$ chart), the range of values in the sample (R chart), the percent of defective units in the sample (p-chart), etc.   See Chapter 3.

Control Group — An experimental group which is not given the treatment under study. The experimental group that is given the treatment is compared to the control group to ensure any changes are due to the treatment applied.

Control Limits — The limits within which the product of a process is expected (or required) to remain. If the process leaves the limits, it is said to be out of control. This is a signal See p. 3-12. that action should be taken to identify the cause and eliminate it if possible. Note: control limits are not the same as tolerance limits.

$$\sigma_{xy} = \frac{\sum\limits_{i=1}^{n} (x_i - \bar{x})(y_i - \bar{y})}{n - 1}$$

Covariance Formula

Covariance — A measure of whether two variables (x and y) are related (correlated). It is given by the formula. "n" is the number of elements in the sample.

$$CP = \frac{Tolerance}{6\sigma}$$

CP Formula

CP — For process capability studies. CP is a capability index defined by the formula. CP shows the process capability potential but does not consider how centered the process is. CP may range in value from 0 to infinity, with a large value indicating greater potential capability. A value of 1.33 or greater is usually desired.

$$CpK = \text{The lesser of:}$$
$$\frac{(USL - Mean)}{3\sigma} \text{ or}$$

$$\frac{(Mean - LSL)}{3\sigma}$$

CpK Formula

CpK — For process capability studies. An index combining CP and K to indicate whether the process will produce units within the tolerance limits. CpK has a value equal to CP if the process is centered on the nominal; if CpK is negative, the process mean is outside the specification limits; if CpK is between 0 and 1 then some of the 6 sigma spread falls outside the tolerance limits. If CpK is larger than 1, the 6 sigma spread is completely within the tolerance limits. A value of 1.33 or greater is usually desired. Also known as Zmin/3.

$$CR = \frac{6\sigma}{Tolerance}$$

CR Formula

CR — For process capability studies: The inverse of CP, CR can range from 0 to infinity in value, with a smaller value indicating a more capable process.

See p. 3-5. *Cumulative Sum Chart (CuSum)* — A cumulative sum chart plots the cumulative deviation of each subgroup's average from the nominal value. If the process consistently produces parts near the nominal, the CuSum chart shows a line which is essentially horizontal. If the process begins to shift,

Definitions

the line will show an upward or downward trend. The CuSum chart is sensitive to small shifts in process level.

Cycle — A recurring pattern.

See p. 3-19.

DataTruck™ — The DataTruck™ harvests data from several 760/860 data collectors located throughout a plant's un-wired FAN™ (Factory Area Network) system. The Data-Truck™ transfers the data to an IBM PC and it can also transmit setup information stored on the PC to the 760/860 data collectors.

See p. 19-22.

Defect — Departure of a quality characteristic from its acceptable level or state. The defects may also be classified by degree of seriousness. See also Nonconformity.

Defective Unit — A sample (part) which contains one or more defects, making the sample unacceptable for its intended, normal usage. See also Nonconforming Unit.

Deformation — The bending or distorting of an object due to forces applied to it. Deformation can contribute to errors in measurement if the measuring instrument applies enough force.

See p. 6-11.

Degrees of Freedom (DF) — The number of unconstrained parameters in a statistical determination. For example, in determining $\bar{x}$ (the mean value of a sample of n measurements), the number of degrees of freedom, DF, is n. In determining the standard deviation (STD) of the same population, $DF = n-1$ because one parameter entering the determination is eliminated. The STD is obtained from a sum of terms based on the individual measurements — which are unconstrained — but the nth measurement must now be considered "constrained" by the requirement that the values add up to make $\bar{x}$. An equivalent statement is that one degree of freedom is "factored out" because the STD is mathematically indifferent to the value of $\bar{x}$.

$$\sigma = \sqrt{\frac{\Sigma \, (x - \bar{x})^2}{DF}}$$

Standard Deviation Formula

Discrete Variable — A variable which assumes only integer values; for example, the number of people in a room is a discrete variable.

Dispersion (of a statistical sample) — The tendency of the values of the elements in a sample to differ from each other.

See p. 1-13.

Dispersion is commonly expressed in terms of the range of the sample (difference between the lowest and highest values) or by the standard deviation.

See p. 2-5. *Dispersion Analysis Diagram* — A cause and effect diagram for analysis of the various contributions to variability of a process or product. The main factors contributing to the process are first listed, then the specific causes of variability from each factor are enumerated. A systematic study of each cause can then be performed.

English System —The system of measurement units based on the foot, the pound, and the second.

Evolutionary Operations (EVOP) — A procedure to optimize the performance of a process by making small, known variations in the parameters entering the process and observing the effects of the variation on the product. This seems identical to Response Surface Methodology, except that it is done in a production situation rather than during process development, and the variations must therefore be kept small enough to meet product tolerances.

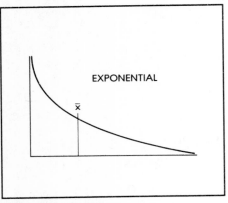

Exponential Distribution

Exponential Distribution — A probability distribution mathematically described by an exponential function. Used to describe the probability that a product survives a length of time *t* in service, under the assumption that the probability of a product failing in any small time interval is independent of time.

$$P(x) = \frac{1}{\mu} e^{-x/\mu}$$

Exponential Distribution Formula

See p. 5-14.

F Distribution — The distribution of F, the ratios of variances for pairs of samples. Used to determine whether or not the populations from which two samples were taken have the same standard deviation. The F distribution is usually expressed as a table of the upper limit below which F can be expected to lie with some confidence level, for samples of a specified number of degrees of freedom.

$$F = \frac{\sigma_1^2}{\sigma_2^2}$$

F Distribution Formula

F Test — Test of whether two samples are drawn from populations with the same standard deviation, with some specified confidence level. The test is performed by determining whether F, as defined above, falls below the upper limit given by the F distribution table.

Factory Area Network (FAN™) — A flexible unwired system for quality control, consisting of DataMyte data collectors and DataTrucks™.

See Chapter 17.

Failure Rate — The average number of failures per unit time. Used for assessing reliability of a product in service.

False x̄ Causes — For x̄ control charts: changes in the x̄ control chart which are not due to changes in the process mean, but to changes in the corresponding R-chart.

See p. 3-22.

Fault Tree Analysis — A technique for evaluating the possible causes which might lead to the failure of a product. For each possible failure, the possible causes of the failure are determined; then the situations leading to those causes are determined; and so forth, until all paths leading to possible failures have been traced. The result is a flow chart for the failure process. Plans to deal with each path can then be made.

Feedback — Using the results of a process to control it. The feedback principle has wide application. An example would be using control charts to keep production personnel informed on the results of a process. This allows them to make suitable adjustments to the process. Some form of feedback on the results of a process is essential in order to keep the process under control.

Figure of Merit — Generic term for any of several measures of product reliability, such as MTBF, mean life, etc.

Fixed Station Data Acquisition Equipment — Data collection equipment used at a fixed location, for example permanently attached to a particular manufacturing process or residing in a test lab.

See Chapter 19.

Flow Chart (for programs, decision making, process development) — A pictorial representation of a process indicating the main steps, branches, and eventual outcomes of the process.

Frequency Distribution — For a sample drawn from a statistical population, the number of times each outcome was observed.

See p. 1-12.

Goodness-of-Fit — Any measure of how well a set of data matches a proposed distribution. Chi-square is the most common measure for frequency distributions. Simple visual inspection of a histogram is a less quantitative, but equally valid, way to determine goodness-of-fit.

See p. 1-11. *Grand Average* — Overall average of data represented on an x̄ chart at the time the control limits were calculated. See also Centerline.

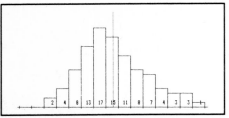

Histogram

See p. 2-12.

Handheld Data Collector — A portable electronic device for recording data on a process or its product. The data collector serves in place of a check sheet for recording data.

Histogram — A graphic representation of a frequency distribution. The range of the variable is divided into a number of intervals of equal size (called cells) and an accumulation is made of the number of observations falling into each cell. The histogram is essentially a bar graph of the results of this accumulation.

$$P(X = r) = \frac{\binom{d}{r}\binom{N-d}{n-r}}{\binom{N}{n}}$$

Hypergeometric Distribution Formula

See p. 7-6.

Hypergeometric Distribution — A probability distribution for the probability of drawing exactly n objects of a given type from a sample of N objects of which r are of the desired type.

See p. 4-5. *Individuals Chart with Moving Range* — A control chart used when working with one sample per subgroup. The individual samples are plotted on the x̄ chart rather than subgroup averages. The individuals chart is always accompanied by a moving range chart, usually using two subgroups (two individual readings) to calculate the moving range points.

Infant Mortality — High failure rate which shows up early in product usage. Normally caused by poor design, manufacture, or other identifiable cause.

Inspection Accuracy — The percentage of defective units which are correctly identified by an inspector. The percentage is determined by having a second inspector review both the accepted and rejected units.

Instability (of a process) — A process is said to show instability if it exhibits variations larger than its control limits, or shows a systematic pattern of variation.

Ishikawa Diagram — See Cause and Effect Diagram.

See p. 2-5.

Just-In-Time (JIT) Manufacturing — Time manufacturing co-ordinates inventory and production to get away from the batch mode of production in order to improve quality.

See p. 8-2.

K — For process capability studies: a measure of difference between the process mean and the specification mean (nominal).

$$K = \frac{(\text{Mean} - \text{Midpoint})}{(\text{Tolerance}/2)}$$

K Index Formula

Kurtosis — A measure of the shape of a distribution. If the distribution has longer tails than a normal distribution of the same standard deviation, then it is said to have positive kurtosis (playkurtosis); if it has shorter tails, then it has negative kurtosis (leptokurtosis).

See p. 5-14.

LCL (Lower Control Limit) — For control charts: the limit above which the process subgroup statistics ($\bar{x}$, R, sigma) remain when the process is in control.

Leptokurtosis — For frequency distributions: a distribution which shows a higher peak and shorter "tails" than a normal distribution with the same standard deviation.

See p. 5-14.

Limit Checking — For handheld data collectors: alerting the operator concerning any entries outside the expected range.

Linearity — The extent to which a measuring instrument's response varies with the measured quantity.

See p. 6-7.

Lot Formation — The process of collecting units into lots for the purpose of acceptance sampling. The lots are chosen to ensure, as much as possible, that the units have identical properties, i.e., that they were produced by the same process operating under the same conditions.

See p. 7-9.

LSL (Lower Specification Limit) — The lowest value of a product dimension or measurement which is acceptable.

See p. 7-11. *Lot Tolerance Percent Defective (LTPD)* — For acceptance sampling: expressed in percent defective units; the poorest quality in an individual lot that should be accepted. Commonly associated with a small consumer's risk.

See p. 2-25. *Matrix* — An array of data arranged in rows and columns.

MTBF (Mean Time Between Failures) — Mean time between successive failures of a repairable product. This is a measure of product reliability.

$$\bar{x} = \frac{x_1 + x_2 + \ldots x_n}{n - 1}$$

Mean Formula

Mean (of a statistical sample) ($\bar{x}$) — The arithmetic average value of some variable. The mean is given by the formula, where "x" is the value of each measurement in the sample. All x's are added together and divided by the number of elements (n) in the sample.

See p. 1-11. *Mean (of a population) (μ)* — The true arithmetic average of all elements in a population. $\bar{x}$ approximates the true value of the population mean.

Measurement Accuracy — The extent to which the average result of a repeated measurement tends toward the true value of the measured quantity. The difference between the true value and the average measured value is called the instrument bias, and may be due to such things as improper zero-adjustment, nonlinear instrument response, or even improper use of the instrument.

Measurement Error — The difference between the actual and measured value of a measured quantity.

Measurement Precision — The extent to which a repeated measurement gives the same result. Variations may arise from the inherent capabilities of the instrument, from variations of the operator's use of the instrument, from changes in operating conditions, etc.

See p. 1-12. *Median (of a statistical sample)* — For a sample of a specific variable, the median is the point X such that half the sample elements are below X, and half above X.

Median Chart — For variables data: a control chart of the median of subgroups.

Metrology — The science of measurement. See p. 6-2.

MIL-STD-105D — A set of specifications for acceptance sampling plans based on acceptable quality level (AQL). For a given AQL, lot size, and level of inspection, the specification lists the number of defective units which is acceptable, and the number which requires rejection of the lot. The specifications allow for tightening or loosening of inspection requirements based on previous inspection results. See p. 7-12.

MIL-STD-414 — A set of specifications for acceptance sampling plans based on acceptable quality level (AQL) for variables data, using the assumption that the variable is normally distributed. See p. 7-12.

MIL-STD-2000 — A set of standard requirements for soldered electrical and electronic assemblies. Standards address design, production, process control and inspection of these assemblies. See p. 7-24.

Mixture — A combination of two distinct populations. On control charts, a mixture is indicated by an absence of points near the centerline. See p. 3-21.

Mode (of a statistical sample) — The value of the sample variable which occurs most frequently. See p. 1-12.

Modified Control Limits — Control limits calculated from information other than the process's statistical variation, such as tolerances. Must be used cautiously, because the process could be working within its normal variation, but show up on the control chart as out of control if limits do not account for that variation.

Monte Carlo Simulation — A computer modeling technique to predict the behavior of a system from the known random behaviors and interactions of the system's component parts. A mathematical model of the system is constructed in the computer program, and the response of the model to various operating parameters, conditions, etc. can then be investigated. The technique is useful for handling systems whose complexity prevents analytical calculation.

See p. 4-7. *Moving Average Moving Range Charts* — A control chart which combines rational subgroups of data and the combined subgroup averages and ranges are plotted. Often used in continuous process industries, such as chemical processing, where single samples are analyzed.

See p. 1-2. *Nominal* — For a product whose size is of concern: the desired mean value for the particular dimension, the target value.

See p. 4-3. *Nominal Chart* — A control chart which plots the deviation from the nominal value. Often used when individual samples are taken in short run, low volume processes. Allows multiple part numbers manufactured by similar processes to be plotted on the same control charts.

Nonconformity — A departure of a quality characteristic from its intended level or state. See also Defect.

Nonconforming unit — A sample (part) which has one or more nonconformities, making the sample unacceptable for its intended use. See also Defective Unit.

See p. 6-7. *Nonlinearity (of a measuring instrument)* — The deviation of the instrument's response from linearity.

$$P(x) = \frac{1}{\sigma \sqrt{2\pi}} e^{-\left(\frac{x-\mu}{2\sigma^2}\right)^2}$$

Normal Distribution Formula

See p. 1-15. *Normal Distribution* — A probability distribution given mathematically by the formula. The normal distribution is a good approximation for a large class of situations. One example is the distribution resulting from the random additions of a large number of small variations. The Central Limits Theorem expresses this for the distribution of means of samples; the distribution of means results from the random additions of a large number of individual measurements, each of which contributes a small variation of its own.

See p. 3-34. *np-Chart* — For attributes data: a control chart of the number of defective units in a subgroup. Assumes a constant subgroup size.

See p. 7-5. *Operating Characteristics Curve* — For acceptance sampling: a curve showing the probability of accepting a lot versus the percentage of defective units in the lot.

Out of Control — A process which exhibits variations larger than the control limits is said to be out of control.

OVERVIEW™ — A computer integrated quality management system providing both real-time SPC and lot control. System consists of DataMyte series 900 data collectors, software running on a VAX computer, and the DataMyte server.

p-Chart (percent defective) — For attributes data: a control chart of the percentage of defective units (or fraction defective) in a subgroup.

Pareto Analysis — An analysis of the frequency of occurrence of various possible concerns. This is a useful way to decide quality control priorities when more than one concern is present. The underlying "Pareto Principle" states that a very small number of concerns is usually responsible for most quality problems.

Pareto Diagram — A "bar graph" showing the frequency of occurrence of various concerns, ordered from the highest to lowest frequency of occurrence.

See p. 2-9.

Percent Defective — For acceptance sampling: the percentage of units in a lot which are defective, i.e., of unacceptable quality.

Performance Study — Analysis of a process to determine the distribution of a run. The process may or may not be in statistical control.

Platykurtosis — For frequency distributions: a distribution which has longer "tails" than a normal distribution with the same standard deviation.

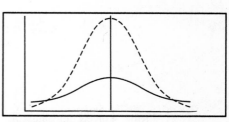

Platykurtic Curve
See p. 5-14.

Point Estimate (Statistics) — A single-value estimate of a population parameter.

Poisson Distribution — A probability distribution for the number of occurrences of an event; n=number of trials; p=probability that the event occurs for a single trial; r=the number of trials for which the event occurred. The Poisson distribution is a good approximation of the binomial distribution for a case where p is small.

See p. 7-6.

$$P(r) = \frac{(np)^r (e)^{-np}}{r!}$$

Poisson Distribution Formula

D-15

Definitions

See p. 1-11. *Population (statistical)* — The set of all possible outcomes of a statistical determination. The population is usually considered as an essentially infinite set from which a subset called a sample is selected to determine the characteristics of the population, i.e., if a process were to run for an infinite length of time, it would produce an infinite number of units. The outcome of measuring the length of each unit would represent a statistical universe, or population. Any subset of the units produced (say, a hundred of them collected in sequence) would represent a sample of the population. Also known as universe.

Post Processing — Processing of data done after it is collected, usually by computer.

See p. 6-8. *Precision (of measurement)* — The extent to which repeated measurement of a standard with a given instrument yields the same result.

Pre-Control — A method of controlling a process based on the specification limits. It is used to prevent the manufacture of defective units, but does not work toward minimizing variation of the process. The area between the specifications are split into zones (green, yellow and red) and adjustments made when a specified number of points fall in the yellow or red zones.

See p. 6-4. *Primary Reference Standard* — For measurements: a standard maintained by the National Bureau of Standards for a particular measuring unit. The primary reference standard duplicates as nearly as possible the international standard and is used to calibrate other (transfer) standards, which in turn are used to calibrate measuring instruments for industrial use.

$$P(x) = \lim_{N \to \infty} \frac{\left(\begin{array}{c} \text{Number of trials} \\ \text{giving outcome X} \end{array} \right)}{\left(\begin{array}{c} N = \text{total} \\ \text{number of trials} \end{array} \right)}$$

Probability Formula

Probability (Mathematical) — The likelihood that a particular occurrence (event) has a particular outcome. In mathematical terms, the probability that outcome X occurs is expressed by the formula. Note that, because of this definition, summing up the probabilities for all values of X always gives a total of 1: this is another way of saying that each trial must have exactly one outcome.

See p. 1-9.

Probability Distribution — A relationship giving the probability of observing each possible outcome of a random event. The relationship may be given by a mathematical expression, or it may be given empirically by drawing a frequency distribution for a large enough sample.

Process Analysis Diagram — A cause and effect diagram for a process. Each step of the process and the factors contributing to it are shown, indicating all cause-and-effect relationships. This allows systematic tracing of any problems that may arise, to identify the source of the problem.

See p. 2-5.

Process Capability — The level of uniformity of product which a process is capable of yielding. Process capability may be expressed by the percent of defective products, the range or standard deviation of some product dimension, etc. Process capability is usually determined by performing measurements on some (or all) of the product units produced by the process.

See p. 5-2.

Process Control — Maintaining the performance of a process at its capability level. Process control involves a range of activities such as sampling the process product, charting its performance, determining causes of any excessive variation and taking corrective actions.

Producer's Risk — The maximum probability of saying a process or lot is unacceptable when, in fact, it is acceptable. See also **α** Risk.

See p. 7-4.

Quality Assurance — The function of assuring that a product or service will satisfy given needs. The function includes necessary verification, audits, and evaluations of quality factors affecting the intended usage and customer satisfaction. This function is normally the responsibility of one or more upper management individuals overseeing the quality assurance program.

Quality Characteristic — A particular aspect of a product which relates to its ability to perform its intended function.

Quality Control — The process of maintaining an acceptable level of product quality.

See p. 1-2.

Quality Function — The function of maintaining product quality levels; i.e., the execution of quality control.

Quality Specifications — Particular specifications of the limits within which each quality characteristic of a product is to be maintained.

See p. 1-13. $\bar{R}$ — Average range value displayed on a range control chart. Value is set at the time control limit(s) are calculated.

See Chapter 3. *R-Chart* — A control chart of the range of variation among the individual elements of a sample — i.e., the difference between the largest and smallest elements — as a function of time, or lot number, or similar chronological variable.

Random — Varying with no discernable pattern.

Random Sample — The process of selecting a sample of size n where each part in the lot or batch has an equal probability of being selected.

See p. 1-13. *Range* — The difference between the highest and lowest of a group of values.

See p. 3-11. *Rational Subgrouping* — For control charting: a subgroup of units selected to minimize the differences due to assignable causes. Usually samples taken consecutively from a process operating under the same conditions will meet this requirement.

Regression Analysis — A technique for determining the mathematical relation between a measured quantity and the variables it depends on. For example, the method might be used to determine the mathematical form of the probability distribution from which a sample was drawn, by determining which form best "fits" the frequency distribution of the sample. The frequency distribution is the "measured quantity" and the probability distribution is a "mathematical relation."

See p. 7-11. *Rejectable Quality Level (RQL)* — For acceptance sampling: expressed as percentage or proportion of defective units; the poorest quality in an individual lot that should be accepted. Commonly associated with a small consumer's risk. See also LTPD.

Reliability — The probability that a product will function properly for some specified period of time, under specified conditions.

Repeatability (of a measurement) — The extent to which repeated measurements of a particular object with a particular instrument produces the same value.

See p. 6-19.

Reproducibility — The variation between individual people taking the same measurement and using the same gaging.

See p. 6-20.

Response Surface Methodology (RSM) — A method of determining the optimum operating conditions and parameters of a process, by varying the process parameters and observing the results on the product. This is the same methodology used in Evolutionary Operations (EVOP), but is used in process development rather than actual production, so that strict adherence to product tolerances need not be maintained. An important aspect of RSM is to consider the relationships among the parameters, and the possibility of simultaneously varying two or more parameters to optimize the process.

Resolution (of a measuring instrument) — The smallest unit of measure which an instrument is capable of indicating.

See p. 6-6.

Route — For inspection, inventory, or other in-plant data collection: the sequence or path that the operator follows in the data collection process.

Run — A set of consecutive units, i.e., sequential in time.

See p. 3-17.

s — Symbol used to represent standard deviation of a sample.

$$s = \sqrt{\frac{\sum_{i=1}^{n}(x_i - \overline{x})}{n - 1}}$$

Sample Standard Deviation Formula

ô — Symbol used to represent the estimated standard deviation given by the formula. The estimated standard deviation may only be used if the data is normally distributed and the process is in control.

$$\hat{\sigma} = \frac{\overline{R}}{d_2}$$

Estimated Standard Deviation Formula

Sample (Statistics) — A representative group selected from a population. The sample is used to determine the properties of the population.

See p. 1-11.

Sample Size — The number of elements, or units, in a sample.

Sampling — The process of selecting a sample of a population and determining the properties of the sample. The sample is chosen in such a way that its properties are representative of the population.

Sampling Variation — The variation of a sample's properties from the properties of the population from which it was drawn.

See p. 2-16. *Scatter Plot* — For a set of measurements of two variables on each unit of a group: a plot on which each unit is represented as a dot at the x,y position corresponding to the measured values for the unit. The scatter plot is a useful tool for investigating the relationship between the two variables.

See p. 6-6. *Sensitivity (of a measuring instrument)* — The smallest change in the measured quantity which the instrument is capable of detecting.

See Chapter 3. *Shewhart Control Chart* — A graphic continuous test of hypothesis. Commonly known as $\bar{x}$ and R charts. See also Control Chart.

Short-run SPC — A set of techniques used for SPC in low-volume, short duration manufacturing.

See p. 6-2. *SI System* — The metric system of units of measure. The basic units of the system are the meter, the kilogram, and the second; the system is sometimes called the MKS system for this reason.

See p. 1-14. *Sigma* — The standard deviation of a statistical population.

Sigma Limits — For histograms: lines marked on the histogram showing the points n standard deviations above and below the mean.

Simulation (modeling) — Using a mathematical model of a system or process to predict the performance of the real system. The model consists of a set of equations or logic rules which operate on numerical values representing the operating parameters of the system. The result of the equations is a prediction of the system's output.

Skewness — A measure of a distribution's symmetry. A skewed distribution shows a longer than normal tail on the right or left side of a distribution. See p. 5-14.

Specification (of a product) — A listing of the required properties of a product. The specifications may include the desired mean and/or tolerances for certain dimensions or other measurements; the color or texture of surface finish; or any other properties which define the product.

Stability (of a process) — A process is said to be stable if it shows no recognizable pattern of change. See also Control, and Constant Cause System. See p. 1-17.

Standard (measurement) — A reference item providing a known value of a quantity to be measured. Standards may be primary — i.e., the standard essentially defines the unit of measure — or secondary (transfer) standards, which have been compared to the primary standard (directly or by way of an intermediate transfer standard). Standards are used to calibrate instruments which are then employed to make routine measurements. See p. 6-10.

$$s = \sqrt{\frac{\sum\limits_{i=1}^{n}(x_i - \bar{x})}{n-1}}$$

Standard Deviation Formula

Standard Deviation — A measure of the variation among the members of a statistical sample. If a sample of n values has a mean of $\bar{x}$, its standard deviation is given by the formula. See p. 1-14.

Statistic — An estimate of a population parameter using a value calculated from a random sample.

Statistical Control (of a process) — A process is said to be in a state of statistical control when it exhibits only random variations.

Statistical Inference — The process of drawing conclusions on the basis of statistics.

Statistical Process Control (SPC) — Statistical methods for analyzing and controlling the variation of a process. See p. 1-13.

Statistical Quality Control (SQC) — The application of statistical methods for measuring and improving the quality of processes. SPC is one method included in SQC.

See p. 3-21. *Stratification (of a sample)* — If a sample is formed by combining units from several lots having different properties, the sample distribution will show a concentration or clumping about the mean value for each lot: this is called stratification. In control charting, if there are changes between subgroups due to stratification, the R-chart points will all tend to be near the centerline.

See p. 1-11. *Subgroup* — For control charts: a sample of units from a given process, all taken at or near the same time.

See p. 3-31. *Systematic Variation (of a process)* — Variations which exhibit a predictable pattern. The pattern may be cyclic (i.e., a recurring pattern) or may progress linearly (trend).

See p. A-5. *t-Distribution* — For a sample with size n, drawn from a normally distributed population, with mean $\bar{x}$ and standard deviation s. The true population parameters are unknown. The t-distribution is expressed as a table for a given number of degrees of freedom and α risk. As the degrees of freedom get very large, it approaches a z-distribution.

$$t = \frac{\bar{X} - \mu}{s_{\bar{x}}}$$

t-Test Formula

t-Test — A test of the statistical hypothesis that two population means are equal. The population standard deviations are unknown, but thought to be the same. The hypothesis is rejected if the t value is outside the acceptable range listed in the t-table for a given α risk and degrees of freedom.

See p. 1-2. *Tolerance* — The permissible range of variation in a particular dimension of a product. Tolerances are often set by engineering requirements to ensure that components will function together properly.

Total Indicator Readout (Runout) — Measurement of a characteristic such as concentricity or flatness for maximum or minimum value as well as range. The part is rotated or moved such that the gage measures the entire area in question. The value(s) of interest such as maximum are either read manually from the gage or determined automatically by automated data collection equipment.

Total Quality Control (TQC) — A management philosophy of integrated controls, including engineering, purchasing, financial administration, marketing and manufacturing, to ensure customer quality satisfaction and economical costs of quality. See p. 8-7.

Transcription — Rewriting; copying.

Trend — A gradual, systematic change with time or other variable. See p. 3-19.

True $\bar{x}$ Causes — For $\bar{x}$ control charts: changes in the $\bar{x}$ control chart which are due to actual changes in the mean produced by the process. True $\bar{x}$ changes are usually accompanied by a stable pattern in the R-chart. See p. 3-22.

TurboSPC™ — TurboSPC software is quality management software for the PC. TurboSPC software receives data from data collectors, other programs, or keyboard inputs. The software produces control charts, performs capability analyses, creates reports, and provides a versatile database and analysis tools for quality management functions. See p. 19-14.

Type I Error — In control chart analysis: concluding that a process is unstable when in fact it is stable. See p. 3-7.

Type II Error — In control chart analysis: concluding that a process is stable when in fact it is unstable.

u-Chart — For attributes data: a control chart of the average number of defects per part in a subgroup. See p. 3-37.

UCL (Upper Control Limit) — For control charts: the upper limit below which a process remains if it is in control.

Uniform Distribution — This distribution means that all outcomes are equally likely.

Unit of Measure — The smallest increment a measurement system can indicate. See also Resolution.

Universe — See Population.

USL (Upper Specification Limit) — The highest value of a product dimension or measurement which is acceptable.

Variability — The property of exhibiting variation, i.e., changes or differences, in particular in the product of a process.

Variables — Quantities which are subject to change or variability.

See p. 2-2. *Variables Data* — Concerning the values of a variable; as opposed to attributes data. A dimensional value can be recorded and is only limited in value by the resolution of the measurement system.

$$\sigma^2 = \frac{\sum\limits_{i=1}^{n} (x_i - \bar{x})^2}{n - 1}$$

Variance Formula

Variance — The square of the standard deviation given by the formula.

x̄ and R Charts — For variables data: control charts for the average and range of subgroups of data. See also Control Chart.

x̄ and Sigma Charts — For variables data: control charts for the average and standard deviation (sigma) of subgroups of data. See also Control Chart.

$$z = \frac{X - \mu}{\sigma}$$

z distribution Formula

z-distribution — For a sample size of n drawn from a normal distribution with mean μ and standard deviation σ. Used to determine the area under the normal curve.

z-test — A test of a statistical hypothesis that the population mean μ is equal to the sample mean x̄ when the population standard deviation is known.

See p. 2-15. $Z_{max}/3$ — The greater result of the formula when calculating CpK. Shows the distance from the tail of the distribution to the specification which shows the greatest capability.

$Z_{min}/3$ — See CpK.

1. INTRODUCTION TO STATISTICAL QUALITY CONTROL

1.1 HOW IT ALL BEGAN

You must have control over an industrial process in order to produce interchangeable parts. Eli Whitney and other forerunners of the industrial revolution discovered this a bit too late. In 1798 Whitney won a contract from the U.S. government to produce 10,000 muskets. He was convinced he could work out a system of making gun parts to a standardized pattern. Nowadays that wouldn't be anything special, but back then it was practically unheard of. The conventional method of manufacturing a gun was to have a skilled craftsman fashion the whole piece, forming and fitting each part. If a part broke, a new one would have to be custom made. For a nation in need of armaments it is easy to see the attraction of the concept of interchangeable parts. Unfortunately, out of Whitney's first manufacturing run of 700 parts only 14 guns could be assembled. The idea temporarily failed for the lack of another.

Quality control was the name for the new idea. Quality control was necessary because mass produced products lacked quality and early manufacturing processes lacked control. The concept of quality control was to ensure that a specification was written and all parts conformed to it. First, a part was evaluated in terms of its function and its meeting the needs and expectations of the customer. Engineering then created a specification (Figure 1.1.1). The specification called out the materials, the dimensions, and the finish. The dimensions were usually expressed as the target dimension, or *nominal,* and the high and low limit, or *tolerance.* The job of quality control then was to inspect parts to make sure they conformed to specification and evaluate bad parts to find out why they didn't. As long as a high percentage of good parts were made in a suitably efficient way the process by which they were made was not that important.

The experience of World War I, however, raised doubts about the adequacy of this kind of quality control. The war was America's first experience with supplying vital materials over long distances. The quality of both finished products and replacement parts had to be guaranteed under conditions that prohibited the on-sight evaluation of failures, and

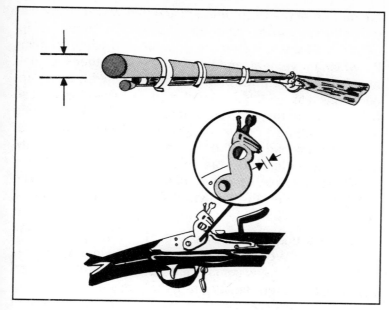

Fig. 1.1.1 A specification is a pattern from which to build parts.

in volumes that made 100% inspection a losing proposition. Quality control was inadequate because it was reactive. It relied too much on the evaluation of parts and parts failures. The logistics of war demanded a more active quality function. The focus had to shift to the home front so to speak — to the manufacturing process. Of course this new emphasis had to have a new name.

Statistical quality control was the new name. It is also called *statistical process control* to emphasize the importance of the process. But the old concept of quality control lacked something else besides a simple shift of emphasis. That something is statistics.

We owe the application of statistics to Dr. Walter A. Shewhart, a physicist at Bell Labs. Shewhart specialized in Brownian movement, the random behavior of small particles in a fluid caused by the collision of molecules. Statistical methods for analyzing large amounts of data were useful to this type of study. It was natural for Shewhart to use his knowledge of statistics when asked to help in the war effort.

Shewhart was assigned the task of designing a standard radio headset for army troops. He began by measuring the head sizes of 10,000 troops (Figure 1.1.2). He arranged the sizes from small to large and marked the frequency of occurrence of individual sizes. What intrigued him was that

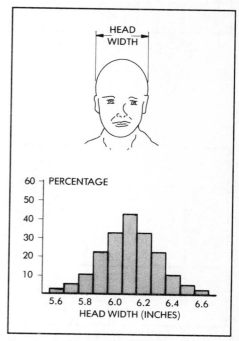

Fig. 1.1.2 Distribution of head sizes among army troops.

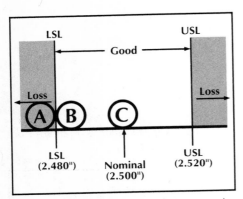

Fig. 1.1.3   Detection based quality control.

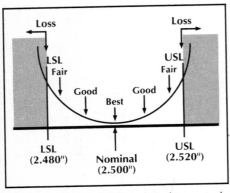

Fig. 1.1.4   Prevention based quality control.

the pattern of the distribution resembled what is known as the *bell-shaped curve*, or normal distribution, shown in Figure 1.1.2. It was a pattern he encountered when studying Brownian movement, and he wondered whether such patterns and the methods of analyzing them had broader applications. He eventually developed some descriptive statistics to aid in manufacturing and wrote a book published in 1931, *Economic Control of Quality of Manufactured Product*. The book had very little impact at the time it was published, but gradually people became aware of the value of statistical methods. One very important technique he developed is called the Shewhart Control Chart, or $\bar{x}$ & R chart.

Statistical quality control saw its first widespread application during World War II. The war department required industries making war materials to implement statistical controls, and hired statisticians to help teach them. One of the statisticians was Dr. W. Edwards Deming, who later helped foster the statistical quality movement in postwar Japan. The Japanese faced much the same problems in guaranteeing quality over long distances that America faced during the war. Lacking in many natural resources, their ability to produce and export manufactured articles was necessary for survival. Fortunately for them, they listened to sound advice and they learned to implement it. The success of their implementation of a basically American methodology has been felt far beyond their shores.

The $\bar{x}$ & R chart shifted the attention of quality control from the detection of defects after the manufacturing process to the prevention of defects during the manufacturing process. Figure 1.1.3 shows this change. In a detection system, the focus is on catching parts or products that fall outside the specification limits (LSL & USL). The drawback to this approach is apparent when we compare the three parts; A, B, and C that are shown in Figure 1.1.3. Part C is right at the nominal or best value. Parts A and B are both undersize. Part B is just over the lower specification and Part A is just under the specification. In a properly functioning detection system parts B and C would be determined to be more alike than parts A and B: B and C both pass inspection, while A fails inspection. A true comparison of the characteristics of each part would reveal A and B to be most alike. In a prevention system using statistical process control the focus is on producing parts or products that are at or near the nominal or best value. Figure 1.1.4 illustrates a prevention based system. While the need for specification limits is still respected, the focus of the system is on producing at the nominal or best value.

1.2 IMPLEMENTING STATISTICAL PROCESS CONTROL

When implementing a statistical process control program it is important to balance and monitor two concerns: applying the tools of SPC and attaining the benefits of SPC. The first does not necessarily lead to the second! The tools of SPC are described in the first one third of this book. These tools include problem identification techniques, process control charts and capability measures. These tools are all used to find problems. The benefit of an SPC program is increasing quality and productivity through *finding and eliminating problems*. Giving the members of an organization the means to find problems without the ability to solve them is the primary cause for SPC failing. Dr. W.E. Deming suggests that when a problem is identified on the manufacturing floor, the person immediately responsible for that process is able to take corrective action and solve the problem only 6% of the time. The other 94% of the time the problem must be solved through the cooperative action of many individuals: operators, supervisors, managers and engineers. See Figure 1.2.1.

For that reason, almost all plans for implementing statistical process control begin with creating an environment within the organization that allows creative and effective problem solving. Creating this environment begins with the commitment of top management to the goals of SPC. Figure 1.2.2 shows a model for implementing statistical process control. The first five chapters of this book are structured around using this model for implementing SPC.

Steps for Implementing SPC

Step 1: Build a system that allows problem solving — The first step is to build a system that allows problem solving. This step is the most difficult to complete and the chosen route will often be different from organization to organization.

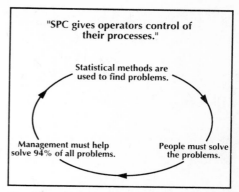

Fig. 1.2.1   *Build a system that allows problem solving.*

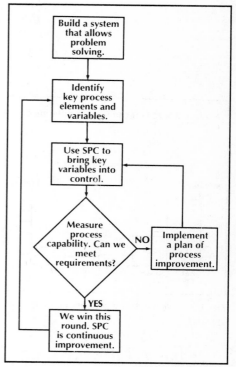

Fig. 1.2.2   *A model for implementing SPC.*

The Tennant Company of Minneapolis cites five elements that they found required for success:

Management commitment — Success improving quality and productivity is tied to the commitment and involvement of everyone in the organization. If some members of top management are not committed, it will be very difficult to secure the commitment and involvement of their employees.

Employee involvement — "Doing it right the first time," "continuous improvement of product and service," and "zero defects" are slogans and themes often used in quality and productivity improvement efforts. They are accomplished in large part through the efforts of the people on the front line who produce and deliver the product or service. Once the commitment of top management is secured, the same commitment must be made throughout the organization.

Cooperative, non-adversarial worker/manager relationships — The goals of improving quality and productivity are shared by everyone in the organization. This creates and requires teamwork. The tools and techniques of SPC will uncover many new problems and challenges that can only be resolved through the cooperative efforts of the team.

Something in it for the people — Tennant found that lasting benefits of improved quality and productivity are only attained when there is something in it for the people — something to secure the long term commitment required for success. The "something" took three forms at Tennant: recognition, rewards and satisfaction.

Time, energy and determination — While improvement may be measured over a few weeks or months, true success requires years of dedicated effort. The example of the Japanese best illustrates this: a 25 year transformation from the perception of poor quality and shoddy workmanship in the 1950s to the perception of high quality and world class manufacturing by the late 1970s.

Accomplishing these elements requires a major transformation for many organizations: the first step is the most difficult step.

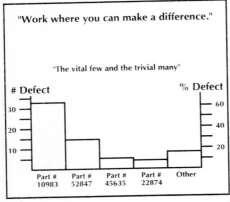

Fig. 1.2.3   *Identify key process elements and variables.*

Step 2: Identify key process elements and variables — The ability to creatively and effectively solve problems is a powerful tool. The second step in implementing SPC is deciding where this tool can most effectively be applied. Chapter 2 describes the common methods used to identify and prioritize problems in SPC. These methods include: brainstorming sessions, cause and effect diagrams, pareto charts, histograms, and scatter diagrams. Properly applying these techniques ensures that subsequent efforts in implementing SPC are focused on selected problems where a difference in quality and productivity can be made and measured. These same problem identification techniques will also be used to further analyze each selected problem to get at its fundamental causes. See Figure 1.2.3.

Step 3: Use SPC to bring key variables into control — Once a problem or process has been identified and analyzed in Step 2 the next step is to make that process do the best that it can with its present set of conditions. These conditions include people, machines, methods, materials, measurement and environment. Chapter 3 describes the application of process control charts to reduce and eliminate causes of variation. Included in this chapter are control charts for variable measurement such as $\bar{x}$ & R charts. See Figure 1.2.4. Applications where characteristics cannot be measured will be addressed by control charts for attributes such as percent defective charts, c-charts, u-charts and np-charts.

Step 4: Measure process capability: can we meet requirements? — Control charts for variables and attributes are proven tools for optimizing a process as it presently exists. Once this optimizing is achieved the remaining question is: "is it good enough?" This question is answered by comparing the characteristics of the individual products against the requirements for each characteristic. These requirements are often stated as engineering limits or product specifications. Figure 1.2.5 clearly shows this relationship. In this case, the individual products (process width) fall well within the upper and lower specification limits (USL & LSL). In other cases the specification limits may fall outside the process width. When this occurs additional decisions and action must be taken (see Step 5). Chapter 5 describes process capability measures. These measures provide a fast and effective means for answering the question "is it good enough?"

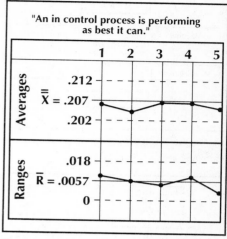

Fig. 1.2.4 Use SPC to bring the key variables into control.

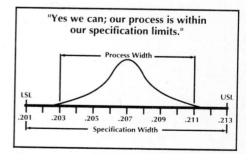

Fig. 1.2.5 Measure process capability. Can we meet requirements?

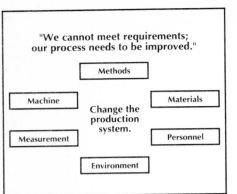

Fig. 1.2.6 *Implement a plan of process improvement.*

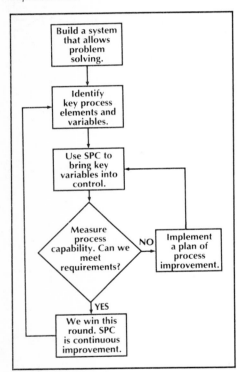

Fig. 1.2.7 *A model for implementing SPC: SPC is continuous improvement.*

Step 5: No we cannot meet requirements. Implement a plan of process improvement — When requirements cannot be met, a decision has to be made on one of five alternatives:

1) Change the process so that it can produce products that meet requirements. See Figure 1.2.6.
2) Determine if the requirements accurately reflect the demands on the product. If they do not, change the requirements.
3) Sort each product produced to insure that the customer only receives products that meet requirements.
4) Let the customer sort good from bad.
5) Quit making the product.

In most cases the final three choices are not desirable or acceptable alternatives. Changing the requirements is the least expensive solution to the problem, but this too is often not an acceptable or allowable option. The focus most often falls on changing the process to enable it to meet the current requirements. These changes can occur in many areas: methods, machines, measurement, materials, people and the environment. Once a change has been made, such as rebuilding a piece of equipment or retraining an operator, the SPC focus moves back to Step 3 and follows through the process again until the question can be answered: "yes, we can meet requirements." See Figure 1.2.7.

Step 6: We win this round, SPC is continuous improvement — Meeting requirements for one characteristic on one process is a measurable improvement and evidence of the impact of SPC; however, improving quality and productivity is a long term commitment that requires solutions to many other problems facing the organization. Fortunately, in Step 2 a clearly defined list of problems and priorities was defined and the next problem area awaits resolution.

Chapters 2 through 5 will detail each step of the SPC implementation process. The remainder of this chapter will describe the statistical concepts on which SPC is based.

1.3 BASIC STATISTICAL CONCEPTS

What Shewhart discovered in the twenties is that variability is as normal to a manufacturing process as it is to natural phenomena like the movement of molecules in a jar of fluid. No two things can ever be made exactly alike, just like no two things are alike in nature. The key to success in manufacturing is to understand the causes of variability and to have a method which recognizes them. Shewhart found two basic causes of variability, *common causes* and *assignable causes*.

Common Causes of Variability

If we flip a coin and count the number of heads versus tails, at first we may get a few more of one than the other but over the long run they will be fairly even (Figure 1.3.1). We say that the *probability* of heads in a coin toss is 50% or 0.5. The probability is a statistic. For a few coin flips this probability may not be a reliable indicator of the outcome, but it tends to be more reliable as larger groups of coin tosses are counted. Coin tosses vary purely by chance, and chance is what is known as a common cause of variability.

Another example of common causes at work is with dice throws. If we repeatedly throw a pair of dice and record the totals, we will get an unequal distribution of results. The possible outcomes are the numbers 2 through 12, but as any craps game player will tell us the frequency of their occurrence varies. Dice pair combinations total some numbers more frequently than others, as shown in Figure 1.3.2. The number 7 will occur in six combinations whereas the number 12 has only one. Over the long run the probability of the number 7 occurring is .167 or about 17%, which is greater than the number 2, which is about 3%. A pair of dice produces an example of an unequal frequency distribution, but it is entirely due to common causes. Whenever the outcomes of a process can be expressed in probabilities, and we are certain about the distribution of outcomes over the long run, we have what is known as a *constant-cause system*.

As you may have guessed, manufacturing processes sometimes behave like constant-cause systems. The causes of variation are commonplace, like dice throws. If left to produce parts continually without change, the variation would remain. It cannot be altered without changing the process itself. Statistics provide us with ways of recognizing

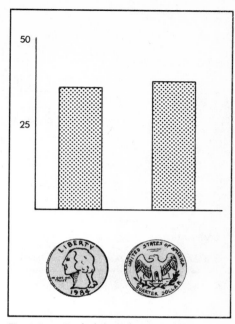

Fig. 1.3.1   *Probability of coin tosses.*

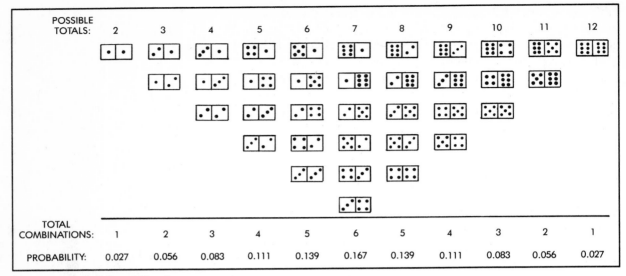

POSSIBLE TOTALS:	2	3	4	5	6	7	8	9	10	11	12
TOTAL COMBINATIONS:	1	2	3	4	5	6	5	4	3	2	1
PROBABILITY:	0.027	0.056	0.083	0.111	0.139	0.167	0.139	0.111	0.083	0.056	0.027

Fig. 1.3.2   Probability of dice throws.

variation due to common causes. The main one is the control chart. By using a control chart we can separate common causes from the second type, which are called assignable causes.

Assignable Causes of Variability

A change of materials, excessive tooling wear, a new operator — these types of things would produce variation in a process that is different from variation due to common causes. They disturb a process so that what it produces seems unnatural. A loaded pair of dice is another example. Since we know what a regular pair of dice produces over a large number of rolls, we can be reasonably sure a pair of dice is loaded if, after a large number of rolls, we have more twelves than sevens.

When we look for problems in a process we are usually just looking for these assignable causes of variability. Assignable causes produce erratic behavior for which a reason can be identified. One might ask why we are going through all the trouble. Why separate assignable causes from common causes when we have to compare parts to a specification anyway? One reason is that we can minimize variability when we know its causes. The less variability we have in our parts and the closer they are to the target, the happier our customers will be. They will be confident in our ability to supply a good consistent product with few or no parts out of specification. On the other hand, large variability may result in parts out of specification. If there are many parts out of specification, we have three choices: 1) continually inspecting all parts and using the good ones, 2) improving the process until most or all parts are good, or 3) scrapping the process and building a better one. Since 100% inspection is expensive and inefficient in most cases, we are better off trying to improve the process and reducing our inspection load. Which brings us back to the process. There is no sense in trying to improve a process that won't do the job for us. Also, we don't want to scrap a process that potentially could work like a charm. Therefore, we need a way to determine whether the process can consistently produce good parts. The only reliable method is through the use of control charts to find and eliminate the assignable causes of variation.

Basic Statistical Terms

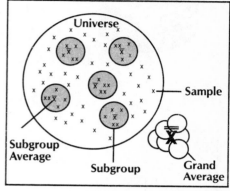

Fig. 1.3.3   Relationship of the basic statistical terms.

Before we go on to the basic statistical concepts presented in the following sections, there are some terms that we need to discuss. Refer to Figure 1.3.3 as these terms are presented.

As we collect data, we pull *samples* from the process. A sample is an individual piece or measurement that we collect for analysis. Samples are usually pulled in rational groups called *subgroups*. Groups of samples that are pulled in a manner that shows little variation between parts within the group, such as consecutive parts off a manufacturing line, are considered rational subgroups.

Once we have collected the samples that make up our subgroup, we can calculate the average of this data, otherwise called the *mean*. The symbol used to represent the mean is $\bar{x}$, pronounced x-bar. We can continue to collect our subgroups at regular intervals. Once we have collected a number of subgroups and calculated the mean of each one, we can also calculate the overall average of the data. This is called the *grand average* and is represented by $\bar{\bar{x}}$.

As we have been collecting data, we have pulled just a few samples from all of the parts we manufacture. All of the parts we make constitute a *population* or *universe*. It would be very difficult and time consuming to measure every part manufactured, so we use the samples and statistical analysis to give us an idea of what all of the parts in our universe look

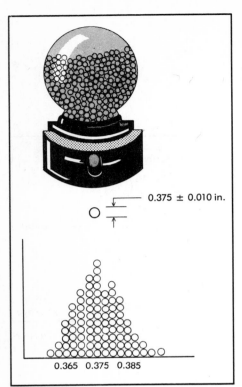

0.375 ± 0.010 in.

0.365 0.375 0.385

Fig. 1.3.4   The central tendency of a process.

like. The statistical concepts that are used to make conclusions about the universe are introduced in the following sections of this chapter and the remaining chapters.

Measures of Central Tendency

Many processes are set up to aim at a target dimension. The parts that come off the process vary of course, but we always hope they are close to the nominal and very few fall outside of the high and low specifications. Parts made in this way exhibit what is called a *central tendency*. That is, they tend to group around a certain dimension (Figure 1.3.4).

The most useful measure of central tendency is the *mean* or average. To find the mean of measurement data, add the data together and divide by the number of measurements taken. The formula would be:

$$\bar{x} = \frac{x_1 + x_2 + \ldots + x_n}{n} \tag{1.3.1}$$

Each 'x' is a measurement and 'n' is the number of measurements. In statistics the average is symbolized by $\bar{x}$, or x-bar. If we use the Greek letter for summation Σ, the formula can be written as:

$$\bar{x} = \frac{1}{n} \sum_{i=1}^{n} x_i \tag{1.3.2}$$

There are two additional measures of central tendency which can be used. The first is the *median*, which is the middle of our data. The median splits our data in half, so 50% of our parts are above the median and 50% are below. The second measure of central tendency is the *mode*, which is the most frequently occurring value in our data.

Figure 1.3.5 is a *histogram* showing one dimension and its variation among 50 parts. The histogram shows the frequency of parts at each dimension by the height of the bars. Notice that the $\bar{x}$ is not at the most frequent value. The most frequent value (mode) is just to the left of it. Now look at Figure 1.3.6. The mean value in this histogram seems to be among the least frequent values that occurred. Obviously, without some kind of statistic that tells us about the spread

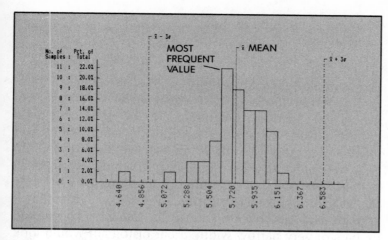

Fig. 1.3.5 *The mean may not be the most frequent value in a distribution.*

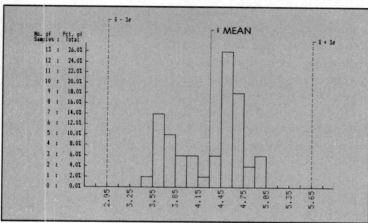

Fig. 1.3.6 *The mean may even be the least frequent value.*

or *dispersion* of our data, the mean does not tell us enough.

Averages are used in many sports, from bowling to baseball. When we know someone's bowling average or baseball hitting average we have some indication of how good that person is, but not really enough information about consistency. Lurking behind a low average could be a lot of great games and a few very bad ones. A good average could merely be the work of an average player with a few lucky games.

Range and Standard Deviation (Sigma)

Two measures of dispersion are used in statistics, the *range* and the *standard deviation*. The range tells us what the overall spread of the data is. To get the range, subtract the lowest from the highest measurement. The symbol for

Fig. 1.3.7 Two histograms having the same mean and range.

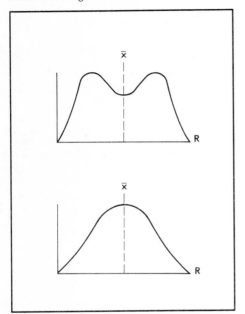

$$R = x_{max} - x_{min} \qquad (1.3.3)$$

This formula tells us to simply subtract the smallest measurement from the largest. Now that we have a measure of spread and a measure of central tendency, why do we need a third statistic? What does standard deviation tell us that range and average do not? To help answer this we need to look at Figure 1.3.7. Two different histograms are pictured. Rather than using bars, continuous lines are used to show the shape of the distributions. We can imagine that if enough data was collected and if the data was represented by bars of very narrow width, a bar histogram would look nearly like the curves shown here. The two histograms have different shapes, yet the ranges and averages are the same. This is what we use the standard deviation for. Standard deviation uses all of the data displayed on the histogram, not just the highest and lowest points and gives us a better idea of what the distribution looks like.

The standard deviation of a population (universe) is called *sigma* in statistics and is symbolized by the Greek letter σ. Sigma can be calculated using this formula:

$$\sigma = \sqrt{\frac{\sum_{i=1}^{N} \left(x_i - \mu\right)^2}{n - 1}} \qquad (1.3.4)$$

μ is the mean of the data, x is an individual measurement and N is the total number of measurements in the universe. It is not very often that we calculate the sigma, because measuring all of the parts in the universe is very time consuming. Instead, we estimate the population standard deviation by pulling samples and calculating the sample standard deviation s. The formula for s is:

$$s = \sqrt{\frac{\sum_{i=1}^{n} \left(x_i - \bar{x}\right)^2}{n - 1}} \qquad (1.3.5)$$

The above formula is called the n-1 formula because of its denominator. The more classic standard deviation formula uses n instead of n-1. The n-1 formula will be used here because it provides a closer approximation of the standard deviation of samples coming from a process that is producing continually. Continuous processes are the most common type used in manufacturing.

Sigma has a special relationship to the distribution shown in Figure 1.3.8. It is called the *normal distribution* and its properties are described in the next paragraph.

The Normal Distribution

There is one type of distribution that can be described entirely by its mean and standard deviation. It is the normal (Gaussian) distribution or bell-shaped curve. It has these characteristics (Figure 1.3.9):

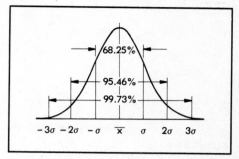

Fig. 1.3.8   The relationship of sigma to a normal distribution.

- The mean equals the mode which equals the median.
- It is symmetrical about the mean.
- It slopes downward on both sides to infinity. In other words, it theoretically has an infinite range.
- 68.25% of all measurements lie between $\bar{x} - \sigma$ and $\bar{x} + \sigma$. See Figure 1.3.8.
- 95.46% of all measurements lie between $\bar{x} - 2\sigma$ and $\bar{x} + 2\sigma$.
- 99.73% of all measurements lie between $\bar{x} - 3\sigma$ and $\bar{x} + 3\sigma$.

The equation for a bell-shaped curve is:

$$P(x) = \frac{1}{\sigma \sqrt{2\pi}}\, e^{\frac{-(x-\mu)^2}{2\sigma^2}} \qquad (1.3.6)$$

The normal distribution is a valuable tool because we can compare the histogram of a process to it and draw some conclusions about the capability of the process. Before making this type of comparison, however, a process must be monitored for evidence of stability over time. This is done by taking small groups of samples at selected inter-

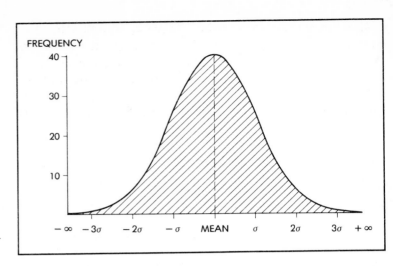

Fig. 1.3.9 *The normal distribution, or bell-shaped curve.*

vals, measuring them, and plotting their averages and ranges on a control chart. The control chart provides us with an indication of whether we have stable variation, in other words, a constant-cause system, or a lack of stability due to some assignable causes. Chapter 2 describes the making and using of x̄ & R charts in more detail. The reason they work has to do with the *central limit theorem* and the normal distribution curve.

Central Limit Theorem

Shewhart found that the normal distribution curve appears when the averages of subgroups from a constant-cause system are plotted in the form of a histogram. The constant-cause system does not itself have to be a normal distribution. It can be triangular, rectangular or even an inverted-pyramid shape like dice combinations, as long as the sample size is reasonably large. The averages of different sized subgroups selected from these distributions, or *universes*, as they are called in statistics, will show a central tendency. The variation of averages will tend to follow the normal curve. This is called the central limit theorem.

Shewhart demonstrated this by using numbered chips and a large bowl. His normal bowl had 998 chips, rectangular bowl had 122, and triangular bowl had 820. The rectangular universe had chips bounded by a certain range and in equal numbers like in Figure 1.3.10. The triangular universe had unequal numbers of various chips as shown in Figure 1.3.11. Shewhart took each chip out of the bowl one

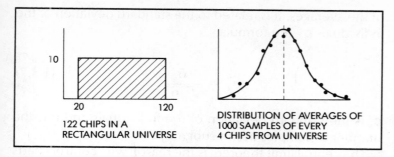

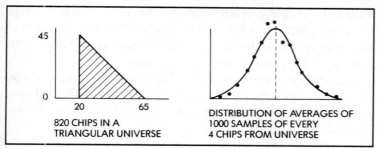

Fig. 1.3.10 Averages of large enough subgroups from a rectangular universe tend to follow a normal distribution.

Fig. 1.3.11 Averages of large enough subgroups from a triangular universe tend to follow a normal distribution.

at a time, recorded the number and put it back. He then mixed the bowl before choosing another. He averaged every four. The points he plotted fell within or along the edges of the bell-shaped curve. What this meant to him is that a process can be monitored over time by measuring and averaging a standard subgroup of parts. The subgroup could be 2, 4, or even 20. The frequency could be once per hour, or once per day depending upon the output. If the process was a constant-cause system, these averages would fall within a normal curve. One could conclude that the process was *stable*. By stable, we mean that the variability was entirely due to common causes. Statisticians also use the phrase *in control* to refer to a process that has stable variability over time.

Frequent checking of the averages of subgroups also provides a way to discover when assignable causes are present in a process. When assignable causes appear they will affect the averages to the point where these averages will probably not fit within a normal curve. Once it is known what the stable variation of the process is, the assignable causes will appear in averages of subgroups taken periodically.

We can calculate the standard deviation of the averages and, if compared to the standard deviation of the individual samples, we will find it is smaller, as shown in Figure 1.3.12. $\sigma_{\bar{x}}$ is the symbol we use to represent the standard deviation

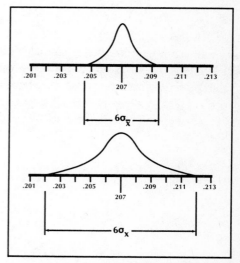

Fig. 1.3.12 Comparisons of the distributions of averages and individuals.

1-17

Intro to SQC

of the averages. It is related to the standard deviation of the individuals by the formula:

$$\sigma_{\bar{x}} = \frac{\sigma_x}{\sqrt{n}}$$

(1.3.7)

σ_x is the standard deviation of the individuals and n is the number of samples in the subgroup.

The central limit theorem is the reason why control charts work. The charting of averages has this particular advantage over the charting of individual data points. The charting of ranges is also used because subgroup ranges will also show stability if a constant-cause system exists.

The SPC concepts presented here will be explained further throughout Part I. Chapters 2 through 5 will deal specifically with some of these concepts and how they are implemented with SPC.

For Further Reference

Burke, James. *Connections*. Boston: Little, Brown and Company, 1978.

Grant, Eugene L., and Leavenworth, Richard S. *Statistical Quality Control*, 5th ed. New York: McGraw-Hill Book Company, 1980.

Green, Constance M. *Eli Whitney and the Birth of American Technology*. Boston: Little, Brown and Company, 1956.

Hale, Roger L., Hoelscher, Douglas R., and Kowal, Ronald E. *Quest for Quality How One Company Put Theory to Work*. Minneapolis, Minnesota: Tennant Company, 1987.

Ishikawa, Kaoru. *Guide to Quality Control*. Hong Kong: Nordica International Limited, 1976.

Quality Control Circles, Inc. *Quality Control Circles*. 2d ed. Saratoga: Quality Control Circles, Inc., 1982.

Shewhart, Walter A. *Economic Control of Quality of Manufactured Product*. Princeton: Van Nostrand Reinhold Company, Inc., 1931.

2. DATA COLLECTION AND ANALYSIS METHODS

2.1 BASIC DATA COLLECTION

The first chapter introduced the concepts of statistical process control. But before we can apply statistics we must first collect data. This chapter explains how to identify characteristics and begin the task of collecting data. The first five sections discuss data collection and data analysis methods and theories. The last section discusses practical data collection methods.

Types of Data

Data can be classified in a very general way by how it is collected. Data that is measured is called *variables* data. Data that is counted or classified is called *attributes* data. Variables data has these characteristics:

- It is measurable, by units of length, diameter, weight, temperature or Newton meters, for example.
- It is continuous (Figure 2.1.1). How we verify it depends entirely on the accuracy and resolution of our gaging. Something could weigh 2 kilograms on our scale but could weigh 1.98 kg on a more precise scale and a slightly different weight on other scales.
- Variables data of the same unit of measure can be compared numerically. We can find the mean, range and standard deviation.

Attributes data has one or more of these characteristics:

- It is countable. Either it exists or it does not, such as with defects on a painted surface (Figure 2.1.2).
- It is classified or graded using a scale, such as small, medium and large eggs. Sometimes an arbitrary scale is used as a substitute for measuring when the exact measurement is not important.
- It can be pass/fail data, such as the picture tube works or does not.

Other types of data such as serial numbers, build sequence numbers and piece counts are used for production control. These types are sometimes used to support the quality control function and very often need to be collected and analyzed for the same purposes. Data collection techniques have their application in many aspects of manufacturing, whether for production control or quality control.

Attributes data is often collected during final inspection. An assembled machine may be composed of parts that can be individually measured, but once assembled it either works or it doesn't. The success of a manufacturing process can be expressed as the percentage of good parts produced. This type of data provides an overall measure of quality improvement when inspection is used, but it does not provide a clue as to how to make the improvement. In general, attributes data is not a good substitute for variable data. Improving a process often depends on the ability to distinguish between minute differences in dimension, weight, or some other quality characteristic. The gaging must be able to detect these differences in order to establish the variability of a process. GO/NO-GO gages based on the low and high limits of a part specification do not provide a means for process improvement (Figure 2.1.3).

Selecting Characteristics

To improve quality we must first identify the characteristics of quality in a part. These characteristics may have to do with fit or finish, or perhaps something very intangible such as product desirability. Selecting characteristics involves a clarification of purpose, and aims at identifying these characteristics as either variable or attribute data. Some of the questions that may need asking are:

- *Purpose* — Is our purpose a general one, or are we addressing a specific quality problem? Can it be defined?

- *Problem clarification* — Where is the problem noticeable? Where does it appear first? Is it a compound problem? Can it be addressed within our factory? Is there currently a method for detecting the characteristics of this problem?

- *Selecting characteristics* — Can we specify the characteristics? Are they measurable? At what point in the process can they be measured or verified? Can we take action on

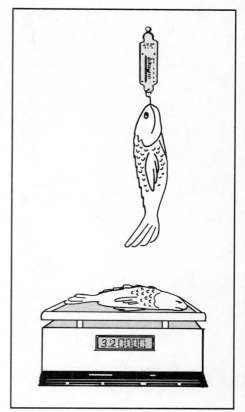

Fig. 2.1.1   Variables data.

Fig. 2.1.2   Attributes data.

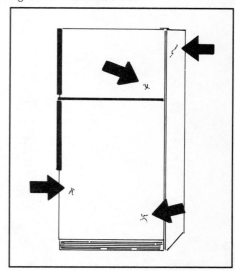

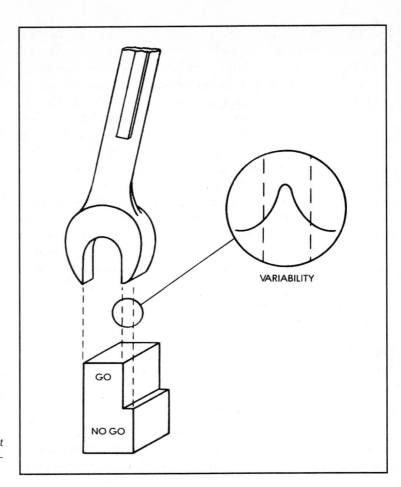

VARIABILITY

GO

NO GO

Fig. 2.1.3 GO/NO-GO gages do not detect the variability in a process — the main indicator of process improvement.

data collected on these characteristics? Are results verifiable?

The importance of selecting characteristics becomes evident when we are faced with the task of collecting data. To be worthwhile, data collection must serve our objectives. The clarification of these objectives and the selection of characteristics that provide evidence and also serve as a basis for action will allow data collection to provide results. If the problem is not clear, or the wrong characteristics are being measured, time will be wasted and we will still have a problem.

Selecting the Means of Analysis

How data is analyzed depends on both the type of data,

Purpose	Type of Data	Means of Analysis	See Chapter
Process Control	Variable	x̄ & R Chart	3
		x̄ & Sigma Chart	3
	Attribute	p, np, c-Charts	3
		Pareto Diagrams	2
Process Capability	Variable	Histograms	2
		Capability Studies	5
	Attribute	c-Charts	3
Acceptance Sampling	Variable	Histograms	2,7
	Attribute	AQL, LTPD, AOQL	7

Table 2.1.1   Means of analysis.

whether variable or attribute, and the purpose for collecting it. There are also several ways of analyzing data for any one purpose. One or all methods may need to be used. Table 2.1.1 classifies several means of analysis described in this book.

2.2   CAUSE AND EFFECT DIAGRAMS (Ishikawa Diagrams)

Dr. Kaoru Ishikawa (1915-1989), a noted Japanese authority on quality and productivity, developed these diagrams for problem solving. A more complete explanation of these diagrams can be found in his book, *Guide to Quality Control,* 1976. A cause and effect diagram is a simple technique for dissecting a problem or process. Making a diagram is possibly the best first step in analyzing a problem prior to data collection. It organizes thinking and provides a plan of attack at the same time. There are three types of cause and effect diagrams:

- *Cause enumeration diagram* — A graphic listing of all the possible causes of a problem.
- *Dispersion analysis diagram* — A diagram used to analyze the causes of variability in a process.
- *Process analysis diagram* — A flow diagram used to study quality problems.

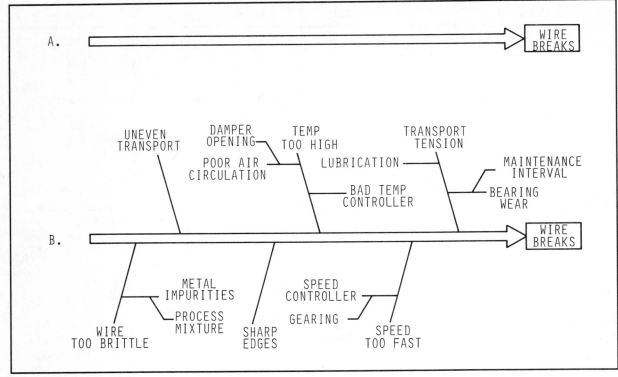

Fig. 2.2.1   How to start a cause enumeration diagram.

All three diagrams look nearly alike. The dispersion analysis and process analysis diagrams are derivatives of the basic cause enumeration diagram and are used to highlight process control problems.

Cause Enumeration Diagram

To create a cause enumeration diagram, first put the problem on the extreme right side of the page as in Figure 2.2.1. Then draw a long-stemmed arrow to it. Along the stem of the arrow draw smaller arrows and label them with the possible causes. It is important to list all possible causes by putting down everything that comes to mind. A good diagram will look extremely complicated and busy; however, if everything is included on the chart, the true causes of the problem should be there.

For a compound problem several diagrams may be necessary. Later we might be able to combine the diagrams and organize the causes, using some criterion such as their relative importance, or the sequence of events. When investigating the causes, we use the diagram to note progress.

We may be able to eliminate most causes, or even add a few that were not described.

Dispersion Analysis Diagram

To make a dispersion analysis diagram, start with the problem on the right and an arrow to it as shown in Figure 2.2.1A. List the main groups that influence the problem by their general category — worker, materials, tools, inspection, and machinery (Figure 2.2.2). The next step is to list all details that would contribute to variability by drawing arrows to the branches and labeling them. This organizes the causes into categories that can be focused on one at a time. We must then keep asking "Why does the dispersion occur?" as we look at each variable on our diagram. This keeps us focused on what we want to accomplish — minimizing the variation.

Process Analysis Diagram

A process analysis diagram is like a process flow diagram. Each step of a process should be labeled and connected by a line going from left to right (Figure 2.2.3). At each step in

Fig. 2.2.2   Dispersion analysis diagram.

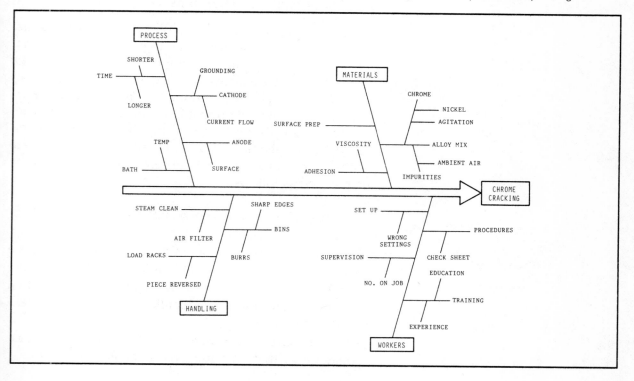

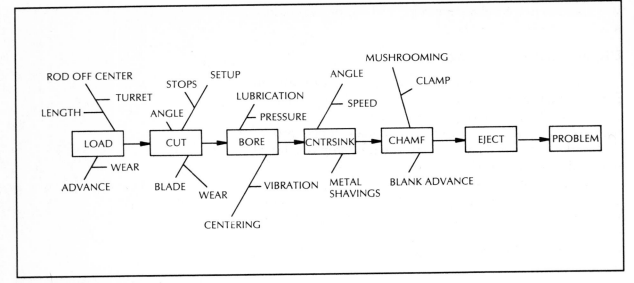

Fig. 2.2.3   *Process analysis diagram.*

the process, draw branches and label everything that could influence the quality of the product at that point in the process. This provides a listing of quality problems arranged by when they appear in the process.

Benefits of Ishikawa Diagrams

These diagrams graphically illustrate all influencing factors of a problem. They force one to put everything on paper so that it all becomes known and can be studied. Of course, the main weakness of a diagram is that its accuracy depends on the person or group who makes the diagram. This is why it is very important to get input from all individuals involved with the process. The different ideas generated by these people will often lead to a more complete picture of the causes actually present. The benefits can be summarized as follows:

- All factors can become known, not just the suspected obvious ones.
- They provide a simple plan of attack. Each cause can be investigated and crossed out if not important.
- We can use the diagram as an ongoing record of when actual relationships were identified and the date corrections were made.
- When the analysis has been completed one can be reasonably sure that everything has been accounted for.

2.3 PARETO DIAGRAMS

Pareto diagrams, named after the Italian economist Vilfredo Pareto (1848-1923), provide a method by which causes of a problem can be arranged by their relative importance. The diagram has as its basis the idea of the "vital few" and the "trivial many." Very often over half of the quality problems are the result of one cause. It is a much better tactic to locate the most important cause and eliminate it than to attempt to eliminate all causes at once. Eliminating the one important cause will result in a dramatic quality improvement with possibly the least amount of effort and greatest payback.

Pareto diagrams can be used with either variable or attribute data, but are used most often with attribute data. Usually the data is expressed in percentages or number of occurrences in each category. For example, inspection data can be broken down by the number rejected due to various causes. Table 2.3.1 lists some sample inspection data. The first step in making the Pareto diagram is to identify the categories of data we wish to display. Often, we will have data previously collected to place on our diagram. If not, we will have to collect data using methods outlined in the following sections.

Next, draw a left vertical axis and label it with a scale going up to the total number rejected. See Figure 2.3.1. Draw a horizontal axis and mark off equal lengths to be labeled with each of the causes. Put the most frequent cause to the left and the rest in descending order. If there is an "other" category, put it on the right side even though it may not be the smallest. The "other" category can be used to group lesser causes and reduce the width of the diagram. Now draw vertical bars of equal width for each cause, each at a height which matches its frequency to the total scale.

An alternative measure would be to put an additional scale on the left side which represents the cost of defectives. On the right side draw a second vertical axis and label it with the percent defective. Finally, if desired, draw a segmented line which represents the cumulative percentage, starting at the bottom left corner and ending at the upper right. The completed diagram will be similar to Figure 2.3.2. The segmented line simply totals each defect, but it provides a way to quickly estimate percentages. Draw two horizontal lines to the percent scale and subtract to get the per-

Defect	Number
Scratch	13
Crack	42
Off center	6
Plating flaw	78
Burr	25
Other	29
	193

Table 2.3.1 Inspection data.

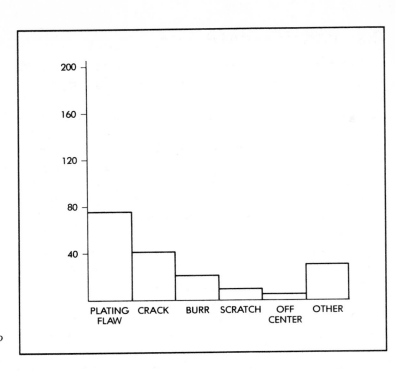

Fig. 2.3.1 Steps for constructing a Pareto diagram.

cent due to any one cause.

Pareto Diagrams have these usages:

- They highlight the few most important causes. If we were to try to reduce scratches by one-half in Figure 2.3.2, it would not have nearly the effect on reducing total defects as reducing plating flaws by one-half. The effort at reducing either of these might be the same even though the "payoff" is much different. Pareto diagrams help focus on effective solutions to problems.

- They highlight the results of improvements. Diagrams drawn side by side will illustrate the overall results of quality improvements in a before/after context. Figure 2.3.3 shows the results of improving the main cause of circuits failing a power-on test.

- Note that it is important to use numbers of defects or costs of defects on the left-hand scale of the diagrams, not the percentages. If percentages were used, the causes in the second diagram would look larger in comparison to the first. Repetitive audits of the causes of this problem should show the main cause shifting to second

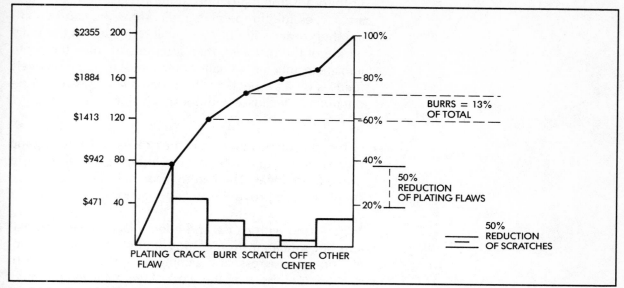

Fig. 2.3.2   Completed Pareto diagram.

Fig. 2.3.3   Side by side diagrams can display the results of improvement.

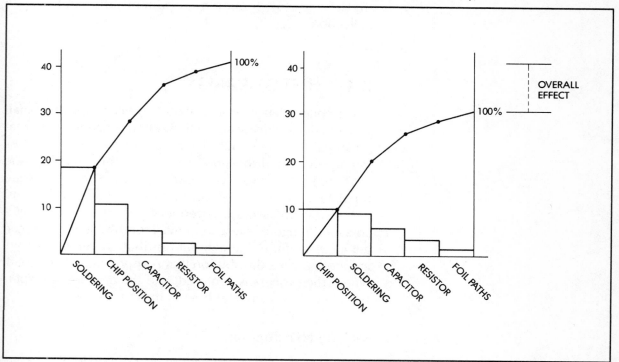

or third position if attempts at improving the main cause are successful. If repetitive audits show the causes shifting their order without an overall reduction in the magnitude of the problem, then attempts to solve the problem are insufficient. A daily control system that effectively focuses on all aspects of the process should cause a reduction of all causes without much shifting of order.

- Often, personnel involved in the process will have a good idea of the major defects or problems, but no data to prove those ideas. The Pareto diagram is an effective, yet simple tool to prove those "gut reactions."

When used with cause and effect diagrams, Pareto diagrams become an instrument for plotting the course of activity, noting progress, and gaining perspective on a problem at any point during the work. After a cause and effect diagram is made, data must be collected to determine the relative importance of causes. The Pareto diagram can be the first useful document produced after initial data collection.

2.4   HISTOGRAMS

We collect data on our process so we can do further analysis and decide if any corrective action is needed. The major questions we should be asking ourselves are "How much variation (dispersion) do I have in this variable?" and "Where is the process centered?" After we have collected our data, we can use a *histogram* to answer these questions.

A histogram shows the frequency of occurrence over a specified range of measurement. From the graph, we can get an idea of the shape of our distribution (remember, we are looking for a distribution that is approximately normal), where the process is centered, and the amount of dispersion. Let's briefly look at how to make a histogram.

Making a Histogram

To get a good idea of what our distribution looks like, we need quite a bit of data, about 50-100 data points. Once the data is collected, follow the remaining steps:

1) Find the smallest and largest data points and write them down.

2) Determine the range of the data by subtracting the smallest data point from the largest: $R = X(max) - X(min)$.

3) Determine the number of *cells* (bars) that we will have on the histogram using Table 2.4.1. The number of cells are based on the number of data points to be included on our histogram.

Number of data points	Number of cells
Under 50	5-7
50-100	6-10
100-250	7-12
Over 250	10-20

Table 2.4.1   Number of cells for histogram.

4) Divide the range of the data (calculated in Step 2) by the number of cells chosen to determine approximately how wide each cell on the histogram should be.

5) Now determine the horizontal scale of the histogram as shown in Figure 2.4.1. It must be large enough to extend out slightly past the smallest and largest data points. If we want to include the specification limits on the histogram, we must make sure the scaling is wide enough to include them as well. Also, we should set the boundaries of each cell at a value such that our data cannot fall on a cell boundary. In other words, if our data went out to 0.01, our cell boundaries should go out to 0.015. Or, if all our values are even numbers, the boundaries could be odd number values. This way we will never have a data point that falls on the cell boundary and will not have to decide in which cell to place the reading.

6) Next, design a frequency table as shown in Figure 2.4.2. Place a tick mark next to the category that each data point falls into. Once all data points have been marked in the frequency table, count the number of ticks in each category and write the total in the frequency column. Adding up the numbers in the frequency column should result in the total number of data points collected.

7) Draw a bar on the histogram for each cell. The height of each bar corresponds to the frequency noted on the frequency table.

Once we have drawn all of the bars on the histogram, we

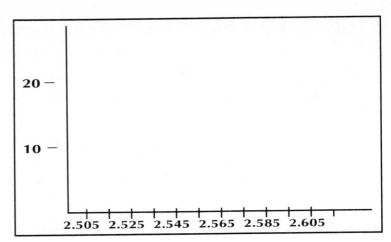

Fig. 2.4.1 Axes of histogram with correct scaling based on the number of data points, data range and number of cells.

Fig. 2.4.2 A frequency table can assist in organizing data in preparation for drawing the histogram.

Class	Cell boundaries	Frequency Tally	Frequency
1	2.505-2.515	//	2
2	2.515-2.525	///	3
3	2.525-2.535	////	5
4	2.535-2.545	//// ///	8
5	2.545-2.555	//// //// ////	14
6	2.555-2.565	//// //// //// ////	20
7	2.565-2.575	//// //// //	12
8	2.575-2.585	//// //	7
9	2.585-2.595	////	4
10	2.595-2.065	/	1

can calculate the mean of the data and draw a vertical line to note its placement on the histogram. We can also place our specification limits on the histogram as shown in Figure 2.4.3. If the distribution is approximately normal, the standard deviation of the data can be calculated and $\bar{x} \pm 3\sigma$ limits added to the histogram. If the data does not result in a normal curve, refer to Chapter 5 and the section on nonnormal distributions for more information.

Parts Outside Specifications

If our histogram looks like a normal distribution, we can estimate the percentage of parts outside our specifications. This is done using z-tables or Student's t-tables. Examples of these tables can be found in the Appendix. What these tables allow us to do is estimate the area under some portion of the normal curve, in this case, outside our specifications. In our example, we will use the z-table. Using the z-table assumes we know our sigma, in other words, that we have

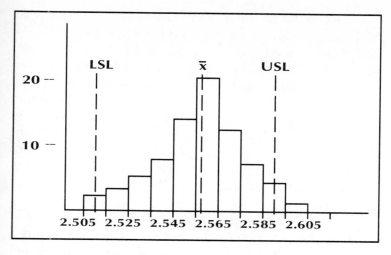

Fig. 2.4.3   Completed histogram.

not estimated standard deviation using a sample of the population. If we do not know sigma, but instead have calculated s (sample standard deviation), we should use the Student's t-tables.

We can calculate the area in the tails of our normal curve that are outside our specification limits using the following formula:

$$Z = \frac{x - \bar{x}}{\sigma}$$

(2.4.1)

We will complete this formula by first putting our upper specification then our lower specification in for x. These two values are then looked up in a z-table, such as the one provided in the Appendix, Table A-3. Multiplying the number from Table A-3 by 100 gives us the percentage of parts we estimate from the normal curve to be outside first the upper, then lower specification limits.

Let's run through an example using the data displayed on our histogram. From the histogram, we know our $\bar{x}$ is 2.557, the lower specification is 2.510 and the upper specification is 2.590. We also know our sigma is 0.0184. Using this data, we can calculate the area under the normal curve that falls outside our specification limits. The calculations are:

$$Z_{lower} = \frac{2.510 - 2.570}{0.0184} = -2.55$$

$$Z_{upper} = \frac{2.590 - 2.557}{0.0184} = 1.79$$

(2.4.2)

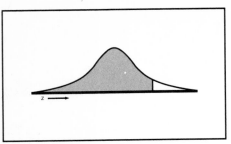

Fig. 2.4.4 Area covered by the z-table always goes from the left tail toward the right. The shaded area is the value displayed in the z-table.

We now refer to Table A-3 to obtain the area in each tail of the curve. The values obtained from the table are .0054 and .9633 for the lower and upper tails, respectively. Refer to Figure 2.4.4. The z calculation always looks at the area from the left side of the curve and moves right (the shaded area of the curve). In order to get the area above the upper specification limit, we must subtract the z value from 1. In our example, this results in a value of .0367 (1 − .9633) for the area above the upper specification. Multiplying each of the areas gives us a result of 0.54% of our parts below the lower specification and 3.67% above the upper specification.

This is just one example of how z-tables can be used to estimate the area under the normal curve. Additional references at the end of this chapter will cover z-tables and Student's t-tables in more detail.

2.5 SCATTER DIAGRAMS

Once we have identified the elements contributing to our problem, we need to determine which causes are truly related to our effect. One tool which can be used to identify a true relationship, or lack of relationship, is the *scatter diagram*.

A scatter diagram is a graphic representation of the relationship between two items. When doing a scatter diagram, we will choose one of our variables that we feel is affecting the end result. We will then change the level of the variable, such as temperature of a plating bath and measure the end result, plating thickness. It is recommended that 50-100 pairs of data be collected. These do not need to be 50-100 different levels of our variable (plating temperature). We can run multiple trials at the same variable level to determine consistency of the end result.

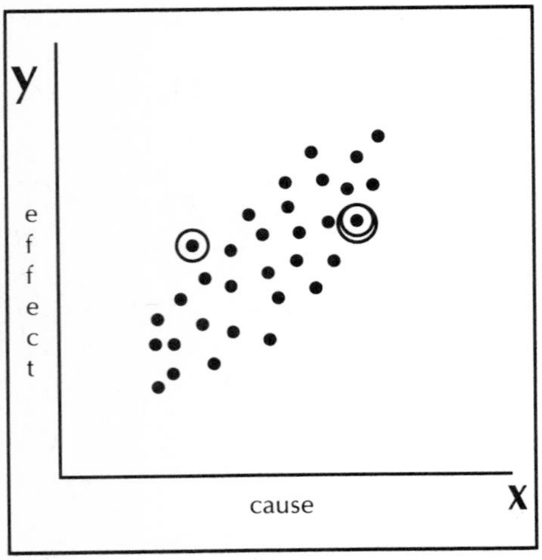

Once we have collected our pairs of data, we can then draw the scatter diagram shown in Figure 2.5.1. The vertical axis of the diagram is the end result (effect) of our test. The horizontal axis is the variable (cause) we were changing (see how this relates back to our cause and effect diagram). The lengths of the two axes should be about the same so the diagram is easier to interpret.

Now each of the data pairs can be plotted on the diagram. If two or three points fall at the same place on the

Fig. 2.5.1 Scatter diagram.

Data Collection and Analysis Methods

diagram, circles can be drawn around the original point to indicate multiple results at this location.

Another method, called a correlation table, can be used to diagram the data, especially if a large amount of data was collected or many of the same data values are obtained (see Table 2.5.1). The correlation table not only gives a graphic representation of the relationship between the variable and the result, but also lists the frequency of occurrence at each test level. With small amounts of data, the correlation and frequency is easily determined from the scatter diagram. With large amounts of data, this becomes more difficult to interpret.

Once the data is plotted, we can analyze the scatter diagram for *correlation* (relationship) between the cause and the result of the tests. The following diagrams show the types of patterns we may observe (see Figure 2.5.2). Note the following information as you look at Figure 2.5.2:

- A positive correlation is found when an increase in the variable (cause) leads to a corresponding increase in the result (effect). A strong positive correlation means that an

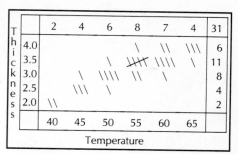

Thickness	2	4	6	8	7	4	31
4.0				\	\\	\\\	6
3.5			\	卌 \\\\	\		11
3.0		\	\\\\	\\	\		8
2.5		\\\	\				4
2.0	\\						2
	40	45	50	55	60	65	

Temperature

Table 2.5.1   Correlation table.

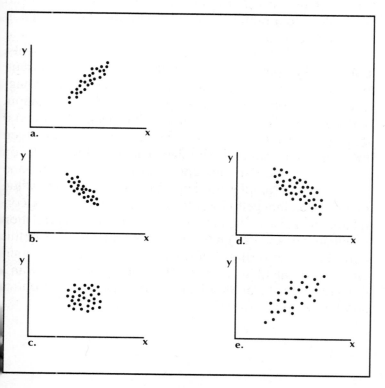

a.

b.

c.

d.

e.

Fig. 2.5.2   Scatter diagrams indicating: a. positive correlation; b. negative correlation; c. no correlation; d. weak negative correlation; e. weak positive correlation.

Data Collection and Analysis Methods

increase in the variable leads to a definite corresponding increase in the result. A weak correlation means that an increase in the variable results in some increase in the result, but it is less dramatic and easy to identify.

- A negative correlation is found when an increase in the variable leads to a decrease in the result. Again, a strong negative correlation indicates that there is definitely a decrease in the result when the variable increases. A weak correlation indicates a weaker relationship between the variable and the result.

- No correlation occurs when there is no real relationship between the variable and the result.

In the example shown in Figure 2.5.1, we see that we have a fairly strong positive correlation. This means that within our testing range, an increase in the plating temperature results in an increase in the plating thickness.

Once we have plotted the points on the scatter diagram and we have visually examined it for correlation, we can also calculate a numeric value to describe the correlation between our items, called a *correlation coefficient (r)*. There are different methods of calculating correlation coefficients, but all of the commonly used methods will result in a value between -1 and $+1$. A value of -1 shows a perfect negative correlation, a $+1$ shows a perfect positive correlation, and a zero shows no correlation. Values falling between 0 and $+1$ or 0 and -1 show a weaker correlation between the items being studied. The significance of the correlation coefficient value depends upon the size of the value and the number of samples used in the calculation. As the value of the coefficient gets closer to ± 1, the strength of the correlation increases. The actual calculation of the correlation coefficient (r) and its significance is beyond the scope of this handbook. For further information on correlation coefficients and calculation of standard error, refer to *Juran's Quality Control Handbook* and Duncan's *Quality Control and Industrial Statistics*.

2.6 RECORDING DATA

Three common ways to record data for process control are:

- Check sheets
- Handheld data collectors
- Fixed station data acquisition equipment

In a factory-wide process control program all three ways would probably be used because each offers certain advantages the others do not have and they are all somewhat complementary.

Check Sheets

Check sheets are the easiest to make and most flexible way to record data. A well designed check sheet enhances all aspects of a study. First, its design allows an auditor to efficiently gather data. Essential information such as the study identification, date, shift, and auditor's name is put on the sheet. If a route is required to collect the data, a description of it is included. Each data collection station is labeled by the type of measurement, number of significant digits, the gaging to be used, frequency of measurement, and possibly the high and low specifications. There must be room on the sheet to record all data and include notes about conditions when necessary. A parts drawing with points of measurement clearly marked can also enhance the check sheet as an effective data recording tool.

Second, a check sheet should provide some method of on-the-spot analysis. The auditor should be able to verify that the data is reasonable while it is being collected to help minimize measurement error. This is not to say that the auditor should make snap judgments about whether the data indicates a problem. But in the case of critical measurements, where parts out of specification cannot be tolerated, the auditor should be able to recognize out of specification data and mark the parts or proceed according to his responsibilities.

The third aspect of a study the check sheet should enhance is the analysis. The sheet should be designed to minimize transcription errors. Such errors occur more often when copying or reading data along a horizontal line than down a vertical column, so columns should be used to record groups of like data. Some sheets, especially those used to mark frequency of occurrence, invite analysis right on the sheet by taking on the appearance of a histogram. If hand calculations will be performed after the data is collected, provide the space and parameters for it on the back of the check sheet. This way the sheet becomes a complete record of the study. If the data needs to be transferred to a computer, then the sheet should accommodate easy transcription. In some situations, where the deviation from nominal is being analyzed or where only the last two digits are significant, the check sheet could be used to record only the deviation number or last two significant digits. The computer can supply the rest.

Check sheets have many usages, including providing information for setting up a complicated piece of machinery, testing all functions of a product, and verifying that all the correct pieces are in a shipping container. The next two sections discuss different methods available for designing variables and attributes check sheets.

Check Sheets for Variables

Variables data can be recorded in two ways. One way is in columns arranged by the route and type of measurement. See Figure 2.6.1. This provides for either manual calculations or transcribing for processing by computer. Another way is to use one check sheet for each type of measurement and label the columns by gradations of the measurement scale as shown in Figure 2.6.2. This allows you to simply mark the sheet where each measurement falls. The result will be a histogram. One drawback of a histogram-type of check sheet is that the data gets categorized and its resolution is limited by the number of columns on the sheet.

Check Sheets for Attributes

There are three basic types of check sheets for attributes data (*Ishikawa, 1976*). The first is *defect-by-item*. See Figure 2.6.3. The types of defects are listed on the left side and checks are made next to them for each type of defect found. The results are then tabulated on the right. This type of sheet translates very easily into a Pareto diagram.

The second type is *defect-by-location*. This sheet is basi-

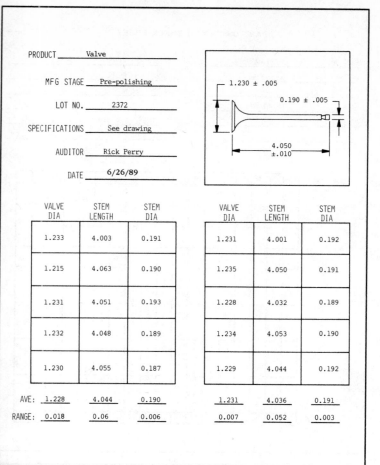

PRODUCT ____Valve____

MFG STAGE ____Pre-polishing____

LOT NO. ____2372____

SPECIFICATIONS ____See drawing____

AUDITOR ____Rick Perry____

DATE ____6/26/89____

1.230 ± .005

0.190 ± .005

4.050
±.010

VALVE DIA	STEM LENGTH	STEM DIA	VALVE DIA	STEM LENGTH	STEM DIA
1.233	4.003	0.191	1.231	4.001	0.192
1.215	4.063	0.190	1.235	4.050	0.191
1.231	4.051	0.193	1.228	4.032	0.189
1.232	4.048	0.189	1.234	4.053	0.190
1.230	4.055	0.187	1.229	4.044	0.192

AVE: 1.228   4.044   0.190   1.231   4.036   0.191

RANGE: 0.018   0.06   0.006   0.007   0.052   0.003

Fig. 2.6.1   Check sheet for variables.

cally a parts drawing with a list of codes for different types of defects. See Figure 2.6.4. The inspector then marks the location on the drawing with the code for the defect. This type of sheet allows quick analysis of problems on large parts or assemblies where the location of a defect provides the key to the cause of a problem. It may be helpful to superimpose a grid on the parts drawing to make the location clearer.

The third type of check sheet is the *defect-by-cause* sheet. See Figure 2.6.5. Defects are recorded against a grid that highlights other variables such as time of day, machine number, and worker. The frequency of defects attributable to these variables can then be checked during data collection. By design, this sheet is very similar to some control

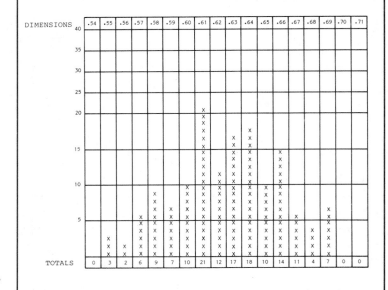

PART DIMENSION CHECK SHEET FILE NO. ___122___

PART NO. OEX250	SPECIFICATION 1.62 ± .05
DESCRIPTION Bearing race	NO. INSPECTED 151
AUDITOR Amy Noyes	NO. BELOW SPEC 5
DATE 6/23/89	NO. ABOVE SPEC 11
ROUTE 101	NOTES Some of yesterday's batch mixed in with today's.

DIMENSIONS

	.54	.55	.56	.57	.58	.59	.60	.61	.62	.63	.64	.65	.66	.67	.68	.69	.70	.71
TOTALS	0	3	2	6	9	7	10	21	12	17	18	10	14	11	4	7	0	0

Fig. 2.6.2   Check sheet for variables in the form of a histogram.

Fig. 2.6.3   Defect-by-item check sheet.

DEFECT TALLY SHEET

PART NO. RD400ES	DATE 5/31/89
	STATION Crating
DESCRIPTION 1/4 H. Pump Assy	INSPECTOR JP
	BIN NO. 3042
REMARKS	STARTING SEQ NO. 4076508
	ENDING SEQ NO. 4076590

DEFECT	NUMBER	TOTAL BY DEFECT
Piece Missing	//// //// //// //	17
Loose Fastener	//// //// //// //// //// //// ///	33
Foiled Power On	//// //// //// //// //// //	27
Scratch	//// //// //// //// //// //// //// //	37
Other	//// //// //// ///	18
	GRAND TOTAL	132
PARTS REJECTED	//// //// //// //// //// //// //// //// //// //// //// //// //// //// //// //// //	82

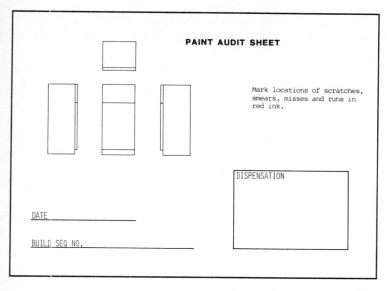

Fig. 2.6.4 Defect-by-location check sheet.

MACHINE	OPER-ATOR	MONDAY SHFT 1	2	TUESDAY 1	2	WEDNESDAY 1	2	THURSDAY 1	2	FRIDAY 1	2	SATURDAY 1	2
FORGE 1	Alex	AA B	CCE AA	CA AA	BA CC	AAAA BB	CCC A	BBA C	CCA BB	CCCC BB	CCCA	AACC CCBB	AAA ACB
FORGE 1	Harv	AA BD	AAB CEE	BBB CCAA	AA	BBD D	AAA	CCCB BCA	BBB AA	DDD BE	DDE	AAB B	BBB A
FORGE 2	Bert	AB	AAA	BBD	E	AAA	CCCA	BBAA AA	AAA BBB	AAAA B	BBAB C	CBAD	CBAD DE
FORGE 2	Ted	AA E	EED	EE DD	BBA	ACAA	AABB	BBAA EE	EEE	DDDE EEE	AAA EEA	AAA BBB	AA

DATE: 2/21/83
CODES
DEFECT ANALYSIS SHEET
INSPECTOR: G.S.
A = TAILINGS   B = MISFORMED   C = OVERSIZE   D = UNDERSIZE   E = BROKEN

Fig. 2.6.5 Defect-by-cause check sheet.

charts. In fact, this sheet could be used for ongoing process control as well as for temporary problem solving.

Handheld Data Collectors

Handheld data collectors have many similarities to check sheets. Because they offer advantages in recording data in computer readable form, and in capturing data directly from electronic gages, they economize with frequent usage. It will help to compare handheld data collectors, such as the

See p. 20-6 for the 2000 series data collector.

DataMyte® 2000 series data collectors, with the qualities of a check sheet to determine its suitability for various data recording tasks.

First are the requirements of data organization, such as identification, date, name of operator and notes about the study. Routing information may be needed to guide the operator to data collection points and indicate the type of measurement. Portions of the data collector's memory are allocated to this type of information. Another block for memory is needed to store interactive prompts for route, gage, and type of measurement. An example of study organization and prompt information is shown in Figure 2.6.6. An operator begins data collection by entering the appropriate study information, following the route, and recording the data. To facilitate data recording, electronic gaging can be used. Otherwise, manual entry of the data is also possible.

Another feature handheld data collectors share with check sheets is on-the-spot analysis. A data collector such as the DataMyte 2003 displays the mean, range, minimum sample, maximum sample, standard deviation of sample group and many other statistical calculations. It also does limit checking, alerting the operator with an audible beep when the data just taken is out of specification.

The third aspect of check sheets, that of enhancing data analysis, is also provided by handheld data collectors. Formatted reports with summary information can be obtained by connecting the data collector to a printer. Charts can also

Fig. 2.6.6   Data collector memory has header and prompt sections.

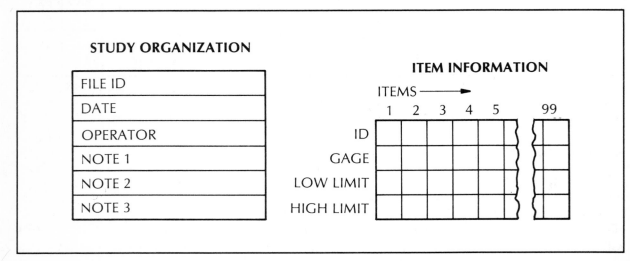

be displayed on the built-in screen or video monitor for real-time feedback at the process. The data can then be transmitted to a computer for archiving and further analysis.

Collecting Variables Data

Handheld data collectors are well suited to collecting variables data. A *matrix* or *file* is used as shown in Figure 2.6.7. The file consists of data cells arranged in columns and rows. One column is used for each type of measurement or *item,* as it is called. The repeated measurements of each item are called *samples.* The width of a data cell is expressed in terms of the number of significant digits sent by the gage or manually entered.

Data can be recorded in a file by moving horizontally or vertically. Horizontal movement allows one to take samples of each dimension of a part before proceeding to the next part. Vertical movement allows one to take several samples of one item and then proceed to the next. An intelligent data collector or a computer program can scale each item and produce a series of histograms or other types of graphs.

Collecting Attributes Data

Attributes data is recorded in a data collector, such as the DataMyte® 769 data collector, by coding it. The coding systems used can be bar codes for wanding information or alphabetical codes for manual entry. Figure 2.6.8 shows an example of a bar code sheet. Required identifying information, such as operator, part number, and batch number can be recorded as well as the defects. We can then simply wand over the bar code for each item as it is required using the bar codes produced by the system. The information can then be sorted by the data collector and pareto or control charts displayed.

The three types of check sheets can be simulated in the data collector. Separate code sections, called fields, identify the defects, locations, machine, worker, and any other information of interest. Each completed data entry is stamped with the time and date of collection by the built-in 24 hour clock. Bar codes allow fast data entry and reduce dependence on the operator's command of the codes and ability to key them correctly. Once collected, the data is then transmitted to a computer for reporting and archiving.

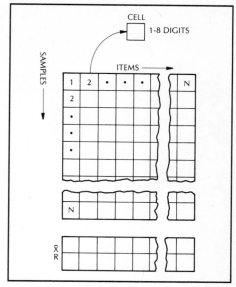

Fig. 2.6.7 *Data is arranged in a matrix-like memory.*

See p. 20-4 for the 769 data collector.

DATA COLLECTION SHEETS FOR 4-AREA CIRCUIT BOARD INSPECTION
DATE -- 01/05/88, 14:46

SERIAL NUMBER

PART NUMBER

848   925   501

BOARD AREA

1   2   3

4

ATTRIBUTES

VOID   PEAK   BALL

OTHER

ENTER

Fig. 2.6.8 Bar codes used for fast entry of attribute data.

Fixed Station Data Acquisition Equipment

Fixed station equipment has several levels of sophistication. At the lowest level, it is dedicated to recording data from a single source. The data analysis is rudimentary and not compatible with data collected from other sources. In some cases it requires an operator to oversee data collection and either record it or transmit it to a computer for analysis.

Some types of fixed station equipment have outputs that are compatible with handheld data collectors so that a periodic linkup with the data collector could be a regular route item for the auditor. An example of this is a weigh scale and an electronic linear gage mounted on a fixture.

When volumes are high enough to require frequent data collection, the process operator could use a system dedicated to collecting data for SPC. Dedicated systems are usually more efficient than check sheets, and allow more frequent recording than what would be obtained with a roving auditor using a handheld data collector. A system such as shown in Figure 2.6.9 allows the operator to record data and also do real-time analysis. In essence, it is both a data collection system and a quality control computer. Summary information, control charts and other graphs are displayed on a monitor, giving the operator a means of identifying a problem as soon as it occurs. Such equipment has a variety of applications, including statistical process control of dimensional characteristics (Figure 2.6.9) and packaging weights (Figure 2.6.10).

The more sophisticated types of fixed station equipment are beyond the scope of this book. In many cases they are a part of test stations or bays. Operation is automatic or semi-automatic and a computer is either resident at the station or directly linked. Their singleness of purpose makes them ideal for high volume tasks critical to process control. In some cases system controllers provide some degree of closed loop data analysis and adjustment. If not, the data available from fixed station sources can be analyzed with the methods of statistical quality control described in this book and others.

Fig. 2.6.9 Fixed station system for dimensional measurements.

Comparing Data Recording Methods

Check sheets, handheld and fixed station data collection systems each have distinct advantages for recording data. Most likely a combination of all three would be used in a factory to suit the types of processes and volume of data to be collected.

Although they are simple to construct and easy to use, check sheets become very inefficient when used to gather large amounts of data (Figure 2.6.11). A 1978 study commissioned by the U.S. government found that the handwritten method of data collection can gather information at a rate of 3.3 times per minute, with an error rate of 11.6 percent. When we add together the time to set up a sheet, record the data, and the time to extract data from a sheet for analysis, we find that the check sheet is very labor intensive. Although paper data sheets and control charts provide good permanent records, they are not the most efficient vehicle for data. The data must be keyed into a calculator to figure averages and other statistics, or keypunched into a computer.

Handwritten methods also contribute significant human error to data. Reading errors occur when a person makes a

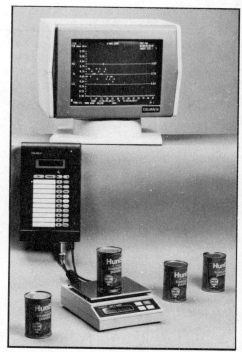
Fig. 2.6.10 Fixed station system for weight measurements.

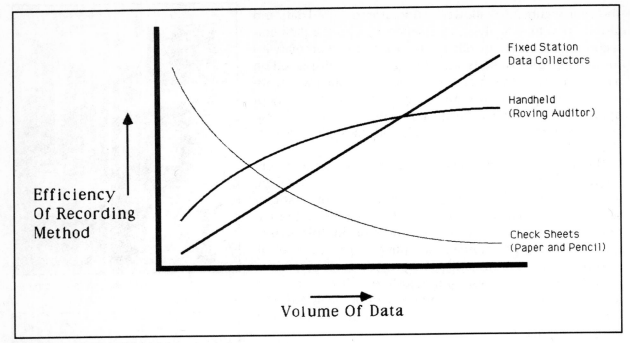

Efficiency
Of Recording
Method

Volume Of Data

Fixed Station
Data Collectors

Handheld
(Roving Auditor)

Check Sheets
(Paper and Pencil)

Fig. 2.6.11   As the volume of data grows, the efficiency of each of the methods shown in the graph changes. At higher volumes some combination of all three methods are needed.

visual observation and wrongly interprets what he sees. Transcription errors occur when a person makes a correct observation but writes it down mistakenly. Keypunching errors occur when a person must enter data from a check sheet into a computer or calculator. Significant error can lead to the false interpretation of data, which nullifies much of the benefit of collecting it in the first place.

It would be better to capture data in a computer readable format right at the source, especially in environments requiring collection of moderate to large amounts of data. This is where handheld and fixed station data collectors are more efficient (Figure 2.6.11). Handheld data collectors can be used by a roving auditor. The auditor is responsible for recording data at several points around a plant. By using electronic gages, data is recorded with greater speed and less error.

A factory having many high volume processes may want to have each process operator record data rather than have a roving auditor record data. A fixed station data collection system provides each operator a way to continually measure the variability of a process. Whenever necessary, adjustments can be made before quality problems occur. The operators thus gain a more effective influence.

For Further Reference

Burke, James. *Connections.* Boston: Little, Brown and Company, 1978.

Duncan, Acheson J. *Quality Control and Industrial Statistics,* 5th ed. Homewood: Richard D. Irwin, Inc., 1986

Grant, Eugene L., and Leavenworth, Richard S. *Statistical Quality Control,* 5th ed. New York: McGraw-Hill Book Company, 1980.

Green, Constance M. *Eli Whitney and the Birth of American Technology.* Boston: Little, Brown and Company, 1956.

Ishikawa, Kaoru. *Guide to Quality Control.* Hong Kong: Nordica International Limited, 1976.

Juran, Joseph M. and Gryna, Frank M. *Juran's Quality Control Handbook,* 4th ed. New York: McGraw-Hill Book Company, 1988.

Quality Control Circles, Inc. *Quality Control Circles.* 2d ed. Saratoga: Quality Control Circles, Inc., 1982.

Rowntree, Derek. *Statistics without Tears.* New York: Charles Scribner's Sons, 1981.

Shewhart, Walter A. *Economic Control of Quality of Manufactured Product.* Princeton: Van Nostrand Reinhold Company, Inc., 1931.

3. CONTROL CHARTS

3.1   INTRODUCTION

The problem identification and analysis tools described in Chapter 2 provide a means to prioritize and analyze individual problems. The result of this analysis will suggest specific problems or processes that require further work using the next step in our SPC model: *Use SPC to bring the key variables into control.* These selected variables or processes may include items such as:

- Outer diameter of a bushing
- Fill weight of a soda bottle
- Torque of a headbolt on an engine
- Hardness of a transmission gear
- Solder defects on a printed circuit board

In each of these cases, the next objective is to reduce variation for the characteristic in question without making significant changes to the production process. The discussion of variation in Chapter 1 identified two types of variation — common causes and assignable causes. Common cause variation is inherent to any process that evidences stability over time. Assignable cause variation results from significant and identifiable changes in the process such as a new lot of raw material entering the process, a new operator at the process, or a change in tooling. It is easy to see that reducing variation through eliminating assignable causes is more practical than reducing variation through eliminating common causes. Process control charts are the best tool available for identifying and eliminating assignable cause variation.

Control charts provide a graphic comparison of a measured characteristic against computed control limits. They plot variation over time, and help us distinguish between the two causes of variation through the use of control limits. These limits are vital guidelines for determining when action should be taken in a process.

Figure 3.1.1 shows a portion of a typical $\bar{x}$ & R chart. The important elements to note are:

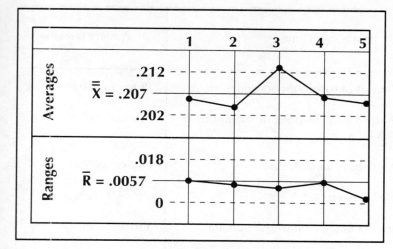

Fig. 3.1.1   A portion of a typical x̄ and R chart.

1) Each of the points on the x̄ portion of the chart represents an average of one subgroup's readings. Because
   each point is an average value, all of the points taken together will tend to be normally distributed if they are
   from a process showing stability (refer to the Central
   Limits Theorem in Section 1.3).

2) The dashed line control limits are drawn at approximately plus and minus 3 standard deviations from the
   central line on the x̄ portion of the chart. Because points
   on the chart from a process showing stability tend to be
   normally distributed, we now expect 99.73% of all the
   points to fall between the two control limits (refer to the
   discussion on standard deviation in Section 1.3).

3) Point #3 on Figure 3.1.1 is beyond the upper control
   limit for x̄. There are four possible explanations for this
   occurrence: a measuring error, a plotting error, an assignable cause, or the possible exception to the probability that 99.73% of all common causes will fall within
   control limits.

4) Because each x̄ value on the chart represents an average
   of individual readings in the subgroup, we must also examine the range chart to determine the variability of the
   individual readings in that subgroup.

Principal Kinds of Control Charts

There are essentially two kinds of control charts — con-

Type Chart	Parameters Plotted	Primary Usages
x̄ & R chart	Averages and ranges of subgroups of variable data.	Process control
x̄ & sigma chart	Averages and standard deviations of subgroups of variable data.	Process control
Median chart	Median of subgroups of variable data.	Process control
Chart for individuals	Individual measurements.	Process control
CuSum chart	Cumulative sum of each x̄ minus the nominal.	Process control
Moving range, Moving average chart	Range or average recalculated for preceeding 3 days.	Process control
p-chart	Ratio of defective items to total number inspected.	Inspection sampling Final inspection
np-chart	Actual number of defective items compared to total inspected.	Inspection sampling Final inspection
c-chart	Number of defects on an item for a constant sample size.	Inspection sampling Final inspection
u-chart	Percent nonconformities on item for varying sample size.	Inspection sampling Final inspection

Table 3.1.1   Principal kinds of control charts.

trol charts for variables data (quantitative data or measurements), and control charts for attributes data (qualitative data or counts). See Table 3.1.1. Variables control charts are more sensitive to changes in measured values and therefore are better for process control. Control charts for attributes data are useful for other reasons; attributes data charts are the easiest to obtain data for, cost the least to use, and often do not require a specialized means of data collection.

x̄ & R Chart — The x̄ & R chart is the most common form of a control chart for variables data, and one of the most powerful for tracking and identifying causes of variation. The x̄ chart is a continuous plot of subgroup averages. The R chart is a continuous plot of subgroup ranges. A subgroup can be from 2 to 20 samples.

x̄ & s Chart — Known as the x̄ and sigma chart, the parts of this chart, like the x̄ & R chart, are always used as a pair. The sigma chart is a somewhat more accurate indicator of process variability than an R chart, especially with a larger size subgroup. Disadvantages of the sigma chart are that it is more difficult to calculate sigma and users often have a more difficult time understanding and interpreting the sigma values.

Median Chart — This chart combines both the x̄ & R information into one graph. The median is the middle value when data is arranged according to size. A median chart yields similar conclusions to the x̄ & R chart, and is easy to use. Typically, median charts are used with subgroup sample sizes of 10 or less.

Control Chart for Individuals — A control chart for individual samples is used in cases where it is necessary for process control to be based on individual readings rather than subgroups, such as when measurements are expensive or destructive or the result of a single daily lab test. Control limits for a chart of individuals should be based on the moving range. Chapter 4 describes control charts for individuals in greater detail.

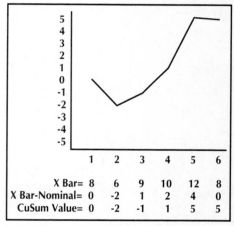

	1	2	3	4	5	6
X Bar=	8	6	9	10	12	8
X Bar-Nominal=	0	-2	1	2	4	0
CuSum Value=	0	-2	-1	1	5	5

Fig. 3.1.2   A typical CuSum chart.

Cumulative Sum (CuSum) Chart — A CuSum chart looks at subgroup averages in a different way than the x̄ & R chart. On a CuSum chart, each point represents the cumulative sum of each x̄ value minus the nominal value. Figure 3.1.2 shows a CuSum chart. In this example the nominal value is 8. The first x̄ value is also 8. Since $8 - 8 = 0$ the first CuSum value is 0. The second x̄ value is 6. To determine the second CuSum value, the nominal value (8) is subtracted from the x̄ value (6) and added to the previous CuSum value (0): $6 - 8 = -2$; $-2 + 0 = -2$. The third x̄ value is 9, The third point on the chart is determined by: $9 - 8 = 1$; $1 + -2 = -1$.

Basically, the CuSum chart exaggerates the shift from nominal. If successive x̄s fall on both sides of the nominal, the chart remains fairly flat. But if two or more successive x̄s fall on the same side of the nominal, the curve begins to rise and fall quite rapidly. The chart, therefore, is much more sensitive to a sustained shift away from the nominal.

Moving average, moving range chart — This chart is used

primarily in industries where the output of a process is very stable from piece to piece, but may vary from day to day or week to week, such as a metal stamping application where the dies wear slowly, or a plating application where the same chemical bath is used each day. In this example a three day moving average chart would plot a point for Friday by averaging Friday's reading with readings from Thursday and Wednesday. Similarly, a moving range chart would plot the range over a three day period each day. These charts tend to dampen what would be a single out of control reading on the x̄ & R chart, and distribute effects over a longer period in time. Moving average and range charts are described in greater detail in Chapter 4.

p-Chart — The p-chart (percent defective chart) is an attributes chart for the percentage of defective items in a subgroup, when the subgroup is not necessarily of a constant size from inspection to inspection. *Fraction defective* is the ratio of defective items to the total number of items inspected, which is another way of expressing the percentage.

np-Chart — The concept of the np-chart is the same as the p-chart except that the np-chart represents the actual number of defective items in the subgroup, rather than the fraction. The np-chart requires a constant subgroup size. It would be better to use the np-chart if the actual number of defectives is more meaningful or simpler to report than the proportion value that is presented by the p-chart.

c-Chart — The c-chart is a special type of attributes control chart which uses the number of defects instead of the number of defectives. In cases where a unit can contain many defects, the c-chart is a practical alternative. Each point on the chart represents the number of defects in the subgroup. The c-chart should be used only when the sample size making up the subgroup remains constant from inspection to inspection. The c-chart is particularly useful where a unit is likely to contain many defects.

u-Chart — The u-chart is similar to the c-chart in that it tracks individual defects. The u-chart requires constant subgroup sizes like the np-chart. A single point on a u-chart represents the average number of defects on each part in a subgroup. For example, if the subgroup size equals 10 and there are five defects in that subgroup, the u value would be 0.5. This signifies an average of 0.5 defects on each part in that subgroup.

3.2 $\bar{x}$ & R CHART

This section describes how to make and analyze an $\bar{x}$ & R chart, also known as a Shewhart Control Chart. The $\bar{x}$ & R chart is the most versatile of control charts for variables. There may be specific situations where sigma charts, median charts and charts of individuals have some advantages over $\bar{x}$ & R charts, but in most applications, an $\bar{x}$ & R chart will do as well or better. Since all are similar in methods of usage and analysis, only the $\bar{x}$ & R chart will be treated in depth.

Rationale for $\bar{x}$ & R Chart

A control chart is used to establish the *operating level* and *variation* of a process. Since the parts coming off a process may be infinite in number, we need a way to establish and monitor this operating level and variation without having to measure every part. The $\bar{x}$ & R chart is extremely efficient at this. It also provides a way to avoid the two types of errors that occur when attempting to control a process:

- *Type I error* — saying a process is unstable when actually it is stable.
- *Type II error* — saying a process is stable when actually it is unstable.

Both of these errors occur when we don't have a method of identifying assignable causes, when we aren't thinking statistically about the data. An example of a Type I error is an operator who adjusts the stops on a turret lathe after taking a single hourly measurement. The data might look like Figure 3.2.1. He starts out at 8:00 a.m. making a batch of 30 parts. The sideways *histogram* represents the distribution of those 30 parts measurements as compared to the nominal and upper and lower limits of the specification. The first batch shows evidence of stable variation (a *normal distri-*

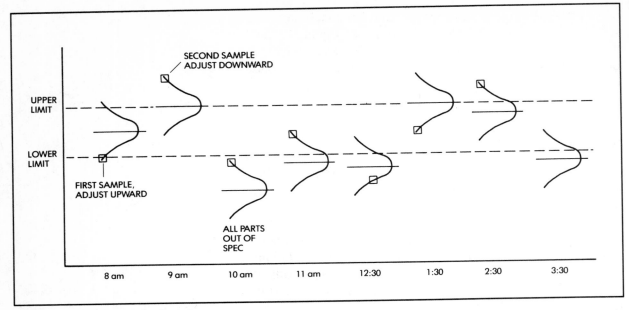

UPPER
LIMIT

SECOND SAMPLE
ADJUST DOWNWARD

LOWER
LIMIT

FIRST SAMPLE,
ADJUST UPWARD

ALL PARTS
OUT OF
SPEC

8 am   9 am   10 am   11 am   12:30   1:30   2:30   3:30

*Fig. 3.2.1   Illustration of a Type I error —
making adjustments to a process when they
are not needed. The curves represent the operating level and dispersion of the process at
various hours during the day.*

bution). The mean is very close to the nominal and nearly every piece falls within the limits. Is there a need to adjust the stops? Certainly not. But at 8:00 a.m. he picks up one part and measures it. It happens to be right at the lower limit, so he moves the stops outward. This shifts the operating level of his process. After his next batch, he takes a second sample and finds he is way above the upper limit, so he moves the stops inward enough to compensate for the shifts. Between 10:00 a.m. and 11:00 a.m., every part he makes is out of spec. In fact, for the whole day, about half of what he produces is out of spec and unusable. Even though the process is stable and capable of making a high proportion of good parts, a large proportion of bad parts are being produced because the operator is taking only one sample and basing his decisions on it. He is making a Type I error. An $\bar{x}$ & R chart could have told him when to adjust the operating level and how much. Just as importantly, it could have told him when to leave it alone.

A Type II error is also easy to illustrate, as shown in Figure 3.2.2. In this case the operating level, or *mean,* is always within specification but the distribution of data shows evidence of a lack of stability. Imagine an operator taking a measurement that happens to fall within spec and, therefore, not making any adjustments. An $\bar{x}$ and R chart could possibly have alerted the operator to an out-of-control

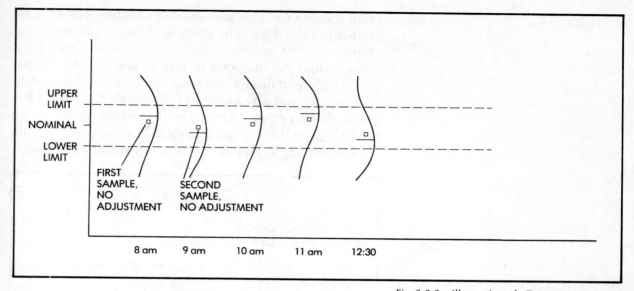

UPPER LIMIT

NOMINAL

LOWER LIMIT

FIRST SAMPLE, NO ADJUSTMENT

SECOND SAMPLE, NO ADJUSTMENT

8 am   9 am   10 am   11 am   12:30

Fig. 3.2.2   *Illustration of a Type II error — not making adjustments when they are needed. The curves show that many of the parts being made are out of specification.*

process after the first hour and steps could have been taken to correct it.

The difference between the chart shown in Figure 3.2.1 and an x̄ & R chart is that an x̄ & R chart is easier to make. Instead of calculating and graphing small histograms of data subgroups, we graph the averages and ranges on separate charts.

An x̄ & R chart can be made manually or by computer. Some of the graphs shown in the following pages were produced by an IBM PC computer. Since the main calculations are finding the average and range of a set of five or so numbers, the charts could also be done manually. The advantages of a computer lie in minimizing errors, improving the readability of the graph, and freeing the auditor up for the more "human" tasks of analysis, judgment and action.

Creating an x̄ & R Chart

To explain the elements of an x̄ & R chart, it helps to create one from scratch as if we were to do it manually. This is an illustration of how a computer would generate a chart, but it can also be done manually.

Vertical Scale — The vertical scale of an x̄ chart should have the grand average of the data at the midpoint (Figure 3.2.3). This could also be thought of as the expected operating level of the process. Extending above and below the midpoint should be evenly spaced scale divisions. The scale increments should be sufficiently wide to graph significant changes in the average. The scale should extend out enough from the midpoint to take in any expected variation. As a rule of thumb, the scale should extend at least 20 percent beyond any element that would be put on the graph, such as control limits. The R chart should be scaled similarly, with the expected range at the midpoint and the scale extending at least 40 percent beyond any element.

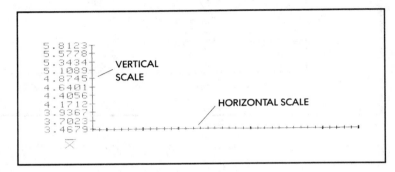

Fig. 3.2.3   Scales used on an x̄ chart.

Horizontal Scale — The horizontal scale should be the same on both charts. The scale divisions correspond to the frequency of sampling and should be labeled by the hour or date when the data was collected, starting at the left. Data should be collected frequently when starting out, perhaps every hour or even more frequently on a high volume process. Once the chart is established and the process has evidence of stability, data can be collected less frequently — every other hour or by shift.

Data Collection — An "x" is the symbol for a single measurement. To start a control chart, at least 100 measurements must be taken. We can begin plotting averages and ranges right away, but we need at least 100 measurements to add the essential elements of the chart — the grand average and the control limits. Several considerations must be made to ensure that the 100 measurements are representative of the process:

- Measurement devices must be accurate enough to record the differences between samples. The devices must have high repeatability. See Chapter 6 for a discussion of measurement systems.

- Measurements must be made in small groupings at selected intervals. Determining the groupings and intervals is called *rational subgrouping* and is explained in more detail later. Normally, subgroup size, symbolized by "n," can be from two to 20 consecutive samples, with n = 5 being the most common.

Plotting the Averages — "x̄," pronounced X-bar, is the symbol of the average of a subgroup. The data should be arranged in columns of five samples each, whether on a sheet or in a data collector (Figure 3.2.4). Each subgroup would then be averaged. The averages would be plotted on the chart in the sequence in which they were collected (Figure 3.2.5). Twenty points on the chart will represent 100 measurements in five sample subgroups.

Plotting the Ranges — "R" is the symbol for a subgroup range. The range of each subgroup in Figure 3.2.4 is plotted against the time intervals that they were collected.

Centerline — "x̄̄," pronounced X-double bar, is the average of the subgroup averages (grand average). See Figure 3.2.4. R-bar is the average of the subgroup ranges. They get plotted as horizontal dotted lines (Figure 3.2.6). They are your

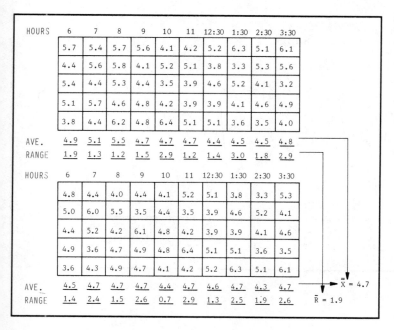

HOURS	6	7	8	9	10	11	12:30	1:30	2:30	3:30
	5.7	5.4	5.7	5.6	4.1	4.2	5.2	6.3	5.1	6.1
	4.4	5.6	5.8	4.1	5.2	5.1	3.8	3.3	5.3	5.6
	5.4	4.4	5.3	4.4	3.5	3.9	4.6	5.2	4.1	3.2
	5.1	5.7	4.6	4.8	4.2	3.9	3.9	4.1	4.6	4.9
	3.8	4.4	6.2	4.8	6.4	5.1	5.1	3.6	3.5	4.0
AVE.	4.9	5.1	5.5	4.7	4.7	4.7	4.4	4.5	4.5	4.8
RANGE	1.9	1.3	1.2	1.5	2.9	1.2	1.4	3.0	1.8	2.9

HOURS	6	7	8	9	10	11	12:30	1:30	2:30	3:30
	4.8	4.4	4.0	4.4	4.1	5.2	5.1	3.8	3.3	5.3
	5.0	6.0	5.5	3.5	4.4	3.5	3.9	4.6	5.2	4.1
	4.4	5.2	4.2	6.1	4.8	4.2	3.9	3.9	4.1	4.6
	4.9	3.6	4.7	4.9	4.8	6.4	5.1	5.1	3.6	3.5
	3.6	4.3	4.9	4.7	4.1	4.2	5.2	6.3	5.1	6.1
AVE.	4.5	4.7	4.7	4.7	4.4	4.7	4.6	4.7	4.3	4.7
RANGE	1.4	2.4	1.5	2.6	0.7	2.9	1.3	2.5	1.9	2.6

$\bar{\bar{X}}$ = 4.7

$\bar{R}$ = 1.9

Fig. 3.2.4   Sample data used to plot an x̄ & R chart.

first estimate of the operating level of the process.

Control Limits — The control limits on the chart are the estimated ± 3 *sigma limits* for the process. Since sigma must be estimated from samples taken from a continuous process, tables of constants have been developed to make these calculations simple and to reduce error (see Appendix Table A-1 — Constants for Calculating Control Limits). The formulas are:

Fig. 3.2.5   *Plotting 50 points on the chart.*

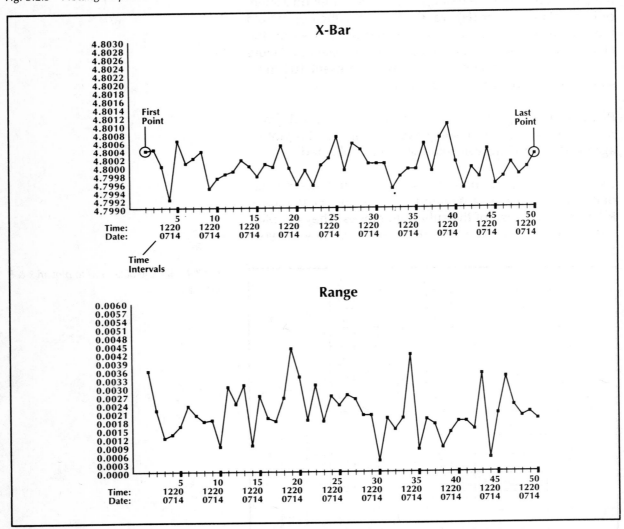

$\bar{x}$ Upper Control Limit (UCL$_{\bar{x}}$) $= \bar{\bar{x}} + A_2 \bar{R}$

$\bar{x}$ Lower Control Limit (LCL$_{\bar{x}}$) $= \bar{\bar{x}} - A_2 \bar{R}$

$$(3.2.1)$$

Range Upper Control Limit (UCL$_R$) $= D_4 \bar{R}$

Range Lower Control Limit (LCL$_R$) $= D_3 \bar{R}$

In the formulas above, A_2, D_4, and D_3 are the constants used for calculating control limits, taken from Table A-1. They vary according to subgroup size (n). For n = 5,

Fig. 3.2.6 The centerline represents the average of the subgroups and is plotted as a dotted line.

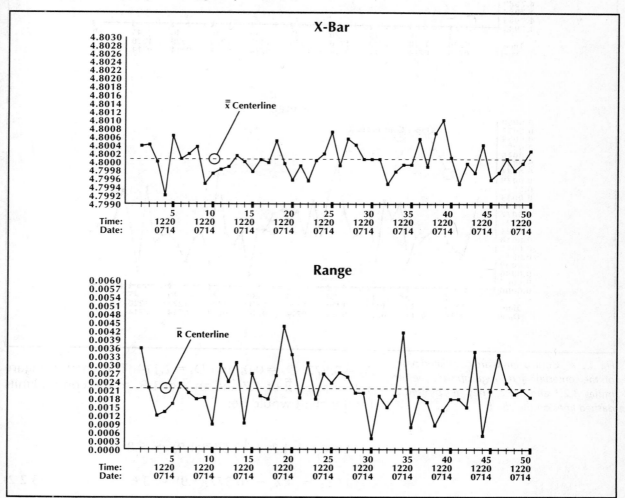

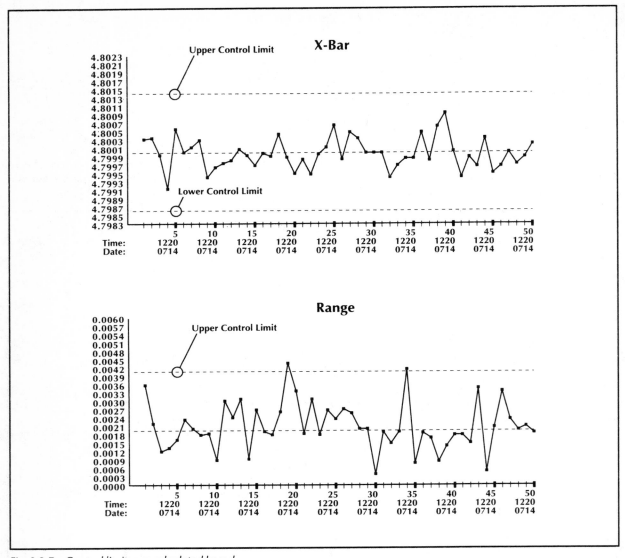

Fig. 3.2.7 Control limits are calculated based on the centerline and subgroup size. See Formulas 3.2.1 and 3.2.2. They get plotted as dashed lines on the chart.

$A_2 = 0.577$, $D_3 = 0.0$, and $D_4 = 2.114$. The chart in Figure 3.2.6 has an $\bar{\bar{x}}$ of 4.7 and an $\bar{R}$ of 1.9, so the control limits for the chart would be:

$$UCL_{\bar{x}} = 4.7 + (0.577)(1.9) = 5.8$$

$$LCL_{\bar{x}} = 4.7 - (0.577)(1.9) = 3.6 \qquad (3.2.2)$$

$$UCL_R = (2.114)(1.9) = 4.0$$

Horizontal dashed lines are used to mark the control limits on the graphs (Figure 3.2.7). *By chance alone,* the subgroup averages and ranges should fall within these control limits 99.73% of the time. The variability that occurs between the control limits can be attributed to common causes for the most part. There are certain patterns that occur between the limits that one should learn to recognize, but the most important characteristic to look for is one or more points falling outside the limits. They indicate the influences of assignable causes of variation.

Some Questions About x̄ & R Charts

How many points are needed before control limits can be calculated? — The general rule is at least 20 points, representing 100 measurements. For control limits to be meaningful they must be based on a representative sample of the population. There must be a high level of confidence that the sample is representative and that is what 100 measurements or more provide. Any less than 100 measurements quickly decreases the confidence level. Since we are basing our work effort at improving quality on the ability to detect assignable causes, we should always aim to make the control limits as statistically significant as possible.

Should engineering limits be put on an x̄ chart? — Engineering limits are a demand on a process that is arbitrary to the process itself. The points on the x̄ chart are averages, so a point at or near an engineering limit can mean many things, including the possibility that one or several data measurements are out of limits. In a pure statistical sense, engineering limits or process objectives do not belong on an x̄ & R chart. However, many industries have to live with these specifications as the criteria for a good process; including them on the chart may serve this goal.

Should the points be connected on the chart? — This is purely a matter of taste. A line connecting the points emphasizes the sequencing of points from left to right. The interpretation of a chart does not depend on this emphasis, so sometimes the sequencing can lead one to the wrong conclusions, like seeing short term "trends" that aren't really there. Lines between points are an ornamentation. If they make the chart more appealing to look at and use, then they may be beneficial.

Rational Subgrouping

To make $\bar{x}$ & R charts truly useful and easy to interpret, they have to be set up that way. This is analogous to a good family portrait. What on the surface is a simple photograph of a group of people usually has to be done by a professional photographer. Quite a bit of expertise and planning must go into portrait photography, and the same is true with rational subgrouping and $\bar{x}$ & R charts.

The main source of information used for this section on rational subgrouping is the *Statistical Quality Control Handbook,* by Western Electric Co., Inc., 1956. A rational subgroup is one where there is the least possibility of assignable causes creating differences between measurements within the subgroup itself. If a subgroup has five measurements, then the opportunities for variation among those measurements must be made deliberately small. This usually means the subgroup should be taken from a batch of pieces made when the process operated under the same settings — one operator and no tooling or material changes. Five consecutive pieces might be the easiest to collect.

The logic behind rational subgrouping is that if we can make variability between pieces within a subgroup entirely due to common causes, then the differences in subgroup averages and ranges will be due to assignable causes. The effects of assignable causes will not be buried within a subgroup and dampened by averaging. They will appear on the chart in the form of points that exceed the control limits or have an identifiable pattern. When planning the data collection, we must have an understanding of what will constitute a subgroup. If we believe that the time of day a piece is produced contributes to variation between pieces, then one subgroup should be of five consecutive pieces from the process. Several subgroups should be collected at selected times throughout the day. If we have a multiple spindle machine and believe that some of the spindles are fine and others not so fine, then the subgroup should be five pieces from one spindle and five from the next, not one piece off each spindle averaged together.

An $\bar{x}$ & R chart is used to plot one system of causes. If several different machines contribute to a single lot of parts, an x̄ & R chart of samples taken from the lot will not reveal nearly as much as separate charts on each machine. Using rational subgrouping and working upstream with an x̄ & R chart, we will find that the chart does indeed help identify assignable causes of variation.

x̄ & R Chart Patterns

Characteristics of a process showing stability — The most common feature of a process showing stability, a constant-cause system, is the absence of any recognizable pattern (*Western Electric*, 1956). Figure 3.2.8 shows such a process. The points on the chart are randomly distributed between the control limits. Since there is a slight chance of a point falling outside the control limits under normal circumstances, it can happen. A rare point out of limits on a process that has shown stability over the long run can probably be ignored. The characteristics can be summarized as follows:

- Most points are near the centerline.
- Some points are spread out and approach the limits.
- No points are beyond the control limits.

Characteristics of a lack of stability — The characteristics of a lack of stability are:

- One or more points are outside the control limits (Figure 3.2.9).
- There is a *run* of seven or more successive points above

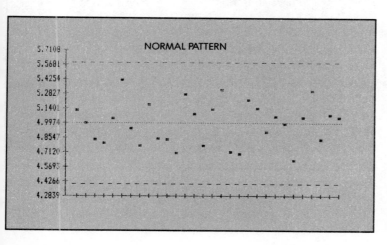

Fig. 3.2.8   *Characteristics of a normal pattern are a random distribution with most points near the centerline, some points near the control limits, but no points beyond the control limits.*

Fig. 3.2.9   A single point outside the control limits is an indication of a lack of stability. Its time of occurrence and possible causes should be investigated.

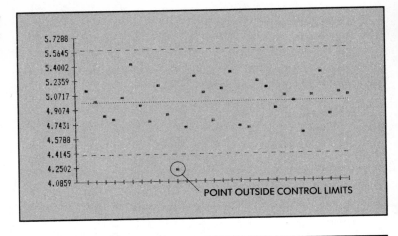

POINT OUTSIDE CONTROL LIMITS

Fig. 3.2.10   This chart has a run of eight points above the centerline, which calls for an investigation.

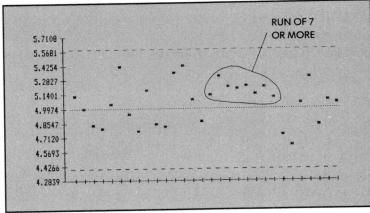

RUN OF 7 OR MORE

Fig. 3.2.11   An upward trend sometimes precedes, and forewarns of, the process going out of control. In this case the trend both precedes and follows points out of control, and is the result of over adjustment by the machine operator.

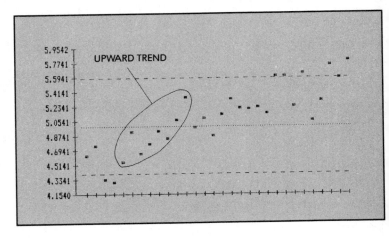

UPWARD TREND

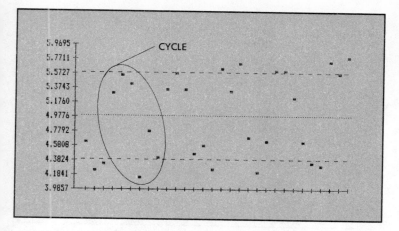

Fig. 3.2.12 A cycle is a pattern that repeats. In this case it preceded points that are out of control and provided an early indication of instability.

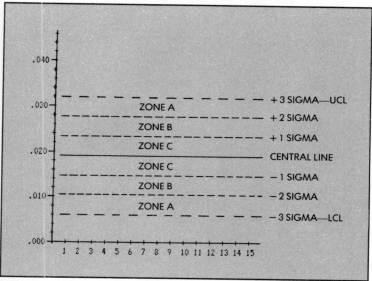

Fig. 3.2.13 Zones used when interpreting a control chart.

or below the centerline (Figure 3.2.10).

• There is a *trend,* downward or upward, of seven or more successive points (Figure 3.2.11).

• There is a *cycle* or pattern that repeats itself (Figure 3.2.12).

Since the control limits are at plus and minus three sigma from the central line, either side of the central line can be divided up into three zones at plus and minus one and two sigma (*Western Electric,* 1956). These zones (labeled C, B, and A going away from the central line), although not plotted on the chart, also are used in reading a control chart. See Figure 3.2.13.

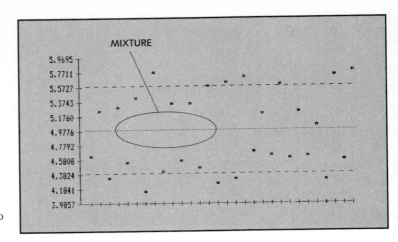

Fig. 3.2.14   Mixtures usually indicate two processes operating at different levels.

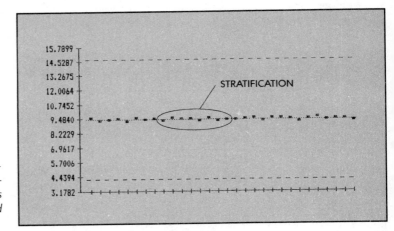

Fig. 3.2.15   Stratification may be an indication of systematic sampling, where the samples consistently offset each other or, as in this case, improperly calculated control limits and chart scales.

Fig. 3.2.16   Clusters are an indication of short duration, assignable causes such as measurement problems, or accidentally sampling from a bad group of parts.

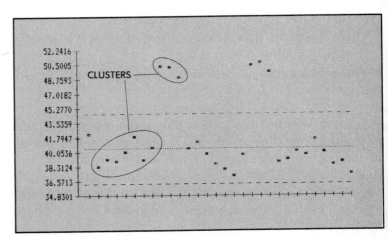

In addition, there are several patterns that may appear which are unnatural and should be investigated (*Western Electric*, 1956):

- a *mixture,* identified by an absence of points near the centerline (Figure 3.2.14)
- *stratification,* identified by 15 or more points consistently hugging the centerline (Figure 3.2.15)
- *clusters,* or the grouping of points in one area of the chart (Figure 3.2.16)

How To Identify Assignable Causes

The first technique to use when looking at an $\bar{x}$ & R chart is to read the R chart first (*Western Electric,* 1956). The R chart is more sensitive to changes in uniformity and consistency. If bad parts start appearing in a process, they will affect the R chart. The variation will increase, so some points will be higher than normal. Generally, the lower the points in the R chart, the more uniform the process. Two machines turning out the same parts can be compared for uniformity by looking at their R charts. Parts mixed together from different processes will also show up in the R chart. Intermittent variation, caused by a switch or relay that strikes occasionally, will cause the R chart to go out of control.

Anything that introduces a new system of causes into the process will show up in the R chart. Any change to the process, such as an inexperienced operator, poor materials, tool wear, or a lack of maintenance will tend to shift points upward. The biggest clues to an assignable cause on an R chart are the time the characteristic occurs and the fact that some parts are affected more than others.

Since the R chart is more sensitive to change, efforts at improving the process will show up first in the R chart. A steady shift downward in points on the R chart is the best evidence of having successfully eliminated assignable causes of variation.

Once the R chart is stable, we can focus on the $\bar{x}$ chart. When the R chart is unstable, the $\bar{x}$ chart can be very misleading. When both charts are stable, the process is said to be *in control,* and the $\bar{x}$ chart indicates the process operating level at various times. Changes in the operating level can be classed by two types: true $\bar{x}$ causes and false $\bar{x}$ causes (*Western Electric*, 1956).

- True $\bar{x}$ causes are causes which change all pieces from a process at pretty much the same rate. This can include a change in materials (thicker or thinner stock), a temperature change, machine calibration or setup, or gradual tooling wear. These things cause a change in level over time that usually can be traced to the moment they occur.

- False $\bar{x}$ causes are causes which show up because the $\bar{x}$ chart reflects changes in the R chart, and those causes are better interpreted on the R chart. They include all the causes mentioned above that create dispersion within a subgroup.

Interpreting the $\bar{x}$ Chart and the R Chart Together

The $\bar{x}$ chart and the R chart must be interpreted together as well as separately (*Western Electric*, 1956). As stated, a stable process will have points randomly distributed between the control limits on the charts. With a stable process, the $\bar{x}$ & R points should tend not to follow each other. A lack of stability will sometimes cause them to move together. For example, a process whose population is skewed in a positive direction (Figure 3.2.17) with a long tail to the high side, will cause a positive correlation between the $\bar{x}$ chart and the R chart. In other words, high $\bar{x}$ points will tend to follow high R points. A process with a negatively skewed population will cause a negative correlation between the charts as shown in Figure 3.2.18. The $\bar{x}$ points will tend to follow the R points, but in the opposite direction.

Changes In Level

A sustained change in level in either chart may call for a recalculation of the centerline and control limits. Control limits must reflect the long term operating characteristics of the process. For that reason, control limits are recalculated only when the sustained change in level appears to be a permanent change. Recalculating limits at that time renews the ability of the control limits to be used to detect assignable causes. It also recognizes a change to the process which, for better or worse, is more or less permanent. A chart can be thought of as a moving window. It should re-

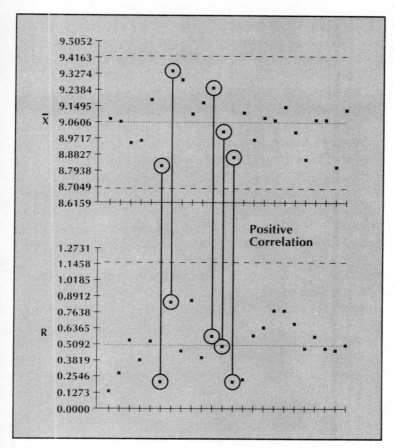

Positive Correlation

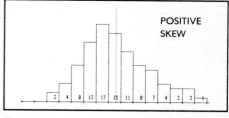

Fig. 3.2.17 The positively skewed distribution shown above can cause positive correlation on the charts, where points tend to follow each other up or down.

flect the actual present conditions as much as possible, since the present is when action based on the chart must take place. Keeping the centerline and control limits constant causes them to eventually become as arbitrary as engineering limits. We must remember, however, that the control limits should be based on at least 20 points. Any sustained change in level should exist for at least 20 points.

Beyond the x̄ & R Chart

An x̄ & R chart is probably the most effective tool for reducing variability. Throughout this section, the emphasis has been on detecting and working to eliminate assignable causes of variation. One might ask, "But what about the common causes?" Common causes are much more difficult to address on an everyday level. Often they are part of the whole manufacturing environment, and addressing them more often than not means addressing decisions on mate-

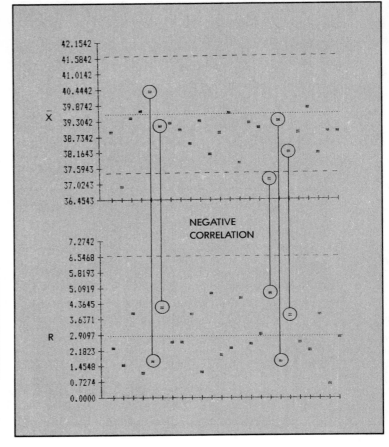

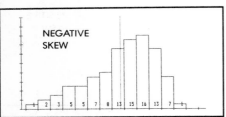

Fig. 3.2.18 A negatively skewed distribution can cause negative correlation on the charts, where the points tend to move in opposite directions.

rials and equipment purchasing, material handling, shop organization, product and process design, employee education, and company orientation. Obviously, eliminating common causes is beyond the scope of this handbook. The task of assessing a process is a quality function, however. It is known generally as *process capability studies,* and is covered in Chapter 5. *Acceptance sampling* is another aspect of quality control that deals with common causes, and it is covered in Chapter 7.

3.3 CONTROL CHARTS FOR ATTRIBUTES

Attributes data on a control chart is simply the count of products or characteristics of a product that do not conform to some established criteria. The chart becomes a line graph showing the variation in the quality of the process over a period of time. Control limits are drawn on the control chart to aid in analyzing the variation of the process. When the variation is shown by the control chart to be due to common causes, the process is said to be *in a state of statistical control*. If the data on the chart shows abnormal points or patterns, then the process has changed due to assignable causes.

Types of Attributes Control Charts and Data

Attributes data is a count of nonconforming units (defective parts), or the number of nonconformities on a unit (number of defects per part). Attributes data does not require actual measurements such as length, width or torque. It is only necessary to count the number of defects or defective units. Many characteristics of quality can only be measured in this way, such as the presence or absence of a required screw or number of bad solder connections on a printed circuit board. Other attributes data may be the result of data that is measurable but recorded as pass or fail, such as that from a GO/NO-GO gage. Control charts for attribute data have a number of advantages:

- Attributes data is easier to collect and, as a result, less costly to acquire. Inspection skills are not complicated and gages, if used, are simple GO/NO-GO gages. In some cases the data is already available from past inspection records. Large amounts of attributes data can frequently be collected at one inspection station.

- Attributes data can be collected from any type of a process. Output from any process can be qualified as conforming or nonconforming.

- Several types of defects can be grouped on one chart. For complex assemblies it would be very impractical to require a separate control chart for each measured characteristic. Attributes charts in this case can indicate problem areas and suggest where more detailed variables control charts may be needed.

- Attributes data is easy to understand by all personnel.

Control charts for attributes are easier to construct and understand.

- Attributes control charts provide an overall picture of the quality of a process and provide useful quality history.

There are also some disadvantages:

- An attributes chart does not always provide detailed data for the analysis of individual characteristics. For attributes charts, a part is defective if it has one defective characteristic or many defective characteristics.

- Attributes data does not indicate different degrees of defectiveness. A nonconforming item may be very defective or slightly defective.

- Because of the above disadvantages of attributes data, control charts for attributes data are less sensitive in indicating changes in the process. The charts for attributes also only indicate when a change in the process occurred and offer little information as to why the change occurred.

Control chart types are defined by the type of data being charted. There are four types of control charts for attributes. See Table 3.3.1. Control charts that are for the count of nonconforming units are called *p-charts* and *np-charts*. A unit is nonconforming if it has one or more defects. If the data is collected in subgroups of constant size then the chart used is an np-chart. If the data is in subgroups of varying size then the p-chart is used. Control charts that count individual

Table 3.3.1 Summary of types of attribute control charts.

TYPE	WHAT IS COUNTED	SAMPLE SIZE
p-chart	defective items	varies
np-chart	defective items	constant
c-chart	defects on an item	constant
u-chart	defects on an item	varies

nonconformities or defects on a product are *c-charts* and *u-charts*. The c-chart involves subgroups of constant size and a u-chart has subgroups that are not of constant size.

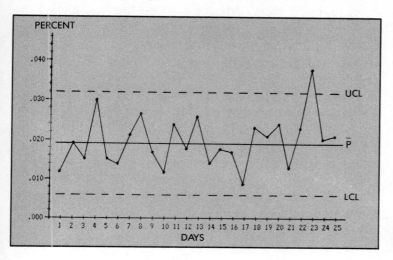

Fig. 3.3.1 *Attributes control chart, with centerline and control limits.*

Elements of a Control Chart for Attributes

A control chart is a two-dimensional line graph (Figure 3.3.1). The plotted points represent the given measure of quality of the process from the data collected at predetermined periods of time. The horizontal axis is divided into rational groups (also called subgroups) — hour to hour, day to day, lot to lot, etc. The vertical scale represents the quantity or percentage of defects per unit or defective units. Three control lines are also drawn horizontally on the control chart — a central line, upper control limit line and lower control limit line. The central line is the average of the number of defects or defective units in the process for the total period of the process being charted. The data used to calculate control limits is also used to calculate the central line.

The control limits are the key to control charts. The control limits are the criteria for analyzing a process for statistical control. Control charts use standard deviation as a measure of variation. The upper control limit is drawn at the central line plus three sigma. The lower control limit is drawn at the central line minus three sigma. The choice of three sigma is somewhat arbitrary but has become fairly standard in the U.S.

The calculation of the three-sigma limits on control charts for attributes is different from that of control charts for variables. For variables it is based on the *normal distribution*. However, for p-charts and np-charts it is based on the *binomial distribution*. For c-charts and u-charts it is based on *Poisson distribution*. When based on the binomial distribution, it is assumed that the possibility of a unit being defective is the same from unit to unit and independent from unit to unit. Each unit has the same chance of being good or bad just as in flipping a coin. When based on the Poisson distribution it is assumed that the possibility for defects occurring on a unit is great but the chances of getting a defect at any one spot is small. Although these distributions differ from the normal distribution, the probabilities of a point falling outside the three sigma control limits on a control chart for attributes is on the same order of magnitude as on a control chart for variables. When only common cause variation is present on a control chart for attributes, the chance of a point being above the upper control limit or below the lower control limit is less than one percent.

Application and Construction of Control Charts for Attributes

The following steps describe how to use a control chart for attributes:

1) *Select the area or process to be charted* — Give high priority to areas where problems are already occurring. Choose characteristics that will provide the type of data needed for finding the problem.

2) *Decide which attributes chart to use* — This will depend on the type of data that may already be available or the type of data that is desired.

3) *Select the frequency of sampling* — In other words, define the subgroups. The subgroup periods may be equal periods of production (hourly, daily, weekly). The periods may also correspond to equal quantities of production (lots, batches). The periods should be chosen to make it easy to find and correct problems. Shorter periods will give faster feedback of problems. It is important that the subgroups be chosen to ensure that there will be minimum variation within a subgroup. This will allow variation to show up on the chart from subgroup to subgroup.

4) *Select the size of subgroup* — The subgroup size should be large enough to ensure that the subgroup has a strong probability of having some nonconformities. Small subgroup sizes tend to result in control limits being wide and being less accurate in depicting an out of control process. The most effective subgroup size for p and np-charts is greater than 50. For the most effective c and u-charts, the subgroup size should be at least one, but is better at five to ten.

5) *Gather the data* — Samples within a subgroup should be collected randomly so each item being inspected has an equal chance of having nonconformities.

6) *Construct the chart* — Be careful not to make the chart too tall vertically. Plot the data. Calculate and draw the control lines. Frequently the value for the lower control limit will be negative for an attributes control chart. In this case there is no lower control limit. At least 20 to 25 subgroup points should be plotted before control limits are calculated.

7) *Analyze the chart for evidence of the process being out of control* — Plotted points indicating a lack of control should be marked with an "x" and investigated. These special causes should be corrected and prevented from recurring.

8) *Eliminate causes and recalculate the chart* — Once the causes of out of control points are corrected, the control limits should be recalculated, excluding those out of control subgroups. The chart then should be re-evaluated with the new limits to look for more out of control conditions.

9) *Extend control limits* — When the process is deemed to be in control, the control limits can be extended forward in time on the chart. Future data can then be plotted on the chart to continue evaluating the process for evidence of the process going out of statistical control. As a process continues to improve, the old control limits may be too wide to gain any further improvement in quality. In this case it may be desirable to recalculate the

control limits using only the most recent data. Quality is improving when the control limits can be made narrower.

Reading a Control Chart for Attributes

The goal in reading a control chart for attributes is defining which points represent evidence that the process is out of control. Points on a chart that indicate only random or chance variation are the result of common causes and do not indicate the need for corrective action. Points that represent non-random variation are due to assignable causes and are the signals for immediate action (*Western Electric*, 1956). Variations above the central line on an attribute chart are called *high spots* and variations below the central line are called *low spots*. The purpose of finding non-random variation is to eliminate the special causes that enter a process causing a change in the quality of the output.

A chart with points that are varying in a random fashion is said to have a *natural pattern*. A natural pattern has most of the points near the central line, a few points spread out and approaching the control limits and no points exceeding the control limits. *Unnatural patterns* are those that are missing one or more of the characteristics of a natural pattern. An unnatural pattern on the chart indicates that something is wrong with the process.

The most important thing to look for in a control chart is points that fall outside of the control limits. Statistically it is extremely unlikely that any point on a control chart will fall outside of the control limits if there are no special causes present in the process. The danger here is finding a point outside of the limits which will signal the need for corrective action when in fact no change has taken place in the process. This is a Type I error. When no corrective action is taken on a point that is within the control limits, but there really was a change in the process, it is a Type II error. In order to reduce the chances of these errors, other tests for unnatural variation can be made.

When reading a control chart for attributes, it is important to keep in mind that evidence of non-randomness near the lower control limit (low spots) may seem to indicate that the process is producing too few defects. This may point to areas of a good process and investigation may lead to ways of improving overall quality. Finding reasons why some subgroups have fewer defects than others can cause action for more permanent improvement in quality. Low spots, however, can also indicate that there has been an error in inspection or the need for tighter standards.

Stratification — The characteristic of this unnatural pattern is the absence of points near the control limits. Some causes of this include: non-random sampling, samples coming from different sources, and samples being screened before inspection.

Systematic variation — In this case the pattern has become predictable. An example of this is a low point always being followed by a high one and vice versa. The cause is usually in collecting the data systematically from different sources.

Trends — A trend is a long series of points that are generally moving up or down without a change in direction. See Section 3.2 for an example. Upward trends indicate more defectives, possibly caused by poorer materials or workmanship or the wearing of a tool. Downward trends may be caused by better materials or workmanship or relaxation of standards.

Sudden shifts in level — This is shown by a number of abnormal points suddenly appearing on one side of the central line or the other. This indicates a major change in the process has taken place. See Section 3.2 for an example. Higher shifts in level may be due to a new batch of poor material, a change to a new operator or poor machine, or a tightening of inspection criteria. Lower level shifts may indicate a change to better operators, machines, methods or a loosening of standards. Gradual changes in level may also occur with similar causes but over a longer period of time.

Cycles — Cycles are short trends occurring in repeating patterns. They may be caused by regular differences in suppliers or non-random data collection.

3.4 p-CHART

When the interest is in determining the quantity of defective units and the data is collected in samples that are not of constant size, then the p-chart is appropriate. The p-chart measures the output of a process as the percentage of nonconforming or defective items in a subgroup being inspected. Each item is recorded as being either conforming or nonconforming even if the item has more than one defect. The following symbols are used on a p-chart:

n — The number of units in a subgroup (the subgroup, sample, or subset size).

k — The number of subgroups in the study period.

np — The number of defective or nonconforming units found in a subgroup.

p — The fraction defective in a subgroup (proportion nonconforming). These are the points that are plotted on the chart. The formula is:

$$p = \frac{np}{n} \qquad (3.4.1)$$

p-bar — The average fraction defective (average proportion nonconforming) for the study period. This is drawn as the central line on the chart. It is found by dividing the total number of defective units found in all subgroups of the study period by the total number of units inspected in the study period. The formula is:

$$\bar{p} = \frac{np_1 + np_2 + \ldots + np_k}{n_1 + n_2 + \ldots + n_k} \qquad (3.4.2)$$

How to Make and Use a p-Chart

Refer to the example in Figure 3.4.1.

1) Gather and record the data.

2) Calculate $\bar{p}$.

3) Calculate upper and lower control limits. Since the value of n varies from subgroup to subgroup, the control limits must be calculated for each subgroup. This makes the p-chart difficult to read and construct. This nuisance can be avoided by keeping the subgroup size the same or within plus or minus 25% of the average sample size and

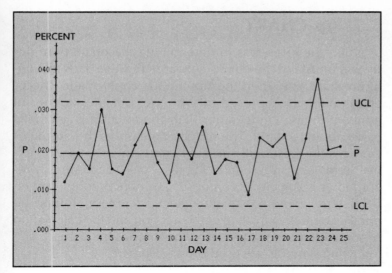

PERCENT

.040

.030 — — — — — — — — — — — — — — — UCL

P .020 — P

.010

.000

1 2 3 4 5 6 7 8 9 10 11 12 13 14 15 16 17 18 19 20 21 22 23 24 25

DAY

— — — — — — — — — — — — — — — LCL

Fig. 3.4.1 The percentage of defective light bulbs from a daily inspection is shown in the table below. Note that an average for n is used. The calculations for the chart at left are:

$$\bar{n} = \frac{25068}{25} = 1003$$

$$\bar{p} = \frac{489}{25068} = .019$$

$$UCL = .019 + 3\sqrt{\frac{.019\,(1 - .019)}{1003}} = .032$$

$$LCL = .019 - 0.13 = .006$$

Note the instability at day 23, indicated by a point out-of-control.

using an average of n. An alternative is showing two sets of control limits using the maximum and minimum subgroup sizes. The formulas for control limits are:

$$UCL_{\bar{p}} = \bar{p} + 3\sqrt{\frac{\bar{p}\,(1 - \bar{p})}{n}} \qquad (3.4.3)$$

$$LCL_{\bar{p}} = \bar{p} - 3\sqrt{\frac{\bar{p}\,(1 - \bar{p})}{n}} \qquad (3.4.4)$$

4) Decide on a scale and put it on the control chart.

5) Plot each p, and add p-bar, and the control limit lines.

6) Analyze the data points for evidence of noncontrol.

7) Find and correct special causes.

8) Recalculate control limits.

day	n	np	p
1	990	12	.012
2	1134	21	.019
3	960	14	.015
4	1000	30	.030
5	1020	15	.015
6	1024	14	.014
7	1040	22	.021
8	1010	26	.026
9	980	17	.017
10	976	12	.012
11	1010	24	.024
12	1010	18	.018
13	900	23	.026
14	900	13	.014
15	1000	18	.018
16	1000	17	.017
17	1020	9	.009
18	1120	26	.023
19	1024	21	.021
20	1060	25	.024
21	920	12	.013
22	970	22	.023
23	980	37	.038
24	1020	20	.020
25	1000	21	.021
	25068	489	

3.5 np-CHART

When the interest is in determining the quantity of defective units and the data is collected in subgroups that are of constant size, then the np-chart is appropriate. The p-chart could also be used in this case, but the number of nonconforming items rather than the fraction is generally easier to understand. The np-chart measures the output of a process as the actual number of nonconforming or defective items in a subgroup being inspected. Each item is recorded as being either conforming or nonconforming even if the item has more than one defect. Except for plotting the number defective instead of the fraction and the calculation of control limits, the np-chart is the same as the p-chart. The following symbols are used on an np-chart:

n — The number of units in a subgroup (the subgroup, sample, or subset size).

k — The number of subgroups in the current calculation.

np — The number of defective or nonconforming units found in a subgroup. These are the points plotted on the chart.

$np\text{-bar}$ — The average number of defective or nonconforming units for the subgroup in the current calculation. This is drawn as the central line on the chart. It is found by dividing the total number of defective units found in all subgroups of the study period by the total number of subgroups in the study period. The formula is:

$$n\,\bar{p} \;=\; \frac{np_1 \,+\, np_2 \,+\, \ldots +\, np_k}{k} \qquad (3.5.1)$$

How to Make and Use an np-Chart

Refer to the example in Figure 3.5.1

1) Gather and record the data.

2) Calculate np-bar.

3) Calculate the upper and lower control limits. The formulas are:

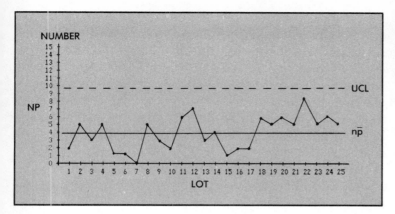

NUMBER

NP

LOT

Fig. 3.5.1   The number of coffee cups manufactured with broken handles is shown in the
table below. Inspection is lot-by-lot with each
lot consisting of 60 cups. The calculations for
the chart at left are:

$$n\bar{p} = \frac{98}{25} = 3.9$$

$$\bar{p} = \frac{3.9}{60} = .065$$

$$UCL = 3.9 + 3 \sqrt{3.9(1 - .065)} = 9.6$$

$$LCL = 3.9 - 5.7 = -1.8 \quad \text{(Assume Zero)}$$

Note the run of eight above the centerline at
lots 18 through 25.

$$UCL_{np} = n\bar{p} + 3 \sqrt{n\bar{p}(1 - \bar{p})} \qquad (3.5.2)$$

$$LCL_{np} = n\bar{p} - 3 \sqrt{n\bar{p}(1 - \bar{p})} \qquad (3.5.3)$$

where $\bar{p} = \dfrac{n\bar{p}}{n}$

4) Decide on a scale for the control chart.

5) Plot each np, and add n$\bar{p}$ and the control limit lines.

6) Analyze the data points for evidence of noncontrol.

7) Find and correct special causes.

8) Recalculate control limits.

3.6   c-CHART

In cases where a single unit of process output is likely to
have many defects, it may be appropriate to use a c-chart.
The c-chart is applied where the defects are scattered continuously throughout the unit output, such as flaws in a roll
of paper or bubbles in glass. The interest here is not only
that the unit is defective but how many defects it has.
Subgroups will be in square yards of cloth, number of glass
containers, an area of sheet metal, etc. The areas or number
of items for a subgroup must be the same from subgroup to

lot	n	np
1	60	2
2	60	5
3	60	3
4	60	5
5	60	1
6	60	1
7	60	0
8	60	5
9	60	3
10	60	2
11	60	6
12	60	7
13	60	3
14	60	4
15	60	1
16	60	2
17	60	2
18	60	6
19	60	5
20	60	6
21	60	5
22	60	8
23	60	5
24	60	6
25	60	5
		98

subgroup. A c-chart usually has one item per subgroup but may have two or more. A c-chart also applies where the defects on a single unit, such as individual automobiles, may come from many sources and where no one such source would produce most of the nonconformities. The c-chart is useful in the final inspection of complicated assemblies. Also, it can be used in non-industrial areas such as bookkeeping errors and accidents. The symbols used on a c-chart are:

n — The number of units in a subgroup (the subgroup, sample, or subset size).

k — The number of subgroups in the current calculation.

c — The number of defects or nonconformities in a subgroup. These are the points that are plotted on the chart.

c-bar — The average number of defects or nonconformities for the study period. This is drawn as the central line on the chart. It is found by dividing the total number of defects found in all subgroups of the study period by the total number of subgroups in the study period. The formula is:

$$\overline{c} = \frac{c_1 + c_2 + \ldots + c_k}{k} \qquad (3.6.1)$$

How to Make and Use a c-Chart

Refer to the example in Figure 3.6.1.

1) Gather and record the data. Remember that the size of the inspection sample must be the same (number of units, area of glass, length of wire, etc.).

2) Calculate c-bar.

3) Calculate upper and lower control limits. The formulas are:

$$UCL_{\overline{c}} = c + 3 \sqrt{\overline{c}} \qquad (3.6.2)$$

$$LCL_{\overline{c}} = c - 3 \sqrt{\overline{c}} \qquad (3.6.3)$$

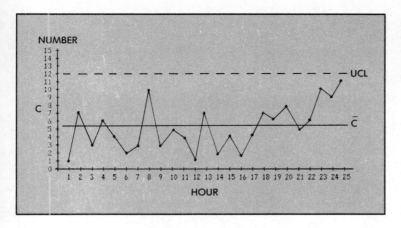

Fig. 3.6.1 The number of assembly errors found in complex circuit board assemblies are shown in the table below. Ten boards were inspected per hour. The calculations for the chart at left are:

$$\bar{c} = \frac{130}{25} = 5.2$$

$$UCL = 5.2 + 3\sqrt{5.2} = 12.0$$

$$LCL = 5.2 - 6.8 = -1.6 \text{ (Assume Zero)}$$

Note the upward trend from hours 16 to 25.

4) Decide on a scale for the control chart.

5) Plot each c, and add c-bar and the control limit lines.

6) Analyze the data points for evidence of noncontrol.

7) Find and correct special causes.

8) Recalculate control limits.

3.7   u-CHART

The u-chart is appropriate in the same situations as the c-chart. The differences are in the calculation of control limits and that the plotted points represent the average number of nonconformities per unit instead of just the number of nonconformities. A u-chart must be used instead of a c-chart if the unit size of a subgroup varies from subgroup to subgroup. The following symbols are used on a u-chart:

n — The number of units in a subgroup (the subgroup, sample, or subset size).

k — The number of subgroups in the current calculation.

c — The number of defects or nonconformities found in a subgroup.

u — The average number of defects or nonconformities per unit in a subgroup. These are the points that are plotted on the chart. The formula is:

$$u = \frac{c}{n} \tag{3.7.1}$$

hour	c
1	1
2	7
3	3
4	6
5	4
6	2
7	3
8	10
9	3
10	5
11	4
12	1
13	7
14	2
15	4
16	2
17	4
18	7
19	6
20	8
21	5
22	6
23	10
24	9
25	11
	130

Fig. 3.7.1 *The number of dents counted on car hoods per day is shown in the table below. The number of hoods inspected varies from day to day. Note that an average for n is used. The calculations for the chart at right are:*

$$\bar{n} = \frac{217}{25} = 8.7$$

$$\bar{u} = \frac{576}{217} = 2.7$$

$$UCL = 2.7 + 3\sqrt{\frac{2.7}{8.7}} = 4.4$$

$$LCL = 2.7 - 1.7 = 1.0$$

Note the sudden shift in level in days 17 through 25.

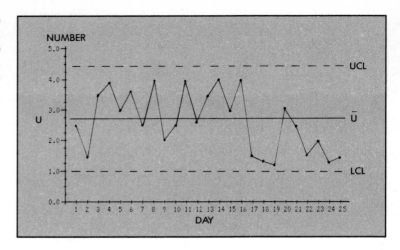

day	n	c	u
1	10	25	2.5
2	9	13	1.4
3	8	28	3.5
4	9	35	3.9
5	9	27	3.0
6	7	25	3.6
7	8	20	2.5
8	8	32	4.0
9	8	16	2.0
10	8	20	2.5
11	10	40	4.0
12	7	18	2.6
13	11	39	3.5
14	9	36	4.0
15	8	24	3.0
16	10	40	4.0
17	8	12	1.5
18	9	12	1.3
19	9	11	1.2
20	10	30	3.0
21	8	20	2.5
22	8	12	1.5
23	9	18	2.0
24	8	10	1.3
25	9	13	1.4
	217	576	

u-bar — The grand average number of defects or nonconformities per unit for the study period. This is drawn as the central line on the chart. It is found by dividing the total number of defects found in all subgroups of the study period by the total number of units inspected in the study period. The formula is:

$$\bar{u} = \frac{c_1 + c_2 + \ldots + c_k}{n_1 + n_2 + \ldots + n_k} \qquad (3.7.2)$$

How to Make and Use a u-Chart

Refer to the example in Figure 3.7.1.

1) Gather and record the data.

2) Calculate u-bar.

3) Calculate upper and lower control limits. As is the case with p-charts, the value of n varies from subgroup to subgroup and the control limits must be calculated for each subgroup. Maintaining the sample sizes the same or within plus or minus 25% of the average sample size, and using an average value for n, simplifies the calculation of control limits. The formulas are:

$$UCL_{\bar{u}} = u + 3\sqrt{\frac{\bar{u}}{n}} \qquad (3.7.3)$$

$$LCL_{\bar{u}} = u - 3 \sqrt{\frac{\bar{u}}{n}} \qquad (3.7.4)$$

4) Decide on a scale for the control chart.

5) Plot each u, and add u-bar and the control limit lines.

6) Analyze the data points for evidence of noncontrol.

7) Find and correct special causes.

For Further Reference

Ford Motor Company. *Process Capability and Continuing Process Control.* Statistical Methods Office, 1983.

Grant, Eugene L., and Leavenworth, Richard S. *Statistical Quality Control,* 5th ed. New York: McGraw-Hill Book Company, 1980.

Ishikawa, Kaoru. *Guide to Quality Control.* Hong Kong: Nordica International Limited, 1976.

Juran, Joseph M., and Gryna, Frank M., *Quality Planning and Analysis.* New York: McGraw-Hill Book Company, 1970.

Quality Control Circles, Inc. *Quality Control Circles,* 2d ed. Saratoga: Quality Control Circles, Inc., 1982.

Western Electric Co., Inc. *Statistical Quality Control Handbook,* 2d ed. Easton: Mack Printing Company, 1956.

4. SPECIAL CHARTING TECHNIQUES

4.1 INTRODUCTION

Many people think that Statistical Process Control (SPC) is only for high volume, repetitive operations. That's simply not true. SPC can be used in processes that make:

- Nuts and bolts in six-minute runs
- One submarine every eighteen months
- Chemicals in batches
- One-of-a-kind prototypes

SPC is not about parts. It's about the process and how to control variability. We see a process with many variables: new shipments of raw material, tool wear, shift changes, and machine updates and replacements. Each of these changes introduces variability. This variability affects the process and must be controlled whether the quantity of parts you produce is large or small.

SPC can be difficult to apply to a short-run process. Types of processes considered to be short-run include those that make:

- A large number of parts in a short period of time
- A large complex part over a long period of time
- Only one part per run
- Parts that are difficult to place into subgroups

Short-run SPC means we have to look at our work in a different way. Conventional $\bar{x}$ & R charts are limited to processes that generate enough subgroups to calculate center lines and control limits. To accommodate short production runs, some modifications are needed. In this chapter, we will deal with four basic charts that can be used for short-run applications. These charts are:

- Nominals chart
- Individuals chart with a moving range
- Moving average moving range chart
- Standardized chart

The information in this chapter assumes that the reader understands $\bar{x}$ & R charting techniques. For more information on $\bar{x}$ & R charts, see Chapter 3.

4.2 NOMINALS CHART

Nominals charts allow more than one part number to be plotted on the same control chart, using the same scale. This reduces the number of charts needed to monitor part numbers and illustrates the effects of the process over time and variation among different part numbers. Because it's the process and not parts we're monitoring, we have to form families of parts. As long as they're made with the same material, method and machine, they can be put into a family. The parts within each family can be different sizes, but if the same process makes these parts, they share common traits.

Limitations of Nominals Charts

Although nominals charts can be useful in some manufacturing processes, there are limitations that should be noted. The first two limitations involve variation (standard deviation) of part numbers. First, the variation of each part number must be similar. Second, if the variation of each part number is not similar, the suspected part number should be plotted on a separate nominals chart. The third limitation to note is that parts must be produced using the same process — machine, method and material.

Before we discuss the steps involved in creating a nominals chart, we need to define *coding*. In coding numbers, two values must be known: the *nominal value* and the *unit of measure*. The nominal value is the desired mean (target) value for the particular dimension. The unit of measure indicates the precision with which the measurements are to be made. This is illustrated in the following example.

Let us suppose we manufacture bearings. The specification for these bearings is 0.500 ± 0.005. The *nominal value* is the 0.500 value. Let us also suppose that we measure these bearings with a caliper and read to the nearest thousandth of an inch. This then would be the *unit of measure*. Now, instead of recording the entire measurement, we can record coded values. The measurements can be recorded as a certain number of increments above or below the nominal value.

If the nominal value is 0.500 with a unit of measure of 0.001, we code the following measurements using this formula:

$$\text{Coded Value} = \frac{\text{Actual Value} - \text{Nominal Value}}{\text{Unit of Measure}} \quad (4.2.1)$$

Table 4.2.1 shows a listing of the actual and coded values for this example. The +1 indicates a dimension of 0.001 above nominal and the −1 indicates a dimension of 0.001 below the nominal of 0.500. These coded values are often multiplied by the unit of measure for graphing purposes.

Steps for Creating a Nominals Chart

1) Select the key variable to monitor for each part.
2) Identify the nominal of the specification.
3) Measure your first sample part.
4) Subtract the nominal (from Step 2) from the actual value (from Step 3).
5) Divide the result from Step 4 by the unit of measure. These coded values are used in place of actual values to make numbers easier to work with.
6) Record the reading from Step 5 as sample measurement #1 in the sample measurement area of the $\bar{x}$ & R chart.
7) Continue with Steps 2 through 5 until subgroup #1 is complete.
8) Determine the average ($\bar{x}$) and the range (R) for the subgroup.
9) Multiply the results from Step 8 by the unit of measure.
10) Plot the $\bar{x}$ & R on the control chart.
11) Continue with Steps 2 through 7 until this part is done running. Then start all over again with each new part number change. Be sure to document when a new part number starts. On the example control chart shown in Figure 4.2.1, see how new parts were started after subgroups 4, 8 and 11.
12) Continue monitoring the process until 20 subgroups have been collected.
13) Calculate the $\bar{\bar{x}}$ & $\bar{R}$ using the standard formulas.
14) Calculate $UCL_{\bar{x}}$, $LCL_{\bar{x}}$, and UCL_R using standard formulas.

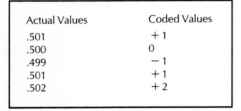

Actual Values	Coded Values
.501	+ 1
.500	0
.499	− 1
.501	+ 1
.502	+ 2

Table 4.2.1   Actual and coded data for nominals charting.

By subtracting each sample measurement from its nominal, we're actually monitoring the process deviation from that point. This enables us to monitor many different sizes of parts that are made by the same process.

Interpretation

The interpretation of nominals charts is the same as Shewhart $\bar{x}$ & R charts. There should be no obvious trends or patterns in the data, beyond what would normally occur by chance. As with $\bar{x}$ & R charts, the goal is to maintain a constant level of statistical control and to take appropriate action when that level deviates.

4.3 INDIVIDUALS CHART WITH MOVING RANGE

In some cases, it is necessary for process control to be based on individual readings, rather than subgroups. This would typically occur when the measurements are expensive — such as in destructive testing — or when the output at any point is relatively homogeneous — the pH of a

Fig. 4.2.1   An example of a nominals control chart.

PART NAME RIVETS		PART # 1-17		CHART # 22		DATE 10/31					
OPERATOR BILL J.		MACHINE 1322		GAGE MITUTOYO MICROMETER							
UNIT OF MEAS. = .0001		SPEC LIMITS ± .003		READING = CODED VALUES							

TIME		8:00	8:15	8:30	8:45	9:00	9:15	9:30	9:45	10:00	10:15	10:30	10:45
SAMPLE MSMTS	1	7	-30	16	1	-15	-3	14	24	14	0	17	21
	2	-7	-14	-1	-24	6	8	-6	-1	-8	-8	2	-7
CODED VALUES	3	26	-4	-16	16	-9	-21	10	24	10	14	24	16
	4	-4	-7	0	-9	-13	7	10	-12	12	0	21	16
	5	-15	6	2	14	-14	4	11	26	9	-6	21	18
SUM		-7	-49	1	-2	-45	-5	39	61	37	0	85	56
AVG. (X BAR)		-1	-11	0	0	-9	-1	8	12	7	0	17	13
RANGE (R)		41	36	32	38	21	28	20	38	22	22	22	28
NOTES					11					11		11	

AVERAGES $\bar{\bar{X}} =$ .0028 / .0000 / −.0028

RANGES $\bar{R} =$ .0082 / .0039 / 0

PART #1   PART #16   PART #4   PA

chemical solution, for example. In these cases, control charts for individuals can be constructed as described below. However, four cautions should be noted:

- Charts for individuals are not as sensitive to process changes as $\bar{x}$ & R charts.

- Care must be taken in interpreting charts for individuals if the process distribution is not symmetrical.

- Charts for individuals do not isolate the piece-to-piece repeatability of the process. In many applications, it may be better to use conventional $\bar{x}$ & R charts with small subgroup sample sizes (two to three) even if this requires a longer period between subgroups.

- Since there is only one sample per subgroup, values for $\bar{x}$ and estimated sigma, $\hat{\sigma}$, can have substantial variability (even if the process is in control) until the number of subgroups is 100 or greater.

Steps for Creating an Individuals Chart With Moving Range

1) Select the key variable to monitor.
2) Record individual readings on an individuals control chart.
3) Calculate the moving range between individuals. It is generally best to record the difference between each successive pair of readings — the difference between the first and second reading, the second and third, etc.

There will be one less range than there are individual readings — 25 readings give 24 ranges. In some cases, the range can be based on a larger moving group — threes or fours, or on a fixed subgroup, like all readings taken on a single shift.

Even though the measurements are sampled individually, it is the moving range group size that determines the nominal sample size "n" for calculating control limits. Continuing with the example above, n=24, not 25.

4) Calculate $\bar{\bar{x}}$ & $\bar{R}$. There is one less range value than the number of individual readings.
5) Calculate control limits using the following formulas:

$$UCL_R = D_4\bar{R}$$

$$LCL_R = D_3\bar{R}$$

$$UCL_x = \overline{\overline{X}} + E_2\overline{R}$$

$$LCL_x = \overline{\overline{X}} - E_2\overline{R}$$

There is no lower control limit for ranges for sample sizes below seven. For sample sizes below seven, the lower control limit is less than zero; therefore, it is not plotted. For a sample of an individuals chart with moving range, see Figure 4.3.1.

4.4 MOVING AVERAGE MOVING RANGE CHART

It is often true that making subgroups from a process is impractical. In some situations, the time required to measure a single observation is so great that repeat observations cannot be considered. These are occasions when moving averages and moving ranges are more in accord with manufacturing practice than ordinary averages and ranges. The method of moving averages and ranges is particularly suitable for lines of production in which it takes some time to produce a single part.

To understand what is meant by a *moving average*, consider these ten parts and their measurements: 3, 4, 5, 3, 2, 9, 5, 2, 6, and 8. We can break up these ten parts into two

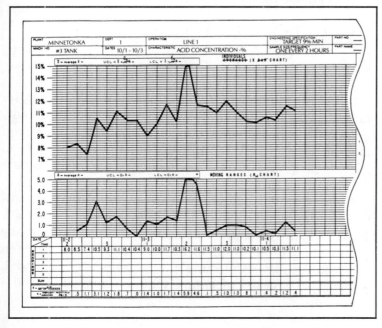

Fig. 4.3.1   *An example of an individuals and moving range chart.*

subgroups of five each and calculate the average of the first group and the average of the second group. These would be ordinary averages. If we applied moving averages, however, we would average the first five, then drop the first sample measurement; add the sixth and average the new set of five; next drop the second sample measurement, add the seventh and average this new set of five, and so on. Numerically, for the first three $\bar{x}$ values calculated with this method, we would have:

$$\bar{x}_1 = \frac{3 + 4 + 5 + 3 + 2}{5} = 3.4$$

$$\bar{x}_2 = \frac{4 + 5 + 3 + 2 + 9}{5} = 4.6 \qquad (4.4.1)$$

$$\bar{x}_3 = \frac{5 + 3 + 2 + 9 + 5}{5} = 4.8$$

The above are moving averages of five samples. *Moving ranges* of the five samples are the ranges of these same subgroups. Here are the range calculations and values:

R_1 is the range of 3, 4, *5* , 3, 2 = 3

R_2 is the range of 4, 5, 3, *2* , 9 = 7

R_3 is the range of 5, 3, *2* , 9 , 5 = 7

Italicized numbers are values used for calculating the range. By adding to our previous sample measurements each new sample measurement (as the part is produced and the oldest sample is dropped), our moving average and moving range will give the usual control *and* an up-to-date picture of the process.

Figure 4.4.1 illustrates one technique for combining sample measurements for moving average moving range on $\bar{x}$ & R charts. In this example, you can see we are using a moving group size of three, grouping three measurements together. Subgroup 1 consists of sample measurements 1, 2, and 3. Subgroup 2 consists of sample measurements 2, 3, and 4 and so on.

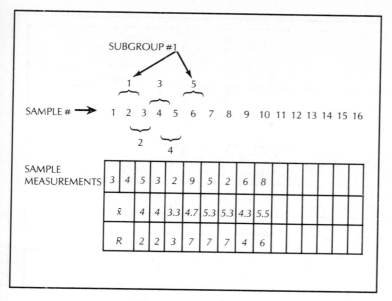

Fig. 4.4.1 Calculating moving averages and moving ranges.

Steps for Creating a Moving Average Moving Range Control Chart

1) Select the key variable to monitor.
2) Select the moving average group size. (We will use three in our example.)
3) Measure your first sample part and record it as sample 1.
4) Measure your second sample part and record it as sample 2.
5) Measure your third sample part and record it as sample 3.
6) Determine the $\bar{x}$ & R values.
7) Plot these values on the chart as subgroup 1.
8) Carry forward samples 2 and 3 to subgroup 2. Place sample 2 where sample 1 was and sample 3 where sample 2 was. (See Figure 4.4.2.) These measurements will be combined with the next sample part measurement to form subgroup 2.
9) Determine the $\bar{x}$ & R values.
10) Plot these values on the chart as subgroup 2.
11) Repeat Steps 8 through 10 until 30 subgroups (minimum) are collected.
12) Calculate $\bar{\bar{x}}$ & $\bar{R}$ values.
13) Calculate control limits using standard $\bar{x}$ & R formulas for the appropriate sample sizes.
14) Continue monitoring the process.

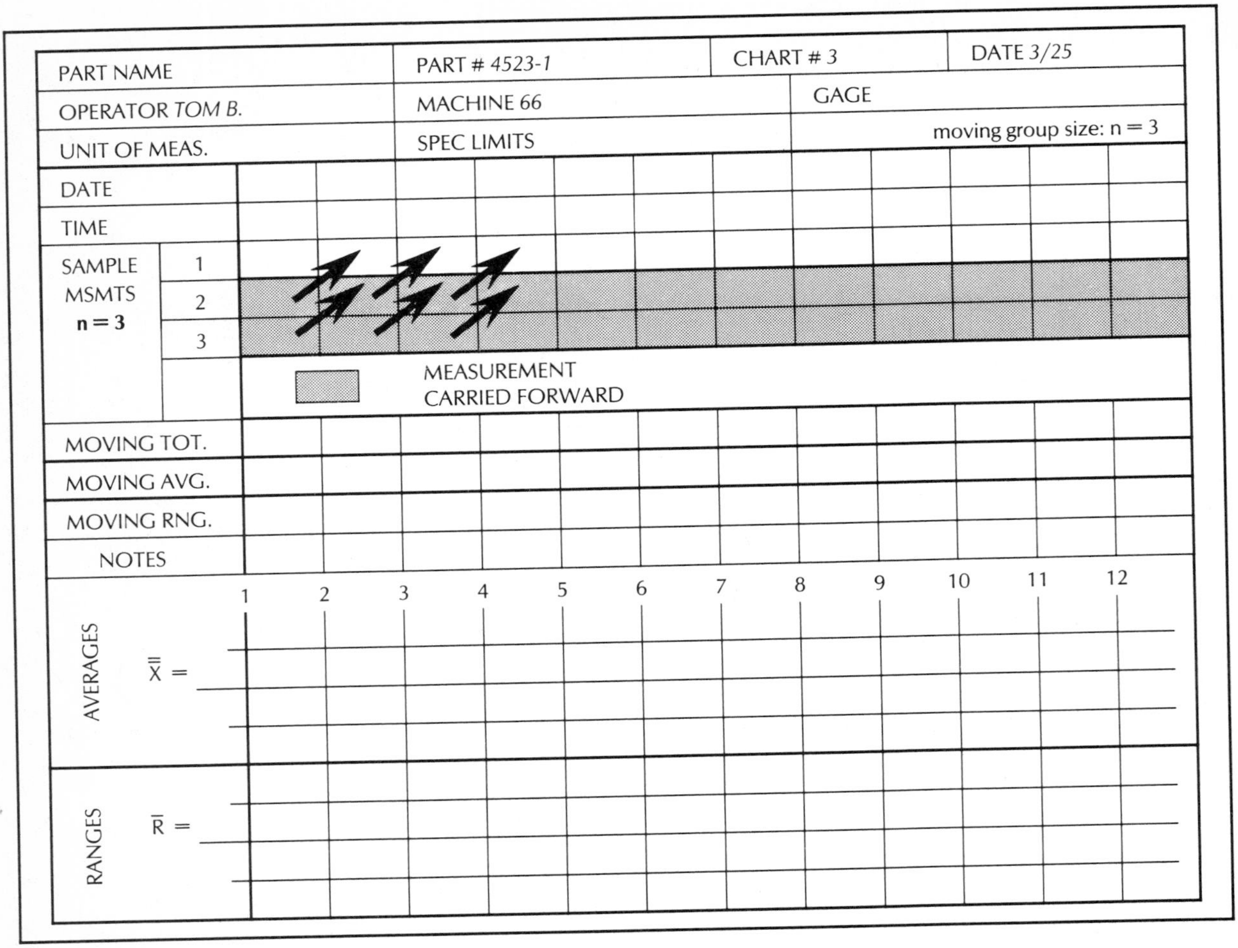

PART NAME		PART # 4523-1			CHART # 3				DATE 3/25		
OPERATOR TOM B.		MACHINE 66				GAGE					
UNIT OF MEAS.		SPEC LIMITS							moving group size: n = 3		
DATE											
TIME											
SAMPLE MSMTS n = 3	1										
	2										
	3										

MEASUREMENT CARRIED FORWARD

MOVING TOT.											
MOVING AVG.											
MOVING RNG.											
NOTES											

AVERAGES $\overline{\overline{X}}$ =

1 2 3 4 5 6 7 8 9 10 11 12

RANGES $\overline{R}$ =

Fig. 4.4.2 *Technique on a moving average moving range chart, where all but the first measurement of a subgroup are shifted to the next subgroup.*

These control charts have some important characteristics. Since we are dealing with moving averages and moving ranges, the plotted points are not independent of each other. Each point reflects the influence of two or more of the same observations. This influence extends over a greater distance when the average involves many observations. Therefore, several points in a row outside or near the control limits do not have the significance that they would on an ordinary chart on which points are independent. Temporary fluctuations in the process are likely to be overlooked also, since the averaging process tends to decrease the degree of variation. Out of control points on the R chart indicate the presence of abnormal short-term variability within the time period during which the data was collected. Therefore, the interpretation of these charts must be

Special Charting Techniques

somewhat different than the regular Shewhart charts and must be more carefully considered.

It should be noted that if we compare individuals with moving averages, we see that the use of moving averages reduces the noise of the process, but the major patterns of the charts remain similar. See Figure 4.4.3. The control limit lines are different, as we would expect since the variability of averages is less than that of individuals, as shown with the central limits theorem (see Chapter 1). This lower degree of variability that moving average charts possess gives them greater power to detect true changes in the process than do individuals charts.

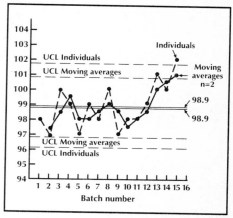

Fig. 4.4.3 Comparison of control chart for individuals with control chart for moving average.

4.5   STANDARDIZED CHART

Nominals charts are one vehicle for short-run charts, but nominals charts have some limitations, as mentioned in section 4.2. Nominals charts must have parts with the same subgroup size. Nominals charts also make the assumption that the standard deviation for each part is similar and that process average equals nominal, even though that is not necessarily the case. There is a type of charting for short-run applications that addresses these issues. These charts are called standardized charts. The standardized chart is the most universal short-run plotting technique. Compared to all other short-run plotting techniques, standardized charting stands out as the pre-eminent method because it is so flexible and easy to interpret. Standardized charts also have the advantage of being easier to learn and compare since all control limits are at ± 3.00 sigma limits.

Standardized charts are based on techniques similar to normal $\bar{x}$ & R charts. The difference is in how you plot values.

Standardized Formula for $\bar{x}$ Charts

Let's start out with the basic $\bar{x}$ chart and work our way to the standardized chart. When charting $\bar{x}$, the goal is to have an $\bar{x}$ value that falls between the upper and lower control limits, which could be expressed by:

$$LCL_{\bar{x}} < \bar{x} < UCL_{\bar{x}} \qquad (4.5.1)$$

where the control limits are defined as:

$$UCL_{\bar{x}} = \bar{\bar{x}} + 3\sigma_{\bar{x}} \qquad (4.5.2)$$

and

$$LCL_{\bar{x}} = \bar{\bar{x}} - 3\sigma_{\bar{x}} \qquad (4.5.3)$$

The standardized $\bar{x}$ chart is based on the principle that control limits for any part on any process are always at ± 3 standard errors of the mean. The standardized $\bar{x}$ chart simply plots points relative to $\pm 3\sigma_{\bar{x}}$ so that part size is not relevant to plot value. Upper and lower control limits on a standardized chart are always at ± 3 respectively with the centerline at zero. A plot point on a standardized chart is determined from the formula:

$$\begin{array}{l}\text{Plot Value} \\ \text{(for } \bar{x} \text{ Chart)}\end{array} = \frac{\bar{x} - \bar{\bar{x}}}{\sigma_{\bar{x}}} \qquad (4.5.4)$$

For example, on a given part on a given process the $UCL_{\bar{x}} = 17.00$ and $\bar{\bar{x}} = 16.00$, knowing this we can find $\sigma_{\bar{x}}$ by solving the $UCL_{\bar{x}}$ equation for $\sigma_{\bar{x}}$ as shown.

$$\sigma_{\bar{x}} = \frac{UCL_{\bar{x}} - \bar{\bar{x}}}{3} \qquad (4.5.5)$$

So $\sigma_{\bar{x}}$ becomes 0.333 and an $\bar{x}$ value of 16.50 will plot at 1.50. Note that on either an $\bar{x}$ or standardized chart, the plotted point will fall exactly half way between the centerline and the upper control limit.

Standardized Formula for Range Charts

Finding the plot value for the range chart is a similar derivation to that for the $\bar{x}$ plot value.

$$\begin{array}{l}\text{Plot Value} \\ \text{(for Range Chart)}\end{array} = \frac{R - \bar{R}}{\sigma_R} \qquad (4.5.6)$$

The Plot Value

The plot value is the actual number that will appear on the chart; it is determined by plugging values into the plot value formula. Calculating the plot value is obviously not as simple as just calculating the average size. The plot value takes into account the subgroup size, the grand average of the data and process standard deviation. The advantage is that all parts can be put on one chart since the chart's control limits range from ± 3. Every part uses the same control limits.

Variable Subgroup Sizes

The standardized chart can handle a variable subgroup size easily. As the subgroup size varies, so does the $\sigma_{\bar{x}}$ in the equation but the physical chart stays the same. When using standardized charts, the standard deviation of different parts on the same chart can be different. Since the standard deviation becomes a factor in the plot value but makes no difference to the control limits on the actual chart.

Natural Boundaries

Another problem with the nominals chart but solved with the standardized occurs when charting a characteristic with a natural boundary, e.g. TIR for roundness. When measuring roundness, the target size is 0 but it is not possible to have parts with a reading smaller than 0. This is shown in the nominals $\bar{x}$ chart of Figure 4.5.1a where all the plotted points are above the centerline. Those unfamiliar with TIR and natural barriers might think that a trend (based on the number of consecutive points above $\bar{\bar{x}}$) was occurring. Looking at Figure 4.5.1b we see a plot for the same type of information that is done in a standardized fashion. Note that the points on the standardized chart are more evenly distributed about the centerline of the chart.

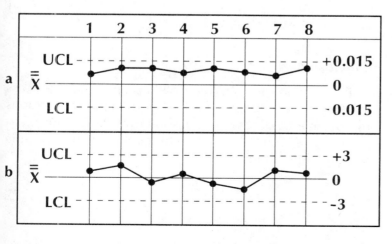

Fig. 4.5.1a&b   Nominals $\bar{x}$ chart

Special Charting Techniques

Standardized Chart Centerline

Notice that the centerline for the standardized chart shown in Figure 4.5.2 represents $\bar{x}$ as well as a plot value of 0. This does not mean that $\bar{\bar{x}}$ equals 0 but that for a given value of $\bar{x}$ equal to $\bar{\bar{x}}$, the plot value will be 0 because there is no deviation from the mean. Also notice that the control limits for the standardized $\bar{x}$ & R chart are fixed at $\pm 3\sigma$.

Using standard statistical techniques we now have a method for low volume, short-run manufacturers. Standardized $\bar{x}$ & R charts can be used to plot multiple parts of the same process on one chart. You can now chart multiple parts with different standard deviations and even with different sample sizes on a single chart. Interpretation of trends, patterns, and out of control points is the same for standardized charts as it is for $\bar{x}$ & R charts (Chapter 3).

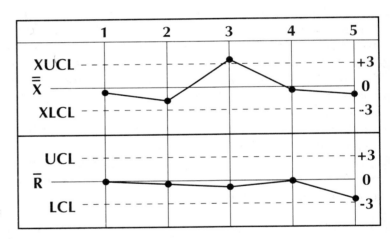

Fig. 4.5.2   Centerline represents $\bar{x}$ as well as a plot value of zero.

Conclusion

SPC is the only tool we know of that can measure success and failure and tell us what must be done for continuous improvement. Because SPC for short runs can be difficult to apply, the special charting techniques explained in this chapter address those needs.

For Further Reference

Duncan, Acheson, J. *Quality Control and Industrial Statistics,* 5th ed. Homewood: Richard D. Irwin Inc., 1986.

Ford Motor Company. *Process Capability and Continuing Process Control.* Statistical Methods Office, 1983.

Grant, Eugene L., and Leavenworth, Richard S. *Statistical Quality Control,* 5th ed. New York: McGraw: Hill Book Company, 1980.

Johnson, Perry. *SPC for Short Runs A Programmed Instruction Workbook.* Southfield, Michigan: Perry Johnson, Inc., 1987.

Juran, Joseph M. and Gryna, Frank M. *Quality Planning and Analysis,* New York: McGraw-Hill Book Company, 1970.

Juran, Joseph M. *Juran's Quality Control Handbook,* 4th ed. New York: McGraw-Hill Book Company, 1988.

5. PROCESS CAPABILITY

5.1 WHAT IS PROCESS CAPABILITY?

Process capability is the measure of process performance. Capability refers to how capable a process is of making parts that are well within engineering specifications. Figure 5.1.1 shows the distribution of parts for a process that is capable. A *capability study* is done to answer the questions, "Does the process need to be improved?" and "How *much* does the process need to improved?"

To define the study of process capability from another perspective, a capability study is a technique for analyzing the random variability found in a production process. In every manufacturing process there is variability. This variability may be large or small, but it is always present. It can be divided into two types:

- Variability due to common (random) causes
- Variability due to assignable (special) causes

The first type of variability can be expected to occur naturally within a process. It is attributed to common causes which behave like a constant system of chances. These chances form a unique and describable distribution. This variability can never be completely eliminated from a process. Variability due to assignable causes, on the other hand, refers to the variation that can be linked to specific or special causes. If these causes or factors are modified or controlled properly, the process variability associated with them can be eliminated. Assignable causes cannot be described by a single distribution.

A capability study measures the performance potential of a process when no assignable causes are present (when it is *in statistical control*). Since the inherent variability of the process can be described by a unique distribution, usually a normal distribution, capability can be evaluated by utilizing the properties of this distribution. Simply put, capability is expressed as the proportion of process output that remains within product specifications.

Capability calculations allow predictions to be made regarding quality, enabling manufacturers to take a preventive approach to defects. This statistical approach is in con-

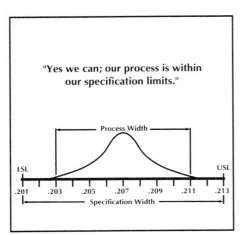

Fig. 5.1.1   *By most measures of capability, this process would be considered capable since its distribution is well inside the specification limits.*

trast with the traditional approach to manufacturing, which is a two-step process: production personnel make the product while quality control personnel inspect and eliminate those products which do not meet specifications. This is wasteful and expensive since it allows time and materials to be invested in products that are not always usable. It is also unreliable since even 100 percent inspection would fail to catch all defective products.

Control Limits are Not an Indication of Capability

Those new to SPC often have the misconception that they don't need capability indices. They think they can compare their control limits to the specification limits instead. This is not true, because control limits look at the distribution of *averages* ($\bar{x}$) and capability indices look at the distribution of *individuals* (x). The distribution of x will always be more spread out than the distribution of $\bar{x}$ values (Figure 5.1.2). Therefore the control limits are often within the specification limits but the plus and minus 3 sigma distribution of parts is not.

Averages of subgroups follow more closely a normal distribution. This is why we can do control charts on processes that are not normally distributed. But averages cannot be used for capability calculation because capability concerns itself with individual parts or samples from a process. After all, parts get shipped, not averages.

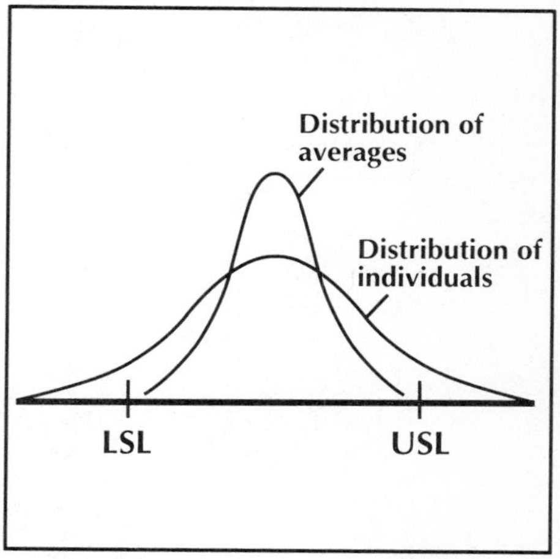

Fig. 5.1.2 *Distribution of averages compared to distribution of individuals, for the same sample data. Control limits (based on averages) would probably be inside specification limits, even though many parts are out of specification. This shows why you should not compare control limits to specification limits.*

5.2   USAGE OF CAPABILITY STUDIES

Capability studies are most often used to determine whether the process can meet specifications or how many parts will exceed specifications. However, there are numerous other practical uses:

- Estimating percentage of defective parts to be expected
- Evaluating new equipment purchases
- Predicting whether design tolerances can be met
- Assigning equipment to production
- Planning process control checks
- Analyzing the interrelationship of sequential processes
- Making adjustments during manufacture

- Setting specifications
- Costing out contracts

Since a capability study determines the inherent reproducibility of parts created in a process, it can be applied to problems outside of manufacturing, such as inspection, administration, and engineering.

There are instances where capability measurements are valuable even when it is not practical to determine in advance if the process is in control. Such an analysis is called a performance study. Performance studies can be useful for examining incoming lots of materials or one time only production runs. In the case of an incoming lot, a performance study cannot tell us that the process that produced the material is in control, but it may tell us by the shape of the distribution what percent of the parts are out of specification or whether the distribution was truncated by the vendor sorting out the obvious bad parts.

5.3 HOW TO SET UP A CAPABILITY STUDY

Before we set up a capability study, we must select the critical dimension or variable to be examined (see Chapter 2 for more information). This dimension is the one that must meet product specifications. In the simplest case, the study dimension is the result of a single, direct process. In more complicated studies, the critical dimension may be the result of several processes. It may become necessary in these cases to perform capability studies on each process. Studies on early processes prove to be more valuable than studies on later processes since early processes lay the foundation which may affect later operations.

Once the critical dimension is selected, data measurements can be collected. This can be accomplished manually or by using automatic gaging and fixturing linked to a data collection device. When collecting measurements on a critical dimension, it is important that the measuring instrument be as precise as possible, preferably one order of magnitude finer than the specification. Otherwise, the measuring process will contribute excess variation to the dimension. Using handheld data collectors with automatic gages will help reduce errors introduced by the process of measurement, data recording, and transcription for post processing by computer.

The ideal situation for data collection is to collect as much data as possible over a defined time period. This will yield a capability study which is very reliable since it is based upon a large sample size.

See Chapter 19 for handheld data collectors for variables data.

In the steps of process improvement, determining process capability is Step 5:

1) Gather process data (this and the following three steps are covered in Chapters 1 through 3).
2) Plot the data on control charts.
3) Find the control limits.
4) Get the process in control (in other words, identify and eliminate assignable causes).
5) Calculate process capability.
6) If process capability is not sufficient, improve the process (reduce the random cause variation), and go back to Step 1.

The Process Must be in Control

The process must be in control and normally distributed before taking samples to measure process capability. All standard capability indices assume that the process is in control and the individuals follow a normal distribution. If the process is not in control, capability indices are not valid, even if they indicate the process is capable.

There are three statistical tools we can use to determine that the process is in control and follows a normal distribution:

- Control charts
- Visual analysis of a histogram
- Mathematical analysis of the distribution to test that the distribution is normal.

No single tool can be used; they must all be used together.

Control charts are the most common method for operators to maintain statistical control of the process. With control charts, such as an $\bar{x}$ & R chart, all points must be inside the control limits with no apparent patterns (trends) present (see Chapter 3 for details).

Chapter 2 explains how to create a histogram. A histogram allows us to see if any parts are outside the specification limits and what the distribution's position is relative to the specifications. If the process is one that is naturally a normal distribution, then the histogram should approximate a bell-shaped curve if the process is in control. Note that a process can be in control but not have its individuals following a normal distribution if the process is inherently non-normal.

See *5.5 Tests for Normality* for statistical calculations that assist in identifying whether or not the process is normal.

The Process Must be Inherently Normal

Many processes naturally follow a bell-shaped curve (a normal distribution) but some do not. Examples of non-normal dimensions are roundness, squareness, flatness and positional tolerances; they have a natural barrier at zero. In these cases, a perfect measurement is zero (for example, no ovality in the roundness measurement). There can never be a value less than zero. The standard capability indices are not valid for such non-normal distributions. See *5.5 Tests for Normality* and *5.6 Capability Measures for Non-normal Distributions* for more information.

5.4   CAPABILITY INDICES

Terminology and Actual Versus Estimated Sigma

Before describing capability indices in detail, several terms must be defined. *USL* stands for Upper Specification Limit and *LSL* for Lower Specification Limit. *Midpoint* is the center of the specification limits. Midpoint is often referred to as the *nominal value* or *target*. *Tolerance* is the distance between the specification limits (TOLERANCE = USL − LSL).

Standard deviation for the distribution of individuals, one important variable in all the capability index calculations, can be determined in either of two ways. It may be calculated or estimated. For an explanation of the concept of standard deviation, see Chapter 1.

Standard deviation may be calculated using the standard statistical formula:

$$\sigma = \sqrt{\frac{\sum\limits_{i=1}^{N} \left(x_i - \mu\right)^2}{N}} \qquad (5.4.1)$$

σ (sigma) symbolizes population standard deviation, N is the population size, x is an individual measurement, and μ is the mean of the population. *Population* means all parts being produced, not just a sample.

Usually we work with a sample of the population, since this is more practical. In this case, the formula for standard deviation is:

$$s = \sqrt{\frac{\sum\limits_{i=1}^{n} \left(x_i - \bar{x}\right)^2}{n - 1}} \qquad (5.4.2)$$

s symbolizes sample standard deviation, and n is the sample size. When over 30 samples are taken, the above formulas for population and sample standard deviation yield virtually the same result. Therefore, we will use the term sigma as we continue our discussion.

Standard deviation may also be estimated using $\bar{R}$ (the average of the subgroup ranges) and a constant that has been developed for this purpose. The formula for estimating sigma is:

$$\hat{\sigma} = \frac{\bar{R}}{d_2} \qquad (5.4.3)$$

$\hat{\sigma}$ (sigma hat or estimated sigma) symbolizes estimated standard deviation. $\bar{R}$ is the average of the subgroup ranges for a sample period when the process was in control. The constant d_2 varies by subgroup size and is listed in Appendix Table A-1.

It is important to remember that the process must be normally distributed *and* in control in order to use the estimated value of sigma. If both of these conditions are not met, the estimated value is not valid. If the process is normally distributed and in control, either method is acceptable and usually yields about the same result. Also, remember that neither actual nor estimated sigma used in calculating capability will be meaningful if the process is inherently non-normal. See *5.6 Capability Measures for Non-normal Distributions* for information on measuring the capability of non-normal processes.

If you have both estimated and actual capability indices available, choose one method and stay with it. Avoid the temptation to look at both and choose the one that is better, since this will introduce variation in results.

CP Index

The most commonly used capability indices are CP and CpK. *CP*, which stands for capability of process, is the ratio of tolerance to 6 sigma. The formula is:

$$CP = \frac{TOLERANCE}{6\sigma} \qquad (5.4.4)$$

The 6σ in the CP formula comes from the fact that, in a normal distribution, 99.73% of the parts will be within a 6σ ($\pm 3\sigma$) spread when only random variation is occurring.

The CP for the sample distribution in Figure 5.4.1 is:

$$CP = \frac{4.0}{3.0} = 1.33 \qquad (5.4.5)$$

As you can see from the CP formula, values for CP can range from near zero to very large positive numbers. When CP is less than 1, tolerance is less than the 6σ spread of the distribution. When CP is greater than 1, tolerance is greater than the 6σ spread of the distribution. The greater the number, the better the CP index is.

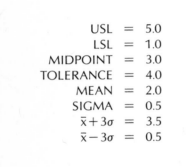

USL	=	5.0
LSL	=	1.0
MIDPOINT	=	3.0
TOLERANCE	=	4.0
MEAN	=	2.0
SIGMA	=	0.5
$\bar{x} + 3\sigma$	=	3.5
$\bar{x} - 3\sigma$	=	0.5

Table 5.4.1   Data for Fig. 5.4.1.

Fig. 5.4.1   *The characteristics of this sample distribution are shown in Table 5.4.1. On the following pages, this distribution is used for the example calculations of capability indices.*

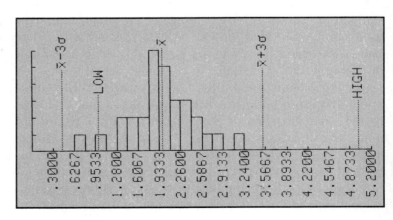

CP is only a measure of the *dispersion* or *spread* of the distribution. It is not a measure of *centeredness* (where the distribution is in relation to the midpoint). Figure 5.4.2 shows two distributions for the same specification limits. Both distributions have a CP of 1.25 but one shows almost all parts being within specification and the other shows a significant number of the parts out of specification. This is why CP is never used alone as a measure of capability. By itself it cannot indicate whether the process is capable. CP only shows how good the process could be if centered. Therefore, CP is usually used with CpK.

CpK Index

While CP is only a measure of dispersion, *CpK* is a measure of both dispersion and centeredness. That is, the formula for CpK takes into account both the spread of the distribution and where that distribution is in regard to the specification midpoint. The formula is:

$$CpK = \text{The lesser of:}$$
$$\frac{(USL - MEAN)}{3\sigma} \quad \text{or} \quad \frac{(MEAN - LSL)}{3\sigma} \qquad (5.4.6)$$

Because we choose the lesser of the two values calculated, we find out how capable our process is on the worst side (the tail closest to the specification limit).

Using the example used in the CP calculation,

$$CpK = \frac{(5.0 - 2.0)}{1.5} \quad \text{or} \quad \frac{(2.0 - 1.0)}{1.5}$$

$$= 2.0 \text{ or } 0.67$$
$$= 0.67 \qquad (5.4.7)$$

The greater the CpK value the better. A CpK greater than 1.0 means that the 6σ (± 3σ) spread of the data falls completely within the specification limits. A CpK of 1.0 means that one end of the 6σ spread falls on a specification limit. A CpK between 0 and 1 means that part of the 6σ spread falls outside the specification limits. A negative CpK indicates that the mean of the data is not between the specification limits. To help you visualize what CpK indicates, Figure 5.4.3 shows the distributions for 4 different CpK values. A small CpK value is due to a process being off center

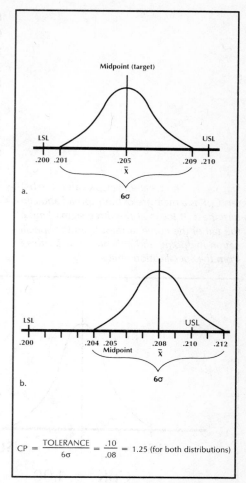

$$CP = \frac{TOLERANCE}{6\sigma} = \frac{.10}{.08} = 1.25 \text{ (for both distributions)}$$

Fig. 5.4.2   Two distributions for the same part, with the same 6σ spread and therefore the same CP. Note, however, that distribution b has few parts within specification. This illustrates that CP is a measure of spread only, not centeredness about the specification midpoint.

from the specification midpoint, spread out, or both. The CpK index will not indicate whether the process is centered above or below target; to see this, look at the mean of the samples compared to the midpoint (target or nominal). Or look at the histogram relative to the specification limits.

Since a CpK of 1.0 indicates that 99.73% of the parts produced are within specification limits, in this process it is likely that only about three out of a thousand need to be scrapped or rejected. Why bother to improve the process beyond this point, since it will produce virtually no reduction in scrap or reject costs? Improvement beyond just meeting specification may greatly improve product performance, cut warranty costs or avoid assembly problems.

Many companies are demanding CpK indexes of 1.33 or

Fig. 5.4.3   The greater the CpK value, the better. CpK is a measure of both spread and centeredness. It looks at the three sigma limit of the tail of the curve farthest from the specification midpoint, and indicates how far this is from the specification limit.

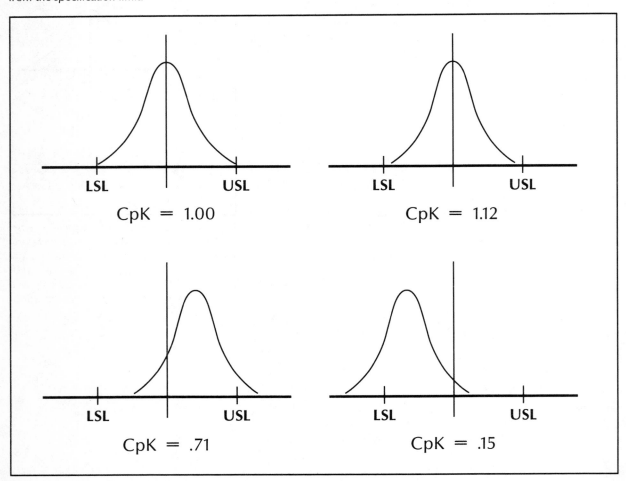

2.0 for the products from their suppliers. A CpK of 1.33 means that the difference between the mean and specification limit is 4σ (since 1.33 is 4/3). With a CpK of 1.33, 99.994% of the product is within specification. Similarly a CpK of 2.0 is 6σ between the mean and specification limit (since 2.0 is 6/3). This improvement from 1.33 to 2.0 or better may be well justified in some cases to produce much more product near the optimal target. Depending on the process or part, this may improve product performance, product life, customer satisfaction, or reduce warranty costs or assembly problems.

Continually higher CpK indexes for *every* part or process is not the goal, since that is almost never economically justifiable. A cost/benefit analysis that includes customer satisfaction and other true costs of quality is recommended to determine which processes should be improved and how much improvement is economically attractive.

Relationship Between CP and CpK

You can see by looking at the formulas the relationship between CP and CpK. CpK cannot be greater than CP for a process. Only when the mean is exactly centered on the specification midpoint is CpK = CP. Thus CP is valuable as an indicator of how much better the CpK could be if the process were setup so that the center of the distribution were closer to the specification midpoint. CP answers whether the distribution of individuals *could* fit within the tolerances (if centered). CpK answers whether the distribution *does*.

CR Index

CR is a less frequently used substitute for CP. It is simply the inverse of CP. Thus the formula is:

$$CR = \frac{6\sigma}{TOLERANCE} \qquad (5.4.8)$$

For our example:

$$CR = \frac{3.0}{4.0} = 0.75 \qquad (5.4.9)$$

Since CR is the inverse of CP, a CP of 1.33 equals a CR

of 0.75. The smaller the CR value the better. CP is more frequently used than CR since it is much easier to compare CP to CpK.

Zmax/3 Index

Another indication of capability, actually an extension of CpK, is *Zmax/3*. CpK looks at the tail of the distribution that is *closer* to the specification. Zmax/3 is a "CpK" for the tail of the distribution that is *further* from the specification. Another way of expressing this is that CpK is Zmin/3 (see the formula for CpK).

5.5   TESTS FOR NORMALITY

Non-normal distributions are any distribution other than the normal distribution described in Chapter 1. Figures 5.5.1, 5.5.2 and 5.5.3 show examples of three common non-normal distributions. Non-normal distributions may or may not be characteristic of the process itself depending on the size of the sample, the probability of the sample being an accurate reflection of the process, and other factors. A process can be in control and still have a non-normal distribution. An example would be a part machined using stops which prohibit making it too small. The distribution would have a sharp cutoff at the low end of the specification and more of a tail toward the high end, yet the machining process could be stable under these circumstances.

In the strict sense of a capability study, the shape of a distribution is not as important as where it lies in comparison to the engineering specifications. However, if we need to express capability as a numeric value, such as the capability indices covered previously, these calculations assume a normal distribution. Remember, all of these indices use the standard deviation calculated using the normal distribution rules. In some cases it may help to understand the characteristics of non-normal distributions and underlying causes. For further reading on this subject refer to the *Statistical Quality Control Handbook,* by Western Electric Company or other standard statistical texts.

Tests for Normality

Many distributions are symmetrical and unimodal even if

they are not normal (Gaussian). Certain statistical calculations can be completed which help determine if a distribution is truly normal. A brief description of one such calculation follows.

One calculation we can complete is *chi-square* (χ^2). This calculation determines how well the collected data follows the expected distribution, in this case, the normal distribution. This test is also called a *goodness-of-fit* test because we are trying to determine how well the normal curve fits our data. To complete this calculation, the data is placed into classes (cells). A class is similar to a cell on a histogram. It has an upper and lower boundary and we determine which data points fall between the boundaries of each class. The frequency for each class is determined and if there are not at least five samples in the class, it is combined with the next class. Once this is completed, the chi-square calculation is completed for each class. The chi-square formula is:

$$\chi^2 = \Sigma \frac{(O - E)^2}{E}$$

(5.5.1)

O is the observed frequency and E is the expected frequency based on the normal distribution. The Σ means that we add all the class results to obtain the chi-square value. This value is then compared to the value from the chi-square table found in the Appendix, Table A-5. The number of degrees of freedom is determined by the number of final classes minus three. We are estimating three statistics (mean, standard deviation and frequency) to complete this test, which is why we subtract three from the number of classes. You begin to see the reason we need a lot of data to determine if a process is normally distributed and capable. If we do not have sufficient data, we will not obtain enough classes to complete the goodness-of-fit test. Normally, we will select an $\propto$ (Producer's) risk of 0.95. The value from the chi-square table is called the critical chi-square. If the calculated value is smaller than the critical value, we can say that the distribution is normal with a high degree of confidence. Further information concerning chi-square and other tests for goodness-of-fit can be found in most statistical texts.

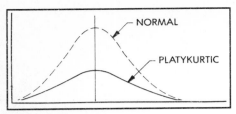

Fig. 5.5.1   Platykurtic curve.

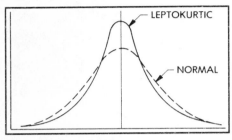

Fig. 5.5.2   Leptokurtic curve.

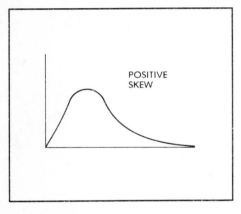

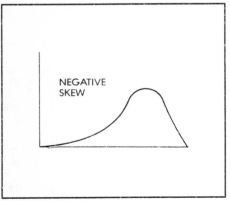

Fig. 5.5.3a & b   Skewed distributions.

There are many other methods for determining if a distribution is normally distributed. This has been just a brief introduction to the topic. Many of the references at the end of this chapter cover this topic in greater detail.

Other symmetrical distributions — Kurtosis is a measure of the flatness of a curve and it can be used to describe curves that are symmetrical but are not normal. Kurtosis is symbolized by γ_2.

- If $\gamma_2 = 0$, the curve is normal
- If $\gamma_2 > 0$, the curve is *platykurtic*
- If $\gamma_2 < 0$, the curve is *leptokurtic*

A platykurtic curve has a low, flat peak and a large dispersion with larger, longer tails than a normal distribution (Figure 5.5.1). A leptokurtic curve has a very high peak and shorter tails than a normal distribution (Figure 5.5.2).

Skewness (γ_1) — Skewness is a measure of a distribution's symmetry. Distributions can be skewed in either a positive or negative direction if one tail extends considerably beyond the other (Figure 5.5.3).

Multi-peaked or bimodal — These distributions have more than one peak (Figure 5.5.4a). A bimodal distribution is characteristic of two mixed distributions each with a different mean. It is possible that two machines produced the lot, or two operators, vendors or materials were involved. When separated, each distribution may be normal, but with a different mean.

Exponential — Exponential curves are encountered often with electronic parts testing (Figure 5.5.4b). The characteristic is having more observations that occur below the mean than above it.

5.6   CAPABILITY MEASURES FOR NON-NORMAL DISTRIBUTIONS

In the previous section, we covered methods for determining if a set of data is normally distributed. If we find our

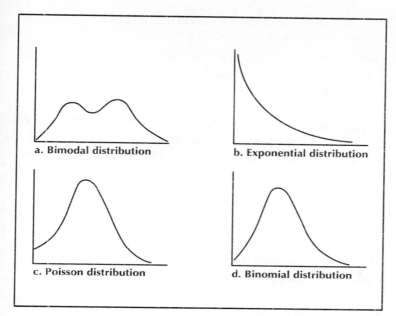

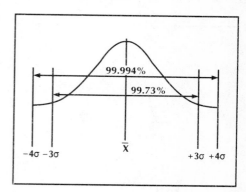

a. Bimodal distribution

b. Exponential distribution

c. Poisson distribution

d. Binomial distribution

Fig. 5.5.4a-d   Non-normal distributions.

Fig. 5.6.1   99.994% of a normal distribution is within the ± 4σ limits.

99.994%

99.73%

−4σ −3σ

X̄

+3σ +4σ

data does not at least approximate a normal distribution, the standard capability indices will not give us valid information because they are based on the normal distribution. How then do we measure process capability if the data is not normally distributed? That is the subject of this section. Note that this section assumes the reader has a more in-depth understanding of statistical techniques than do previous sections. Much of the analysis of non-normal distributions is beyond the scope of this handbook.

The first tool for determining the capability of a non-normal process is visual inspection of a histogram. Specification limits may be placed on the histogram and the data examined for parts outside specification limits. If all parts are well within the specification limits, we have shown graphically that our process is capable. Remember, if the data is non-normal we cannot use either the estimated or standard formula for standard deviation because they are both based on the normal distribution.

Let's look back for a moment at what we were trying to accomplish when we expressed capability in terms of capability indices such as CP and CpK. With those indices, we aimed for values of 1.33 or greater. This correlated to ± 4σ capability or 99.994% of all parts within the specifications (see Figure 5.6.1). If this was our target when working with normal distributions, why not carry the target of 99.994%

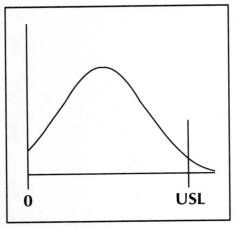

Fig. 5.6.2   *An example of a process distribution with a natural barrier at zero.*

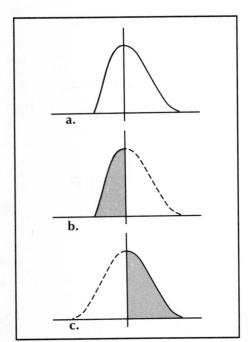

Fig. 5.6.3   *A unimodal but asymmetrical distribution.*

over to our analysis of non-normal distributions? If we can show that 99.994% or more of our parts are within specification, the process is capable.

One statistical method of showing capability requires first transforming the data to approximate a normal distribution. The data can be transformed by taking the natural logarithm (ln) of the data. After calculating the average and standard deviation of the transformed data, we can then use z tables (Chapter 2). The natural log of the specifications must also be calculated. Using the z values calculated for the area outside of the specification limits, we can also calculate the percentage of parts inside the specification limits. In this way, we can express the process capability numerically.

If we are working with a distribution where there is a natural barrier at zero (Figure 5.6.2), we are only concerned with the area above the upper specification limit. The capability of the process can then be expressed as the percentage of parts between zero and the upper specification limit.

Another option in this case is to calculate the CpK for just the upper half of the distribution. Because we know no parts can be below zero, only the upper half of the curve is of concern. We cannot calculate CP, however, because there is no lower tail to this type of distribution.

If we are working with a distribution which is not symmetrical but is unimodal, we may consider looking at the curve as two halves (Figure 5.6.3). The mode should be used as the determination of central tendency. Each half is drawn separately and its mirror image added to give two normal distributions. Each curve can then be analyzed for the percentage outside the appropriate specification limit. In Figure 5.6.3, we would calculate the area below the lower specification limit using curve b and the area above the upper specification limit using curve c. The average and standard deviation of each curve, b and c, must be calculated as well as the z values for the area outside the specification. Once the area outside the specification limit is calculated for each curve, the area within the specification is found by subtracting the area outside the appropriate specification from 0.50 (remember, we are only looking at the shaded half of b and c). The percentage within the specification limits for the original curve is obtained by adding the two results for percentage within the specification for b and c. This then gives us a numerical result for process capability.

These are just a couple of methods for dealing with non-normal distributions. Many other methods are available, such as utilizing Weibull distributions, but involve a higher level of statistical knowledge. Further information can be obtained from standard statistical texts and computer software packages. The method chosen should be appropriate for the process distribution in question and be understandable to the end user.

5.7 ESTIMATING SCRAP AND REWORK

One of the most valuable applications of statistical quality control is that of reducing scrap and rework. The reduction of scrap and rework alone can pay for the installation of an SPC system in a very short period of time. This section covers calculating the amount and cost of scrap and rework, and possible methods of reducing scrap and rework.

Estimating scrap and rework can be done by analyzing the data that was collected in the capability study. It is done most reliably if the distribution of the measurements is reasonably normal, by calculating the area under the normal curve that falls outside the specification limits. An area under the normal curve z-table (Appendix Table A-3) is needed for this calculation. Z-tables were introduced in Chapter 2. Let's review how to use z-tables to find the area outside the specification limits (USL and LSL stand for upper and lower specification limits):

1) Calculate (LSL − Mean) ÷ sigma.
2) Look up the calculated value from areas of normal curve table (Table A-2).
3) Calculate (USL − Mean) ÷ sigma.
4) Look up calculated value from areas of normal curve table (Table A-3).
5) The value calculated in Step 4 gives the area from the left tail to the USL. Subtract this value from 1 to get the area above the USL.
6) Add the values from Steps 2 and 5 together to obtain the total area outside the specification limits.

As an example, let's use the data from Table 5.7.1. Calculations are:

1) $(LSL - Mean) \div sigma = (1.0 - 2.0) \div 0.5 = -2.0$.
2) Table value $= 0.0228$.
3) $(USL - Mean) \div sigma = (5.0 - 2.0) \div 0.5 = 6.0$.
4) Table value $= 1.0$.
5) $1 - 1.0 = 0$ (no area outside the USL).
6) $0.0228 + 0 = 0.0228$.

Total scrap/rework is thus about 2.28%. In addition to calculating the amount of scrap and rework, one can also arrive at a cost. To estimate the cost of scrap and rework we first must estimate the cost of scrap per part and rework per part. If rework costs more than scrap (this condition should be rare) it is more cost efficient to scrap those parts that require rework. The cost of scrap and rework is equal to:

		(5.7.1)
Rework cost =	Scrap cost =	
(Rework cost/part)	(Scrap cost/part)	
(% rework)	(% scrap)	
(number of parts)	(number of parts)	

USL	=	5.0
LSL	=	1.0
MEAN	=	2.0
SIGMA	=	0.5

Table 5.7.1   Sample data.

Fig. 5.7.1   Sample distribution for estimating scrap and rework. The table below summarizes the characteristics.

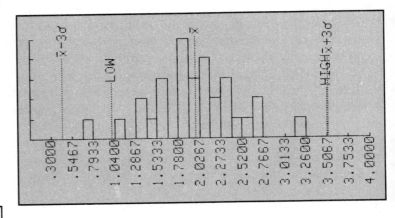

Scrap cost	=	$5.00
Rework cost	=	$2.00
USL	=	3.5
LSL	=	1.0
MEAN	=	2.0
SIGMA	=	0.5

Table 5.7.2   Data for Fig. 5.7.1.

As an example, let's examine the conditions shown in Figure 5.7.1 and Table 5.7.2. Samples below LSL are scrapped, while samples above USL are reworked. First we must calculate the amount of material above USL and below LSL. This is similar to our example above:

1) $(LSL - Mean) \div sigma = -2.0$.
2) Table value is 0.0228.

3) (USL − Mean) ÷ sigma = 3.0.
4) Table value is 0.9987 so area above USL is
   1 − 0.9987 = 0.0013.

Next we calculate the cost:

Cost of scrap = $5.00/part × 0.0228 = $0.114/part
Cost of rework = $2.00/part × 0.0013 = $0.0026/part
Total cost = $0.114 + $0.0026 = $0.1166/part

Calculated Value	Normal	Camp-Meidell	Other
2.00 – 2.49	2.28	5.6	12.5
2.50 – 2.99	0.62	3.6	8.0
3.00 – 3.49	0.14	2.5	5.6
3.50 – 3.99	0.025	1.8	4.1
4.00 – 4.49	0.003	1.4	3.2
4.50 – 4.99	0.00035	1.1	2.5
5.00 or above	0.00003	0.9	2.0

Table 5.7.3 This table is derived from an area under the normal curve table. Use the "other" column if data is not normal and does not conform to the Camp-Meidell conditions.

Estimation Using Non-Normal Data

Estimating scrap and rework on non-normal data is less reliable than similar calculations on normal data. This section will separate data into three categories: data that is roughly normal, data that conforms to the *Camp-Meidell conditions*, and other data. The Camp-Meidell conditions are: mode is equal to mean (mode is in the tallest bar of a histogram) and frequency declines continuously on both sides of the mode (Table 5.7.3). Use the following procedure for this:

1) Calculate (Mean − LSL) ÷ sigma.
2) Look up the calculated value in Table 5.7.3.
3) Calculate (USL − Mean) ÷ sigma.
4) Look up the calculated value in Table 5.7.3.
5) Sum of the two values looked up is the estimated amount of scrap or rework.

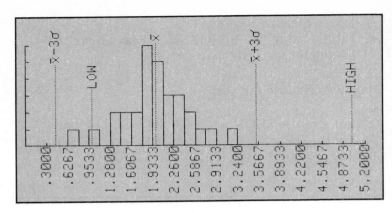

Fig. 5.7.2 Sample distribution having the characteristics shown in the table below.

USL	=	5.0
LSL	=	1.0
MEAN	=	2.0
SIGMA	=	0.5

Table 5.7.4 Data for Fig. 5.7.2.

As an example, let's examine the condition shown in Figure 5.7.2 and Table 5.7.4. The data does not meet the Camp-Meidell conditions; therefore, the "other" column is used on the chart.

1) Index for area under normal curve table (Table A-2) is:
$(2.0 - 1.0) \div 0.5 = 2.0$
2) Area below LSL is 12.5% (from Table 5.7.3).
3) Index for area under normal curve table is:
$(5.0 - 2.0) \div 0.5 = 6.0$
4) Area above USL is 2.0% (from Table 5.7.3).
5) Total scrap and rework is: 12.5% + 2.0% = 14.5%.

The best method for reducing scrap and rework is to improve the manufacturing process to make better parts. This option is not always available due to limitations in materials or equipment. A second option that may reduce the cost of scrap and rework involves shifting the mean of a process. We conclude with an example of how shifting a process mean can minimize cost of scrap and rework. Rework costs $0.20 per part and scrap costs $2.00 per part.

Figure 5.7.3a shows a process where scrap accounts for 25% of production and rework accounts for 10% of production. Cost of scrap and rework is $0.52 per part.

After shifting the process to the midpoint of the tolerance, scrap accounts for 15% of production and rework accounts for 15% of production (Figure 5.7.3b). Cost of scrap and rework is $0.33 per part.

When the process is shifted to the extreme of having no scrap, rework accounts for 84% of production (Figure 5.7.3c). The cost of scrap and rework is $0.17 per part. Al-

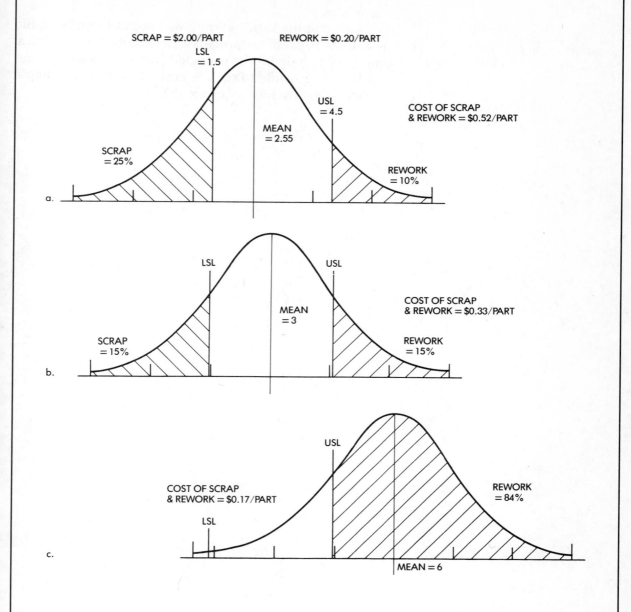

Fig. 5.7.3a, b, & c   An example of maximizing
profits by shifting the mean of the distribution.

though the cost of scrap and rework has been reduced significantly, the cost of quality (which includes the labor of checking parts) may actually be more than in the other examples.

Note that the distribution is assumed to remain stable when the mean of the process is shifted. By using statistical methods to monitor the expected amount of scrap and rework, it is possible to reduce costs through fairly simple means (e.g., machine adjustments).

For Further Reference:

Dovitch, Robert A. "CPI, CpK, Capability Ratio — Measures of Performance or Contributors to Confusion?" *Machine and Tool Blue Book,* pp.10-14, October, 1987.

Downing, Douglas and Clark, Jeff. *Statistics the Easy Way.* New York: Barrons Educational Series, Inc., 1983.

Ford Motor Company. *Quality System Standard.* Product Quality Office, 1983.

Ford Motor Company. *Process Capability and Continuing Process Control.* Statistical Methods Office, 1983.

Ford Motor Company. *Supplier Five Day Seminar on Statistical Thinking.* Ford Motor Body & Assembly Operations Div., 1982.

Grant, Eugene L., and Leavenworth, Richard S. *Statistical Quality Control,* 5th ed. New York: McGraw-Hill Book Company, 1980.

Juran, Joseph M., and Gryna, Frank M. *Juran's Quality Control Handbook,* 4th ed. New York: McGraw-Hill Book Company, 1988.

Juran, Joseph M., and Gryna, Frank M. *Quality Planning and Analysis.* New York: McGraw-Hill Book Company, 1970.

Ott, Ellis R. *Process Quality Control.* New York: McGraw-Hill Book Company, 1975.

Western Electric Co., Inc. *Statistical Quality Control Handbook,* 2d ed. Easton: Mack Printing Company, 1956.

6. MEASUREMENT

6.1 INTRODUCTION TO MEASUREMENT

In modern industry, uniformity for interchangeability of parts is vital for cost effective manufacturing. If a machine part fails or wears out, a replacement must be available and must fit. It will if the parts are made according to specific and accurate measurements. Data must be analyzed on parts and processes to determine conformance to product specifications, and data must be fed back to the manufacturing process to prevent production problems. The quantification of data on parts and processes involves the defining of standard units, calibrating instruments to these standard units, and using these instruments to quantify parts and processes. This quantifying is called *measurement*.

What is Measurement?

The term measurement has several meanings. It can be defined as the process of quantification, comparing an unknown magnitude to a known magnitude. Measurement can also mean the resulting number. In the case of quantification (comparing an unknown magnitude to a known magnitude), it is the act of obtaining specific data about a characteristic of a part or a process; i.e., "The measurement was done in the assembly plant." A resulting number is the specific data on a part or a process; i.e., "The measurements on part A all fell within the specified tolerances." The science of measurement is called *metrology*.

6.2 UNITS OF MEASURE

To allow the quantifying of data on parts or processes, defined standard units must be used. These standard units are called *units of measure*. They are definitions of standardized units which are used to quantify characteristics about which we are interested. Metrologists have developed systems of international units of measure for the purpose of international commerce. The primary systems which are in use today are the English, the metric system, and the Systeme International d'Unites (SI). The more preferred system is now the SI system. Table 6.2.1 gives examples of units of mea-

Characteristic	English	Metric	SI
Length	Foot	Meter	Meter
Force	Pound	Kilogram	Newton
Time	Second	Second	Second
Mass	Slug	Kg − Sec²/Meter	Kilogram

Table 6.2.1   *Examples of units of measure in the three measuring systems.*

sure for the three systems.

The English system was retained by the American colonies when separating from England in 1776. This system consists of units of measures which were developed before the industrial revolution, i.e., a foot is 12 inches, a yard is 3 feet, etc. Only part of the English system is based on decimal multiples.

In 1799 a committee of French scientists, under the direction of the French government, established a system of measures and weights. A basic unit, called the meter, was defined and is the basis for the metric system. This system of measures, primarily concerned with length, area, volume, and mass, is based entirely on a decimal system. See Table A-6 in the Appendix for a list of the units of measure used in this system.

By the 1970s all industrialized countries adopted the metric system with the exception of the United States. However, most U.S. companies who have international involvement tend to use the metric system. The metric system is preferred to the English system on a purely technical basis and has been widely accepted by the scientific community. Most of the world is now adopting the SI system, including the United States and the United Kingdom.

The Systeme International d'Unites (SI) system has evolved more recently from the metric system and is an international system. It consists of seven basic units of measure: length, mass, time, temperature, light intensity, electric current and amount of substance. All are fully compatible with the metric system. Two supplemental units are used for solid angles and planes. There is a long list of units of measure derived from the seven basic units of measure, and standardized terminology for subdivisions and

Prefix	Symbol	Multiple or Subdivision
tera	T	1 000 000 000 000 $= 10^{12}$
giga	G	1 000 000 000 $= 10^{9}$
mega	M	1 000 000 $= 10^{6}$
kilo	k	1 000 $= 10^{3}$
hecto*	h	100 $= 10^{2}$
deka*	da	10 $= 10^{1}$
deci*	d	0.1 $= 10^{-1}$
centi*	c	0.01 $= 10^{-2}$
milli	m	0.001 $= 10^{-3}$
micro	μ	0.000 001 $= 10^{-6}$
nano	n	0.000 000 001 $= 10^{-9}$
pico	p	0.000 000 000 001 $= 10^{-12}$
femto	f	0.000 000 000 000 001 $= 10^{-15}$
atto	a	0.000 000 000 000 000 001 $= 10^{-18}$

Table 6.2.2   Terminology for subdivisions and multiples of units in the SI system.

*Use is discouraged

multiples of units of measure. Latin prefixes are used to indicate subdivisions, and Greek prefixes to indicate multiples of any standard unit. Table 6.2.2 lists the SI system multiples and subdivisions. Tables A-7 and A-8 in the Appendix are listings of the SI system units of measure and conversion charts for the different systems.

6.3   MEASUREMENT STANDARDS AND TRACEABILITY

Primary Reference Standards

For any standards system to be usable, the system must be based on units that are unchangeable. The SI system defines most of its units based on natural phenomena that are unchangeable. An example of this is the definition of a meter. The definitions of a meter and other main units of the SI system can be found in Table 6.3.1. One meter can be reproduced with an accuracy of about ten to the minus eight, which is 0.01mm by definition. These definitions of units are called *standards*.

All countries maintain *primary reference standards* through a "Bureau of Standards" whose purpose is to con-

UNIT	DEFINITION
Meter (m)	1650763.73 wavelengths (in a vacuum) of the uninterrupted transition $2p_{10}$ to $5d_5$ in Kr^{86}.
Kilogram (kg)	Mass of the international kilogram at Sevres, France.
Seconds (s)	1/315,569,259,747 of the tropical year at 12^h ET, 0 January 1900, supplementally defined 1964 in terms to the cesium F,4; M,0 to F,3: M,0, transition, the frequency assigned being 9,192,931,770 hertz.
Kelvin (K)	Defined in the thermodynamic scale by assigning 273.15K to the triple point of water (freezing point, at one standard atmosphere). $1K = 1/273.16$ of the thermodynamic temperature of the triple point of water.
Ampere (A)	The constant current which, if maintained in two straight parallel conductors of infinite length, of negligible circular sections, and placed 1 meter apart in a vacuum, will produce between these conductors a force equal to 2×10^{-7} M.K.S. unit of force per meter of length.
Candela (cd)	1/60 of the intensity of one square centimeter of a perfect radiator at the temperature of freezing platinum.
Mole (mol)	The amount of substance of a system which contains as many elementary entities as there are atoms on 0.012 Kilograms of carbon-12. The elementary entities must be specified.

Table 6.3.1   Definitions of fundamental units of the SI system.

struct and maintain these standards. The standards consist of copies of the international kilogram and measuring systems which can verify the units and subunits of the defined standards. These standards are then used as a basis for the calibration of equipment. It is not, however, practical for the bureaus of standards and standards laboratories to calibrate all equipment, so secondary and tertiary standards were developed to transfer the primary standards for the calibration of instruments used in general laboratories and manufacturing areas.

Standards Hierarchy

Equipment used by the technicians and inspectors is calibrated against a set of *working standards*. The working standards are referred back to the primary standards through the use of one or more *transfer standards*. This reference from one standard to a higher, more accurate standard is known as *calibration*. Some specialists suggest that a precision of ten to one be used to transfer from one standard to the next. This may not be necessary because the combination of many levels can be represented by the square root of the sum of the squares instead of by the sums of the precisions of the levels. A precision ratio of five to one has been accepted among transfer standards to allow for a longer hierarchy of transfer standards.

When a piece of equipment is calibrated and can be related back to a primary standard through transfer standards, this is known as *traceability*. A system of documented certification of accuracy allows for calibration to be traceable to the National Bureau of Standards, which maintains the primary reference standards. Figure 6.3.1 shows the hierarchy of standards used for traceability.

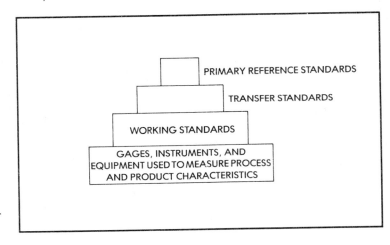

Fig. 6.3.1 Hierarchy of standards used for traceability.

6.4 MEASURING INSTRUMENTS

There are several terms often used when referring to characteristics of measuring instruments. These terms describe how sensitive instruments are to measured quantities, how bias affects the measurements, and how repeatable an instrument is when measuring.

Resolution

Resolution (or sensitivity) is the measure to which an instrument can sense the variation of a quantity to be measured. It is the maximum incremental change in the instrument's output with a change in any specified portion of its measuring range. Figure 6.4.1 shows that there is a discrete change in output for a large enough change in the input measuring range.

Accuracy

Accuracy is the condition or quality of conforming exactly to a standard. The accuracy of an instrument is the extent to which the average of many measurements made by the instrument agrees with the true value or standard being measured. The difference between the average and the true value is the error or inaccuracy. A lack of accuracy is sometimes referred to as a *bias*. This condition, when a result of the measuring instrument, is known as *out of calibration*.

A measuring instrument's accuracy must be considered over the whole range of the measuring instrument. This is often expressed as *linearity*. Linearity is the maximum deviation of the actual measurements from a defined theoretical straight line characteristic. It is expressed as a percentage of the theoretical output and measured output over the total theoretical output characteristic. The ratio can be expressed as follows:

$$\text{Linearity} = 1 - \frac{\theta - E}{\theta_T}$$

(6.4.1)

where: θ is the theoretical output
E is the measured output
θ_T is the total theoretical output

Often the linearity of an instrument is expressed in terms of *nonlinearity* ($1 -$ linearity ratio). Nonlinearity can be expressed as a percentage of deviation from the theoretical output and measured output over the total theoretical output.

$$\% \text{ Nonlinearity} = \left(\frac{\theta - E}{\theta_T}\right)(100)$$

(6.4.2)

Figure 6.4.2 illustrates graphically the concept of linearity.

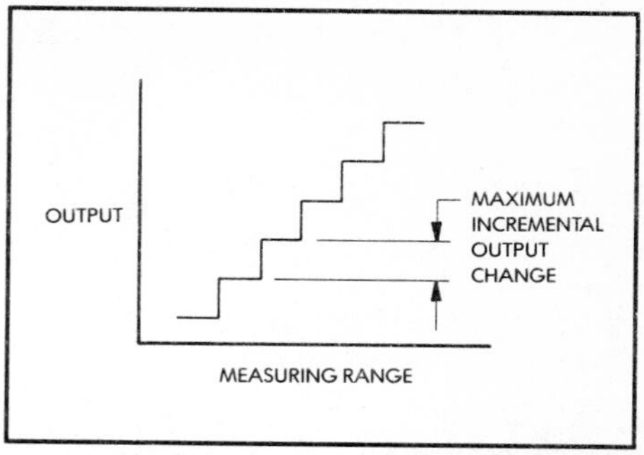

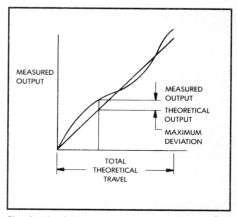

Fig. 6.4.1 *Instrument resolution is a measure of the sensitivity, or discrete change in output, for a change in measuring range.*

Fig. 6.4.2 *Instrument accuracy expressed as a measure of linearity, or deviation from a theoretical straightline characteristic.*

Precision

Precision (also known as *repeatability*) is the variation in readings obtained when repeating exactly the same measurement. The precision of an instrument is the ability to repeat a series of measurements on the same piece and obtain the same results for each measured value. The variation in measured values can be expressed in terms of a standard deviation of the measuring error. The smaller the standard deviation the more precise the instrument.

Accuracy Versus Precision

Confusion often exists between the terms accuracy and precision. The confusion exists because the terms are often interchanged in their usage. Accuracy and precision are two different concepts. The accuracy of an instrument can be improved by recalibrating to reduce its error, but recalibration generally does not improve the instrument's precision. The difference between the two terms will be further clarified in the following examples.

Figure 6.4.3 represents a set of 28 measurements made

Fig. 6.4.3 Example of accuracy without precision.

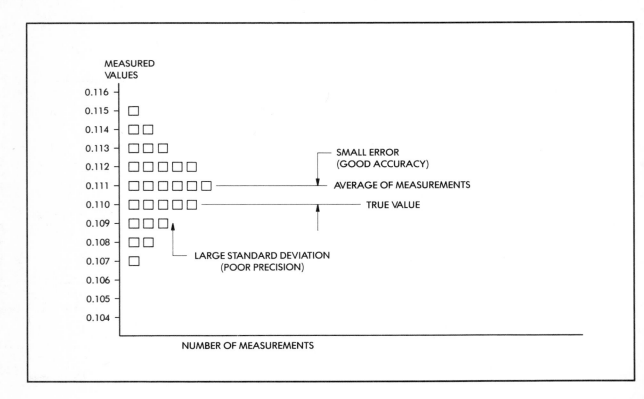

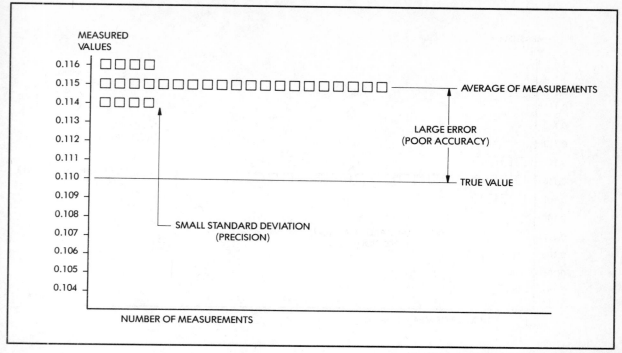

MEASURED
VALUES

NUMBER OF MEASUREMENTS

AVERAGE OF MEASUREMENTS

LARGE ERROR
(POOR ACCURACY)

TRUE VALUE

SMALL STANDARD DEVIATION
(PRECISION)

Fig. 6.4.4   Example of precision without accuracy.

with the same instrument on the same part which shows good accuracy with little precision. The accuracy is represented by the small difference (error) between the true value of 0.110 and the average of the measurements of 0.111, which is 0.001. The precision in this case is poor because of the wide distribution of measurements (ranging from 0.107 to 0.115), as shown by the bar graph (each box represents a measurement). This variation can be expressed in terms of a large *standard deviation of the measurements* error.

Figure 6.4.4 shows 28 measurements taken with a different instrument on the same part as in Figure 6.4.3. It shows that there is precision or good repeatability but that the accuracy is poor. The precision can be seen in the diagram by noting that the distribution of the measurements (ranging from 0.114 to 0.116) is closely grouped around the average (0.115) of the measurements. The standard deviation of the measurements is small in this case. The large error between the true value (0.110) and the average (0.115) of the measurements is 0.005 and represents poor accuracy.

Figure 6.4.5 shows 28 measurements taken with a different instrument on the same part as the two previous ex-

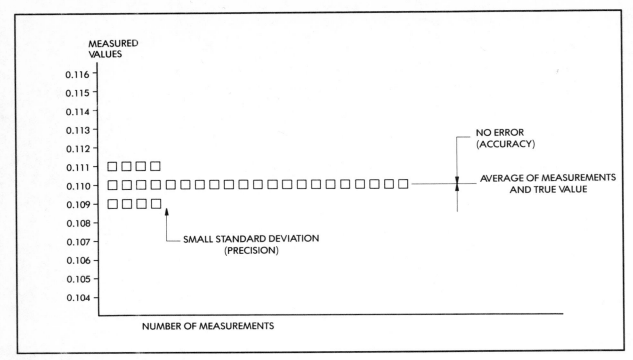

MEASURED VALUES

NUMBER OF MEASUREMENTS

NO ERROR (ACCURACY)

AVERAGE OF MEASUREMENTS AND TRUE VALUE

SMALL STANDARD DEVIATION (PRECISION)

Fig. 6.4.5 Example of accuracy with precision.

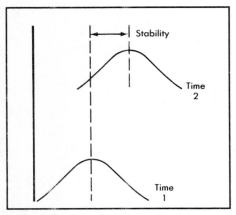

Fig. 6.4.6 Example of lack of stability over time.

amples. It shows that the precision or repeatability is good as well as the accuracy. Note in Figure 6.4.5 that the true value (0.110) and the average value of the measurements (0.110) are the same, indicating that the accuracy is very good. It can also be noted that the variation of the measurements is quite small (ranging from 0.109 to 0.111) which indicates precision or good repeatability.

Stability

Stability refers to the difference in the average of at least two sets of measurements obtained with the same measuring device on the same parts taken at different times. See Figure 6.4.6.

6.5   SOURCES OF ERROR IN MEASUREMENT

Error in measurement can result even with the best of equipment and measuring techniques. A measurement error is the difference between the measured value of a magnitude and the true value. This difference from the true value can be a problem of accuracy or precision.

Several sources of error exist in most instruments — non-linearity, hysteresis (e.g., gear back-lash), and sensitivity to environmental factors such as temperature, magnetic or electrical fields are a few examples. Instrument error is often magnified by the fixturing required in the measuring process. Poor electrical connection, improper fastening of mechanical linkages, and loose clamps are examples of fixturing problems. Temperature induced error, in addition to its effects on measuring instruments, affects the specimen being measured. As temperature changes, the length (L) of a specimen changes as follows:

$$\Delta L = (L)(\propto)(\Delta T)$$

where: L = Original length of the specimen   (6.5.1)
       $\propto$ = Thermal expansion coefficient
       ΔT = Temperature variation

Table 6.5.1 gives the thermal expansion coefficients for several industrial materials.

Deformation

Deformation is the second largest source of error following temperature. Deformation can be caused by the following:

- Force exerted on a specimen by the measuring instrument.
- Placement of the specimen supports.
- Placement of the instrument supports.

A compression force (within the elastic limit) will cause deformation. The deformation (DL) can be calculated as follows by Hook's law:

$$\Delta L = \frac{(F)(L)}{(E)(A)}$$

where: F = measuring force, Kgf   (6.5.2)
       L = length of test piece in mm
       E = Young's modulus, Kgf/mm^2
       A = cross sectional area in mm^2

Material	Thermal Expansion Coefficient (Per°Celsius)
Aluminum	23.8×10^{-6}
Brass	18.5×10^{-6}
Bronze	17.5×10^{-6}
Carbon Steel	$11.7 - (0.9 \times C\%) \times 10^{-6}$
Cast Iron	9.2 to 11.8×10^{-6}
Ceramics	3.0×10^{-6}
Chromium Steel	11 to 13×10^{-6}
Copper	18.5×10^{-6}
Crown Glass	8.9×10^{-6}
Duralumin	22.6×10^{-6}
Flint Glass	7.9×10^{-6}
Gold	14.2×10^{-6}
Gunmetal	18.0×10^{-6}
Invar (36% nickel)	1.5×10^{-6}
Iron	12.2×10^{-6}
Nickel	13.0×10^{-6}
Nickel-Chromium Steel	13 to 15×10^{-6}
Nickel Steel (58% nickel)	12.0×10^{-6}
Nylon	10 to 15×10^{-6}
Phenol	3 to 4.5×10^{-6}
Plutonium	9.0×10^{-6}
Polyethylene	0.5 to 5.5×10^{-6}
Quartz	0.5×10^{-6}
Silver	19.5×10^{-6}
Steel	11.5×10^{-6}
Tin	23.0×10^{-6}
Vinyl Chloride	0.7 to 2.5×10^{-6}
Zinc	26.7×10^{-6}

Table 6.5.1   Thermal expansion coefficients.

An example using Hook's law is as follows: a gage block (A = 9mm × 35mm = 315mm² and L = 1000mm) is measured with a measuring force of F = 1kg. Young's Modulus for steel is E = 2 × 10⁴ kg/mm². The deformation would be as follows:

$$\Delta L = \frac{(1 \text{ Kgf}) (1000 \text{ mm})}{(2 \times 10^4 \text{ Kgf/mm}^2) (315 \text{ mm}^2)} \qquad (6.5.3)$$

$$= 0.0001587 \text{ mm}$$
$$= 0.16 \text{ } \mu m$$

Operator Error

There are several sources of operator error which will cause variation in the data that is recorded. An operator of a measuring system can get different results even when measuring the same product and with the same measuring system. This is due to the fact that even with the same operator there will be slight differences in measuring techniques from one measurement to the next. When two operators are measuring the same product with the same measuring instrument, differences will occur in the recorded measurements. These differences can be even greater than with the same operator due to even greater differences in measuring techniques. Depending on the measuring technique, variation in data can be more dispersed or show up as a systematic error.

Variations in recorded measurements can also be evident if different test procedures are used to measure the same product. Different test procedures can introduce variation in measurement technique if they are not carefully designed. Errors in data often occur when the operator takes a reading from the instrument and then enters it on a form or similar media. These errors can consist of transposition of numbers, recording incorrect numbers, or writing them illegibly so transcription errors occur later when the data is analyzed.

6.6   REDUCING MEASUREMENT ERROR

Tolerances and Measurement Error

It is recommended that the ratio of the product tolerance to the precision of the measuring instrument be a 10:1 ratio for an ideal condition. In a worst case condition a 5:1 ratio can be used. These are rules of thumb; the actual ratio should be based on the level of confidence required for each situation. When the tolerance is mixed with the measurement error, a good component may be diagnosed as being bad, or a bad component may be accepted. See Figure 6.6.1 for an example of a 10:1 precision versus part tolerance distribution.

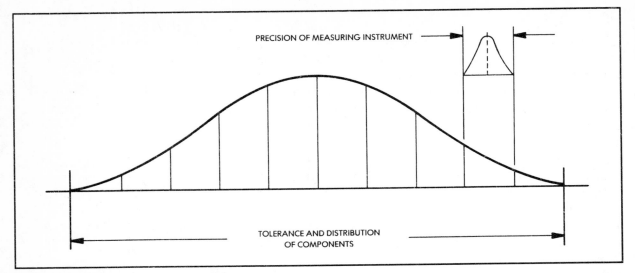

PRECISION OF MEASURING INSTRUMENT

TOLERANCE AND DISTRIBUTION
OF COMPONENTS

Fig. 6.6.1   The part tolerance compared to
the precision of the measuring instrument
should be at 10:1 ratio ideally.

Analysis of Measurement Error (Capability)

Section 6.5 described many sources of measurement error. The error causes variations in the observed values (measurements). Conclusions can be drawn about measurement error based on the formula below. The relationship of various causes is assumed to be independent.

$$\sigma_{obs} = \sqrt{\sigma^2_{cause(a)} + \sigma^2_{cause(b)} + \ldots + \sigma^2_{cause(n)}} \qquad (6.6.1)$$

where σ_{obs} = observed variation
$\sigma^2_{cause(n)}$ = variance of a cause

The causes are not always easy to identify and may be interrelated. Often it is feasible to identify causes by experiment. Generally, measurement variation occurs because of variation in the system of measurement and the variation of the product being measured. This can be expressed as follows:

$$\sigma_{obs} = \sqrt{\sigma^2_{meas} + \sigma^2_{prod}} \qquad (6.6.2)$$

where σ^2_{meas} = variance due to measurement
σ^2_{prod} = variance due to product

If the variation in the system of measurement is less than 10% of the observed variation, then the effect upon the variation in product will be less than 1%. The rule of thumb of 10% variation is based on the statement below.

$$\sigma_{prod} = \sqrt{\sigma_{obs}^2 - \sigma_{meas}^2} \qquad (6.6.3)$$

The following examples will illustrate this concept. To determine the amount of variation within a group of product, samples were selected and measured using the same measuring device. The observed variation, or standard deviation was found to be 12. Next, a number of repeat measurements were taken on the same sample, using the same measuring device, and the observed variation on standard deviation was found to be 3. Product variation was then calculated to be:

$$
\begin{aligned}
\sigma_{prod} &= \sqrt{12^2 - 3^2} \\
&= \sqrt{135} \qquad (6.6.4) \\
&= 11.6
\end{aligned}
$$

It can be seen that the measurement variation has a small impact on the product variation by comparing the observed and the calculated product variation. There is about a 3% difference in the observed and product variation.

In another example, the observed variation was 12.6 and the variation of measurement was 9. The product variation was calculated at 8.8.

$$
\begin{aligned}
\sigma_{prod} &= \sqrt{12.6^2 - 9^2} \\
&= \sqrt{77.8} \qquad (6.6.5) \\
&= 8.8
\end{aligned}
$$

Where product variation and measurement variation are almost the same, further evaluation is needed to determine if the measurement variation can be reduced to a more acceptable level.

Minimizing Instrument Error

Accuracy and precision can be controlled if appropriate steps are taken. If a systematic error is evident, then a correction can be applied to the data. If the accuracy is low, then a correction factor can be added to each measurement to adjust the data to the proper reading. An adjustment could also be made to the instrument to bring it back into calibration.

Calibration programs are a means of checking equipment that is used for quality inspection. Calibration control would include provisions for periodic audits on instruments to check their accuracy, precision, and general condition. New equipment should be included in a calibration program to ensure it is properly functioning before use in an inspection process.

Inventory control of instruments is often used and records kept to keep track of the use and calibration of instruments. Calibration schedules are kept on instruments as a means to monitor instrument conditions. Elapsed calendar time is the most widely used method. Checks are made at calendar intervals to check the instrument's performance. Another method is to schedule calibration by the actual usage. This is done by counting the number of measurements an instrument has made. Metering the hours of use is also a technique to monitor calibration intervals.

Adherence to calibration schedules is probably the most important aspect of calibration control. Without this, the calibration schedule's value would be greatly reduced. Many different systems can be used to accomplish this adherence.

Tool Control Charts

To minimize the effects of calibration intervals on the observed variation of a process, it is recommended that control charts be kept on all auditing tools. The charts can be placed on the gaging fixture or taken on the audit route. An extra reading, the tool control reading, should be included with each subgroup. For example, if five samples are taken on each item in an audit route, then the sixth should be the tool control reading, and it should be plotted on a chart.

Tool control charts are the only reliable method of corroborating data from the current process with historical data. The tool control chart can be used to compensate for any bias due to measurement error when trying to equate the data. The charts will also provide a time independent indication of a need for calibration or repair.

Reducing Operator Errors

Adequate training is required for the operators of instruments to be able to properly utilize them. Many errors can be introduced into measurement by variation in operator technique. With the use of training, these types of errors can be reduced. Proper test procedures and fixturing will also help reduce errors in measurement. Procedures will give a documented systematic approach to the way a test or inspection is to be performed. These procedures should include enough detail to prevent differences in tests or inspections from one operator to the next. Test procedures themselves should be checked to ensure that they are fully useful and properly designed. It does no good for operators to follow a procedure that does not provide the desired results.

Automation of measurement can be an effective means to reduce operator errors in the recording of data. An instrument that automatically records its measurements will eliminate transposition errors and other types of recording errors. Not only do automated instruments reduce errors, but in many cases the time required to record measurements is markedly reduced.

6.7   GAGE CAPABILITY STUDIES

All gages and test equipment have inherent variability. Whether or not this measurement error (capability) precludes using a given gage for a statistical analysis in a specific application can be readily determined using an established methodology.

Consider a sample study: five parts, three inspectors, one gage, a single control dimension and two sets of readings.

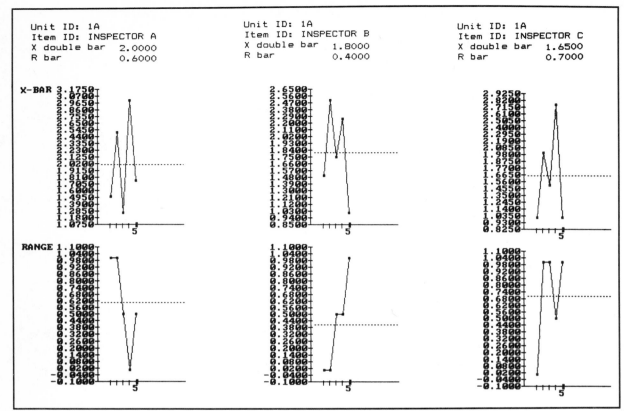

```
Unit ID: 1A                    Unit ID: 1A                    Unit ID: 1A
Item ID: INSPECTOR A           Item ID: INSPECTOR B           Item ID: INSPECTOR C
X double bar   2.0000          X double bar   1.8000          X double bar   1.6500
R bar          0.6000          R bar          0.4000          R bar          0.7000
```

Fig. 6.7.1 *The Results of readings made by three inspectors plotted on x̄ & R charts, on five parts of two readings on each part.*

(The following is reproduced with permission from *Industrial Quality Control*. See the reference at the end of the chapter.)

Five parts are selected and a single dimension specified for measurement. The parts are then numbered sequentially, one through five. Three inspectors are selected, each uses the same gaging instrument, and measures the parts in a random order, to assure that any drift or change will be spread randomly throughout the study. When the first set of readings are obtained, the inspectors again measure a second set in a random order. To eliminate the possibility that one inspector could bias another one's reading, the individual conducting the study should be certain that no information is exchanged. In Figure 6.7.1, the results are plotted on an average and range chart. The readings and results will be carried forward and used as a sample capability study.

Gage Repeatability

Repeatability is the variation obtained when one person, using the same measuring instrument, measures the same dimension two or more times. See Figure 6.7.2. In this example only two measurements were made on each piece, or a sample size of two. The standard deviation for these values can be estimated using the average range. In control chart applications this is done using:

$$\hat{\sigma}_x = 1/d_2 \cdot \overline{R} \qquad (6.7.1)$$

The factor, d_2 is essentially independent when the number of samples, k, is larger than 10 or 15. For smaller k values, Table 6.7.1 gives corrected $1/d_2$ factors.

Calculating the estimated standard deviation within parts for each inspector in Figure 6.7.1 (repeatability) gives the following results:

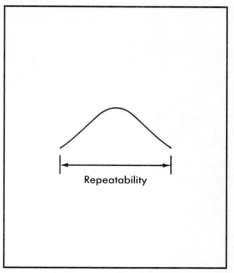

Fig. 6.7.2 Gage repeatability.

Table 6.7.1 Factors for calculating standard deviation.

Factors $1/d_2$* for Converting the Average Range, $\overline{R}$, into a Standard Deviation $\hat{\sigma}_x$.

	k = 1	2	3	4	5	8	10	∞
n = 2	0.709	0.781	0.813	0.826	0.840	0.855	0.862	0.885
3	0.524	0.552	0.565	0.571	0.575	0.581	0.581	0.592
4	0.446	0.465	0.472	0.474	0.476	0.481	0.481	0.485
5	0.403	0.417	0.420	0.422	0.424	0.426	0.427	0.429
6	0.375	0.385	0.388	0.389	0.391	0.392	0.392	0.395
7	0.353	0.361	0.364	0.365	0.366	0.368	0.368	0.370
8	0.338	0.344	0.346	0.347	0.348	0.348	0.350	0.351
9	0.325	0.331	0.332	0.333	0.334	0.334	0.336	0.337
10	0.314	0.319	0.322	0.323	0.323	0.324	0.324	0.325

k = Number of Samples (Number of parts measured)
n = Sample Size (Number of times each part was measured)

*Based on d_2 factors, Table D3, p. 910, *Quality Control and Industrial Statistics*, A.J. Duncan.

$$\text{Inspector A}$$
$$R = 0.6$$
$$1/d_2 = 0.840$$
$$\hat{\sigma} \text{ (within parts)} = (0.840)(0.6) = 0.504$$

(d_2 is taken from Table 6.7.1 — Sample size is 2 and 5 parts were measured.)

$$\text{Inspector B}$$
$$R = 0.4$$
$$1/d_2 = 0.840$$
$$\hat{\sigma} \text{ (within parts)} = (0.840)(0.4) = 0.336$$

$$\text{Inspector C}$$
$$R = 0.7$$
$$1/d_2 = 0.840$$
$$\hat{\sigma} \text{ (within parts)} = (0.840)(0.7) = 0.588$$

Assessing these results individually, Inspector B has the least variation and the best repeatability. Assuming, however, that all three inspectors normally perform this gaging operation, a standard deviation can be calculated using the average range for all three:

$$R = (0.6 + 0.4 + 0.7) \div 3 = 0.567$$

$$\hat{\sigma} \text{ (within parts)} = 1/d_2 \times \overline{R} =$$
$$(0.855)(0.567) = 0.5018 = 0.502$$

Note that for d_2, k has changed from 5 to 15. From Table 6.7.1, $1/d_2 = 0.885$.

Gage Reproducibility

Reproducibility is the variation in measurement averages (between inspector variation), where:

$$\overline{R}_3 = \overline{X}_{max} - \overline{X}_{min} \qquad (6.7.2)$$

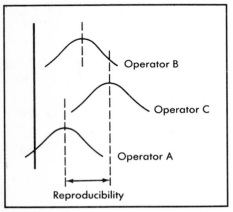

Fig. 6.7.3 Gage reproducibility among three operators.

See Figure 6.7.3. The $1/d_2$ factor is based on one sample, with a sample size equal to three. From Table 6.7.1 this value is 0.524. Then,

$$\hat{\sigma}_{(Between\ Inspectors)} = 1/d_2 \cdot \overline{R}_3$$
$$= (0.524)(2.0 - 1.65) = 0.183 \qquad (6.7.3)$$

Statistically, variances can be combined to give a single value according to the formula:

$$\sigma_A^2 = \sigma_B^2 + \sigma_C^2 \tag{6.7.4}$$

This resultant value would be used to measure the repeatability and reproducibility, or

$$\tag{6.7.5}$$

$$\hat{\sigma}_{\text{(Repeatability \& Reproducibility)}} = \sqrt{(0.502)^2 + (0.183)^2} = 0.534$$

It is immediately apparent that reduced variability could be attained by inspector training, which could minimize the differences in averages, or by obtaining a more precise gaging device.

Assume the total tolerance for the parts used in this study is 3.0. A 99 percent spread factor is chosen to provide a high confidence level for the gage repeatability and reproducibility. The tolerance consumed by the measuring system is calculated using this formula:

$$\tag{6.7.6}$$

$$\frac{(5.15)\,(\hat{\sigma}_{R\&R})}{\text{Tolerance}_{(Parts)}}\,(100)$$

$$\frac{(5.15)\,(0.534)}{3.0}\,(100) = 91.67\%$$

The constant, 5.15, is derived from the "Table of Areas Under the Normal Curve" (Table A-3). The constant represents a 99 percent confidence level that any gage R&R readings with these parts, equipments and operators would fall within the same range of variability.

This percentage (91.73) cannot be directly associated with the percent good parts rejected or the percent bad parts accepted. The situation is undesirable, however, since the accepted standard for gage capability is approximately ten percent or less of the total tolerance.

The same information is shown in Figure 6.7.4, using a standardized form which simplifies the calculations required. (This is a form developed by a major automotive manufacturer.) All of the calculations given above are trans-

REPEATABILITY AND REPRODUCIBILITY

GAGE TYPE _____ DATE _____

B/P SPEC. _____ CHARACTERISTIC _____ MACIL NO. _____

PART NUMBER _____ PART NAME _____ GAGE NO. _____

COL. NO.	1	2	3	4	5	6	7	8	9
INSPECTOR	A—			B—			C—		
SAMPLE #	1st TRIAL	2nd TRIAL	DIFF.	1st TRIAL	2nd TRIAL	DIFF.	1st TRIAL	2nd TRIAL	DIFF.
1	2.0	1.0	1.0	1.5	1.5	0	1.0	1.0	0
2	2.0	3.0	1.0	2.5	2.5	0	1.5	2.5	1.0
3	1.5	1.0	.5	2.0	1.5	.5	2.0	1.0	1.0
4	3.0	3.0	0	2.0	2.5	.5	2.5	3.0	.5
5	2.0	1.5	.5	1.5	.5	1.0	1.5	.5	1.0
6									
7									
8									
9									
10									
TOTALS	10.5	9.5	3.0	9.5	8.5	2.0	8.5	8.0	3.5
AVERAGES	2.1	1.9	.6	1.9	1.7	.4	1.7	1.6	.7

	2.1	$\bar{R}_A$		1.9	$\bar{R}_B$		1.7	$\bar{R}_C$
SUM-	4.0		SUM-	3.6		SUM-	3.3	
$\bar{X}_A$	2.0		$\bar{X}_B$	1.8	—	$\bar{X}_C$	1.65	

RANGE VARIATION

$\bar{R}_A$ (Col. 3)	0.6
$\bar{R}_B$ (Col. 6)	0.4
$\bar{R}_C$ (Col. 9)	0.7
SUM	1.7
$\bar{R}_1$	0.567

$UCL_R = (3.268)(\bar{R}_1)$

$= (3.268) \boxed{0.567}$

$= \boxed{1.85}$ ①

REPRODUCIBILITY - APPRAISER VARIATION

Difference in Means

$\bar{R}_3 = \bar{X}_L - \bar{X}_5 = (2.0) - (1.65) = \boxed{0.35}$

Standard Deviation $(SDM) = (1/d_2)(\bar{R}_3)$

$= (0.524)(0.35) = \boxed{0.183}$

Variance $= (SDM)^2 = \boxed{0.183}^2 = \underline{0.0335}$ ②

REPEATABILITY - (EQUIPMENT VARIATION)

Difference in Readings

Standard Deviation $(SDR) = (1/d_2)(\bar{R})$

$= (0.885)(0.567) = \boxed{0.502}$

Variance $= (SDR)^2 = \boxed{0.502}^2 = \underline{0.252}$ ③

REPRODUCIBILITY AND REPEATABILITY (COMBINED)

Standard Deviation (R & R) $= \sqrt{(SDM)^2 + (SDR)^2} = \sqrt{\underline{0.0335} + \underline{0.252}}$

$SDRR = \boxed{0.5343}$ ④

PERCENT TOLERANCE CONSUMED BY REPRODUCIBILITY AND REPEATABILITY

P.T.C. $= [(5.15)(SDRR) \div DRAWING\ TOLERANCE](100)$

$= [(5.15)(\underline{0.5343}) \div \underline{3}](100) = \boxed{91.73}$ %

Fig. 6.7.4 Standard Gage Capability Form.

The procedure for estimating gage R & R summarized in the following steps:

1. Select five or more parts and prepare them for gaging (wash, de-burr, and number).

2. Choose two or more appraisers (those who ordinarily use the equipment are preferable). Have each appraiser note any characteristics about the instrument which makes reading difficult. The slightest defect should be corrected before conducting the study.

3. Each appraiser, using the same instrument, measures the parts in a random order and records the values obtained.

a. Appraisers should not see other appraisers' readings.

b. Record readings to one more decimal place than the instrument's least count, i.e., if the least count is 0.001, "read to 0.0001." This requires estimating but in many cases is essential.

4. After the first set of measurements have been recorded, each appraiser repeats the measurements — without referring to his first results. (More than one repeat reading can be taken if deemed advisable.)

5. Record the readings in the appropriate columns on a form similar to Figure 6.7.4.

6. Calculate the range (difference between individual appraiser's readings) for each appraiser. Fill in all totals and calculate the averages indicated. Compute R_1.

7. Calculate the upper control limit (circle 1 in Figure 6.7.2) for the ranges and compare individual ranges to this value. Discard points out-of-control and recalculate R_1. (Follow the procedure suggested previously.)

8. Determine the standard deviation to measure reproducibility. Calculate R_3, the difference between the largest and smallest appraiser means $(R_3 = \bar{x}_L - \bar{x}_S)$; select appropriate $1/d_2$ factor from Table 6.7.1. (For three appraisers, n = 3, k = 1.) Calculate the variance and enter at circle 2 in Figure 6.7.4.

9. Determine the standard deviation to measure repeatability. Select appropriate $1/d_2$ factor from Table 6.7.1. (For three appraisers, each measuring five parts twice, n = 2, k = 15, use k = ∞.) Calculate the standard deviation using R_1, the variance, and enter at circle 3 in Figure 6.7.4.

10. Combine the variances calculated at circle 2 and 3, take the square root and enter at circle 4 to determine the standard deviation for reproducibility and repeatability.

11. To find the percent tolerance consumed by repeatability and reproducibility, multiply the value at circle 4 by 5.15, divide by the drawing tolerance and multiply by 100 to convert to a percent.

ferred and shown on Figure 6.7.4. Figure 6.7.6 is the same type of information as in Figure 6.7.4 but Figure 6.7.6 shows the long method of calculating. What follows is the sequence of operations needed to complete this standardized form.

Measurement Error Effect on Acceptance Decisions

Previous mention was made concerning the effect that measurement error had on accepting defective material or rejecting good material. This aspect was explored by Alan R. Eagle. (See the references at the end of the chapter.) The general concept is shown graphically in Figure 6.7.5. In this

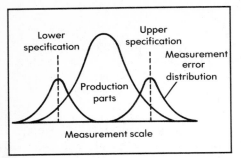

Fig. 6.7.5 *Measurement error schematic.*

illustration, the gage or testing device is zeroed on the upper and lower specifications, and the measurement errors distributed about these points.

It is obvious that, due to these errors, a probability exists that a good part could be rejected or a bad part accepted. Several assumptions were made in calculating the probabilities involved (a normal practice in statistical analysis). These are:

- The distribution of production parts is normal.
- The parts are centered on the blueprint nominal.
- The specifications intersect the parts distribution at plus and minus two standard deviations.
- The measurement error distribution is normal and distributed about the zero setting.

Using curves developed by Eagle, it is shown that a gage which consumes 91 percent of the specified tolerance would only accept 1.45 percent defective parts on the low end and would reject about seven percent of parts on the high side. Since these would logically be re-inspected before final rejection, relatively few errors would be made.

In *Quality Planning and Analysis* (see references), a rule of thumb is given:

If the ratio of three standard deviations of measurement error to product tolerance is less than about 25 percent, then the effect of measurement error on decisions can usually be ignored.

For situations where the specifications include more than plus and minus two standard deviations of the parts, the graphs overestimate the actual probabilities. For the converse situation, i.e., less than two standard deviations included in the specifications, the graphs underestimate, but this latter condition would indicate that the major problem exists in the machine or process, not the gaging.

If the process is not centered or not normal, errors will also exist, and the problem would have to be solved by other methods. Simulation is one technique which could be employed if the conditions warranted such an approach. Also, if the gage is used to obtain data for control chart purposes, the fact that the process mean fluctuates could produce results that would have to be individually analyzed.

This method was presented primarily to reveal that the

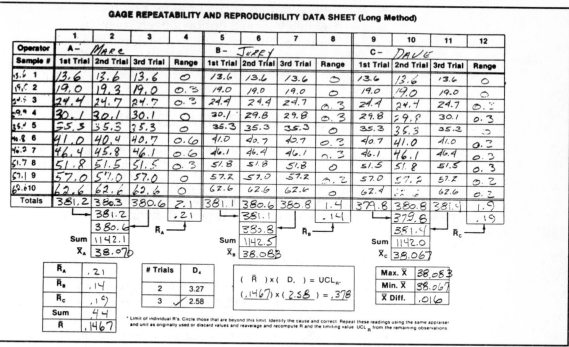

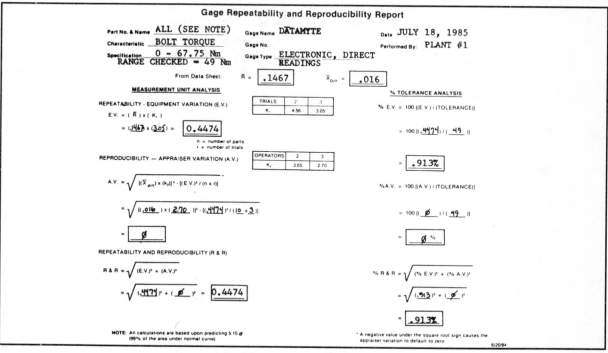

Fig. 6.7.6 Standard Gage Capability Form
showing the long method for calculating R&R.

percent tolerance consumed by repeatability and reproducibility should not be judged solely by its magnitude, but must be further evaluated to determine its effect on potential decision errors.

6.8   TORQUE AUDITING

Measuring fastener torque on an ongoing basis for quality control purposes is called *torque auditing*. To help understand the need for torque auditing, it will help to review some of the basic principles of torque measurement. Some case studies of torque auditing appear in the applications section.

How Fasteners Fasten

A special property of fasteners is their *elasticity*. When a fastener is tightened it actually stretches or elongates like a rubber band. A fastener stretched in one direction, such as when tightening a nut down on the threads of a bolt, will provide tension in the opposite direction. As shown in Figure 6.8.1, this tension (T) is what is used to provide the *clamping force* (C) of a joint.

One way to measure tension is to measure the length (or elongation) of the fastener after it is tightened and compare it to its length at rest. This is not very practical in an industrial environment. A more practical but less accurate method is torque measurement.

What is Torque?

Torque is the force that tends to produce rotation or torsion. Figure 6.8.2 illustrates the application of torque and its basic unit of measure, the *pound-foot* (also called a footpound). One pound-foot is defined as one pound of force applied perpendicular to, and at a distance of one foot away from, the axis of rotation. The wrench acts like a lever to produce the torsion or twisting effect. The unit of measure for torque in the metric and SI systems is the Newton-meter. A Newton is a measure of force and is equivalent to one kilogram-meter per second squared ($kg\text{-}m/sec^2$).

Table A-8 in the Appendix provides conversion factors among various units of measure. In an ideal situation, the amount of torque applied to a fastener is a function of the

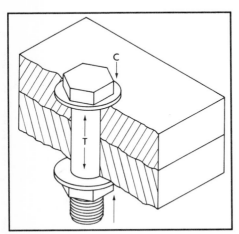

Fig. 6.8.1   *Tension and clamping force.*

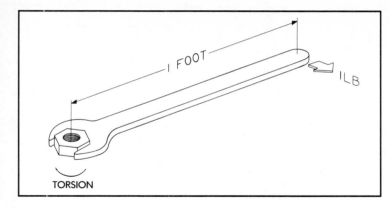

Fig. 6.8.2   Torque applied to a fastener.

amount of force and the distance at which it is applied. For example, looking at Figure 6.8.2, a torque of 2 pound-foot (lb-ft) will result from either 2 pounds of force on a one foot long wrench or 1 pound of force on a two foot long wrench.

Torque and Clamping Force

Since torque is used to put tension in a fastener, the measurement of fastener torque should tell us something about the clamping force of a joint. But torque measurement is by no means a direct indication of clamping force. Other variables are involved.

For one thing, only a small part of the torque applied to a fastener contributes to clamping force. The rest, as much as 90%, is used to overcome friction. Some torque is also absorbed by the fastener shank when it twists slightly.

The major factors that determine clamping force are shown in Figure 6.8.3. The quality of the engineering of a joint includes the physical characteristics and the torque specification. Proper installation includes both assembly and the application of fastening torque. If everything is done correctly, the desired clamping force results. Any variation will affect clamping force. Differences will be due to changes over time, such as metal fatigue, and the ability of the auditor and torque tool to get an accurate reading.

Torque Auditing Methods

The two most common ways to audit torque are:

- *Static* — using handheld torque tools to check breakaway torque.

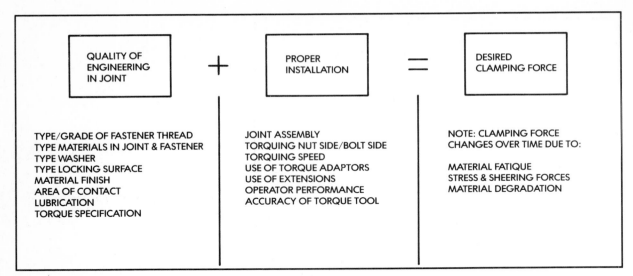

QUALITY OF ENGINEERING IN JOINT	+	PROPER INSTALLATION	=	DESIRED CLAMPING FORCE

TYPE/GRADE OF FASTENER THREAD	JOINT ASSEMBLY	NOTE: CLAMPING FORCE
TYPE MATERIALS IN JOINT & FASTENER	TORQUING NUT SIDE/BOLT SIDE	CHANGES OVER TIME DUE TO:
TYPE WASHER	TORQUING SPEED	
TYPE LOCKING SURFACE	USE OF TORQUE ADAPTORS	MATERIAL FATIGUE
MATERIAL FINISH	USE OF EXTENSIONS	STRESS & SHEERING FORCES
AREA OF CONTACT	OPERATOR PERFORMANCE	MATERIAL DEGRADATION
LUBRICATION	ACCURACY OF TORQUE TOOL	
TORQUE SPECIFICATION		

Fig. 6.8.3   Factors that determine clamping force.

- *Dynamic* — using in-line torque transducers to check peak torque.

There is a good deal of difference between these two methods, and advantages and disadvantages to both. The basic characteristics are listed here briefly, and a detailed description follows.

Static torque auditing has these characteristics:
- It is checked after installation.
- The reading accuracy is operator dependent.
- The differences between how a fastener is installed and how a torque is checked must be taken into account.
- The time between when a fastener is installed and when torque is checked must be taken into account.
- Fairly uncomplicated gaging is used — a torque wrench or torque driver with dial readout, or an electronic (strain gage) torque tool connected to a handheld data collector.
- A variety of joints and fastener types can be audited with different range torque tools and adapters.

Dynamic torque auditing using in-line torque transducers has these characteristics:
- It is checked during installation.
- The readings are operator dependent.
- It is used only with powered torque tools, such as nut runners and air ratchets.

- It must be installed in-line with the production tool.
- Strain-gage type output requires an electronic readout or handheld data collector.

Torque Breakaway (Static Torque)

Torque breakaway is the point at which torque applied to a fastener restarts the fastener in a positive direction. The restarting movement is characterized by a momentary drop-off in torque followed by an increase in torque with further rotation in the positive direction. See Figure 6.8.4.

There are two ways to check torque breakaway. The first is by having an operator apply torque and feel for the breakaway point, releasing the wrench at the moment it occurs. The dial reading of the wrench indicates the breakaway. The second way is with the use of an intelligent data collector, such as the DataMyte model 2003 data collector, which can sense the drop in load on a strain-gage type torque wrench, and record breakaway automatically. The second way does not depend on the sensitivity of the operator, so it provides a more consistent reading of the breakaway point.

As shown in Figure 6.8.5, the breakaway point is a good relative indicator of installation torque, and the clamping force attained, when certain conditions are accounted for. By relative indicator, we mean that there is a positive correlation. Typically, for different types of fasteners and methods of installation, a breakaway reading is either consist-

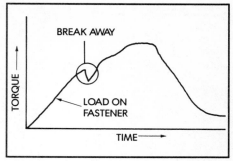

Fig. 6.8.4   A graph of a torque signal.

See p. 20-6 for 2000 series data collectors.

Fig. 6.8.5   Relationship between an audit reading and clamping force.

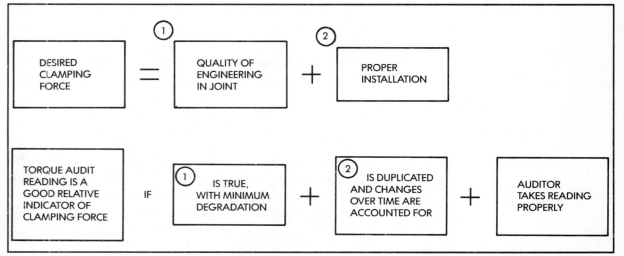

ently higher by a slight amount or consistently lower than installation torque. Since the purpose is not so much to equate installation specifications with audit data, but rather to monitor a process for control and capability, correlative data is adequate. Given a stable process and a sufficient amount of data, the correlative factor can be accounted for.

Sources of Variation Between Installation and Breakaway Torque

Some of the areas that can make a breakaway reading different from the installation torque specification are:

Actual installation versus the specification — The torque achieved at installation may not be the same as the specification. The design of the joint and fastener, its assembly, the performance of the operator, and the accuracy of the torque tool affect the installation. Use of adaptors, torquing the nut side or bolt side, and tightening speed, all greatly affect actual torque.

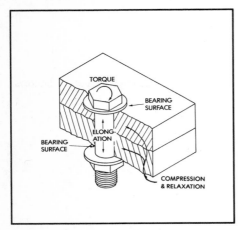

Fig. 6.8.6   Torque fall-off.

Torque fall-off (joint relaxation) — In as little as one hour after installation, the bearing surfaces of a joint will fatigue due to the compression from clamp load. See Figure 6.8.6. This tends to reduce the torque required to achieve breakaway.

Hard joints and soft joints — Hard joints, such as metal-to-metal with a locking fastener, will have a more pronounced (and detectable) breakaway than joints using a soft gasket or holding softer materials. Breakaway may be imperceptible in some types of joints, such as pipe threaded junctions, due to the composition of joint materials and the nature of the threads.

Tightening speed (frictional coefficients) — Speed affects a torque reading because the coefficient of friction varies with it. Since the greatest amount of torque is used to overcome the friction in the joint at rest and restart further tightening motion, friction contributes a great deal to variability among readings. Generally, the tightening speed at installation should be matched during an audit. This is a justification for the use of in-line torque transducers to audit torque on power tools.

Other factors — Other factors that contribute to differences between installation and breakaway torque are:

- Temperature
- Absorption of torque by the joint
- Stress and shearing forces
- Training of the auditor in reading breakaway torque
- Accuracy of the audit method

Dynamic Torque

To control torque on an ongoing basis, we can monitor it dynamically — as the torque is applied to the fastener. Individual readings tell us only the torque currently being applied, which may or may not be within suggested tolerance. To apply statistical analysis, we must be able to average a number of readings. The average torque output of a tool supplies criterion for adjusting the tool, and the range or dispersion of measurements becomes criterion for judging the capability of the tool.

A dynamic torque monitoring system, therefore, consists of:

- the torque tool, such as a pneumatic nut runner, ratchet, or multi-spindle nut runner
- an in-line transducer
- a handheld data collector and analyzer such as the DataMyte 2003 data collector

The term *transducer* means a device that converts energy from one form to another. In this case, the in-line transducer converts mechanical force into electrical energy. The transducer is termed "in-line" because it is installed in the line of force being applied, as an extension between the driving tool and the socket being driven. (See Figure 6.8.7.)

As torque is applied to the fastener, the center shaft of the transducer undergoes a very slight twisting motion. The amount of flex is related to the torque applied. Strain gages are mounted to the shaft in a resistive bridge configuration. When the shaft undergoes a twisting movement, the strain gages twist with it. This motion stretches the resistive gages, causing their resistance to increase.

When the data collector is connected to this transducer,

See p. 20-6 for 2003 data collectors, and p. 19-12 for 753 data collectors.

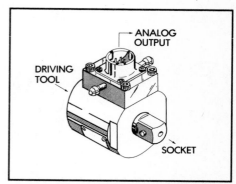

Fig. 6.8.7   In-line torque transducer.

6-31

Measurement

it measures the torque by sensing the change in resistance across the strain gages. This signal is converted into a digital count representative of the torque applied. The data collector is programmed to record the peak value of torque applied to the fastener. It does this by continually sampling the input signal. When a torque signal appears, the data collector continually updates its memory with the highest signal it sees until the torque drops to zero again. It then records the peak value and alerts the operator if the torque recorded falls outside of specified limits.

The resistive bridge is known as a Wheatstone bridge. It consists of four precision resistors configured as shown in Figure 6.8.8.

The resistors used in an in-line transducer are called strain gages. They are high precision resistors made in a variety of alloys for different applications. They are made very thin so that any stretching or compressing will cause a change in resistance. When properly mounted, they act as an adding and subtracting electrical network that allows for compensation with temperature changes and other extraneous signals.

In-line transducers are available in a wide range of capacities and configurations. Most can be used directly with the data collector with no modification. The data collector supplies the signal conversion, data storage, and analysis by displaying the average, range, standard deviation, and other summaries.

6.9 GEOMETRIC DIMENSIONING AND TOLERANCING

Geometric dimensioning and tolerancing provides a uniform method of stating the requirements of a part so that there is a standard interpretation by all users of the requirements. It ensures that design requirements are stated as they relate to the function of a part and relationship to other parts used in an assembly. This, in turn, ensures the interchangeability of parts in a manufacturing environment.

Geometric dimensioning and tolerancing is widely used in European industries and its use is rapidly growing in the United States. A majority of government contracts use geometric tolerancing and other industries are also finding that its use as a universal engineering drawing language and

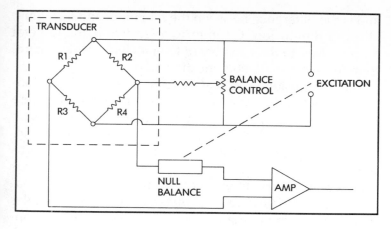

Fig. 6.8.8 Schematic of a Wheatstone bridge.

technique can assist with productivity and quality.

Primary references on the subject of geometric dimensioning and tolerancing include Lowell W. Foster *GEOMETRICS II The Application of Geometric Tolerancing Techniques* published by Addison-Wesley, and the American National Standards Institute (ANSI) Y14.5M-1982 *Dimensioning and Tolerancing* published by the American Society of Mechanical Engineers. These standards are applied throughout the world with some exception to the International Standards Organization (ISO) standards. These exceptions are noted in the ANSI Y14.5M-1982.

Geometric tolerancing is the concept of categorizing tolerances used to control form, profile, orientation, location, and runout. This will clearly define the engineering intent, or in many cases, provide the true position which may result in a bonus tolerance.

Geometric dimensioning and tolerancing provides for the maximum producibility of a part by allowing maximum production tolerances. It can provide "bonus" tolerances in many applications and will more easily fit a particular manufacturing process capability. In the long run, this will save money by reducing the cost to produce a part.

Foster states that geometric dimensioning and tolerancing is the dimensioning and tolerancing of a drawing with respect to the actual function or relationship of part features. Dimensioning is a means to define the size or geometric characteristic of a part or part feature. Tolerance is the total amount by which a specific dimension can vary

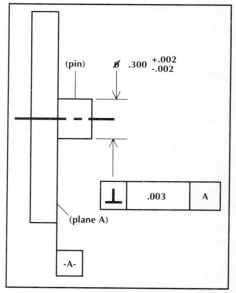

Fig. 6.9.1 Using geometric tolerancing, a manufacturer can specify tolerance and location by referencing a dimension, a plane, and a perpendicularity callout to that plane. In this example, the pin diameter must be within the specified limits (± .002). The specified tolerance of perfect form (.003 max), which actually specifies the pin form, will be perpendicular to plane A. The pin must be the correct size and perpendicular to plane A within .003.

⬭	FLATNESS
—	STRAIGHTNESS
◯	ROUNDNESS (CIRCULARITY)
⌀	CYLINDRICITY
⌒	PROFILE OF A LINE
⌓	PROFILE OF A SURFACE
⊥	PERPENDICULARITY (SQUARENESS)
∠	ANGULARITY
//	PARALLELISM
↗	CIRCULAR RUNOUT
↗	TOTAL RUNOUT
⊕	POSITION
◎	CONCENTRICITY
≡	SYMMETRY

Fig. 6.9.2 Terms and symbols used in geometric dimensioning and tolerancing.

and is the difference between the minimum and maximum allowed dimension. Geometric tolerance is the general term applied to the category of tolerancing used to control form, profile, orientation, location, and runout.

The true position of a feature is the theoretical exact location of a feature established by dimensioning. A feature is a physical portion of a part, such as a surface, hole, or slot. The limits of the part's features must be defined with minimum and maximum tolerances specified for that part's function.

Max material condition (MMC) is the condition of a feature which contains the most material (i.e. minimum holes diameter, maximum shaft diameter) and least material condition is the other extreme condition of a feature which contains the least material (i.e. the maximum hole diameter, the minimum shaft diameter). It is the application of the material condition in relation to other features in geometric tolerancing that will allow for "bonus tolerances" or zero tolerances on a feature because of its material condition. This vital condition is the boundary generated by the collective effects of the specified MMC limit of size of a feature and any applicable geometric tolerances.

The terminology and corresponding symbols used for geometric dimensioning and tolerancing are listed in Figure 6.9.2.

Along with providing interchangeability of mating parts, uniformity and convenience in the drawings, geometric tolerancing provides a common language that accurately and reliably reflects the engineering design requirements. This translates into dollars saved and maximum producibility, a must in this new era of high technology and quality improvement.

For Further Reference

American Society of Mechanical Engineers. *Dimensioning and Tolerancing, ANSI Y14.5M-1982.* New York: American Society of Mechanical Engineers, 1983.

Charbonneau, H.C. and Webster, Gordon. *Industrial Quality Control,* New Jersey: Prentice-Hall, 1978.

Eagle, A. R. "A Method for Handling Errors in Testing and Measuring." *Industrial Quality Control,* pp. 10-14, March, 1954.

Foster, Lowell W. *GEO-METRICS II (The Application of Geometric Tolerancing Techniques).* Reading, Massachusetts: Addison-Wesley Publishing Company, 1979.

Foster, Lowell W. *Modern Geometric Dimensioning and Tolerancing With Workbook Section,* 2d ed. Ft. Washington, MD: National Tooling & Machining Association, 1982.

Grant, Eugene L., and Leavenworth, Richard S. *Statistical Quality Control,* 5th ed. New York: McGraw-Hill Company, 1980.

Juran, Joseph M. and Gryna, Frank M. *Juran's Quality Control Handbook,* 4th ed. New York: McGraw-Hill Book Company, 1980.

Juran, Joseph M., and Gryna, Frank M. *Quality Planning and Analysis.* New York: McGraw-Hill Book Company, 1970.

Mitutoyo Metrology Institute. *Fundamentals of Precision Measurement Textbook.* Tokyo: Mitutoyo Mfg. Co. LTD.

7. INSPECTION

7.1 THEORY OF INSPECTION

Inspection theory is changing nowadays. Inspection used to be a means of weeding out lots that have more than a certain percentage of defectives and accepting lots that have less or no defectives. It is now becoming the supplier's responsibility to supply lots that are 100% good or at most, a couple of parts per thousand, (and in many cases a couple of parts per million) defective.

The only way defect-free shipments can be achieved is for the supplier to have ongoing in-process controls in place. This means that the supplier is continuously monitoring the process and working toward improvement. Inspection is now becoming more a method of reviewing data provided by the supplier to see that the process is still in control and confirming visually that the customer has the right part.

Since it's the supplier's customer asking for the inspection data, the customer may set the standards for what data is collected, how the report is compiled and how the report is formatted. As the use of SPC grows, more and more customers are requiring that suppliers meet these standards. For some industries, suppliers are being asked to use automated data collection systems, such as the DataMyte 769 data collector, and supply customers with the standard reports printed from the data collector. In addition to requiring data and reports, customers may also conduct periodic audit inspections.

Chapter 2 covers methods of collecting data. Chapter 8 examines supplier certification further. Now, let's examine some traditional methods of inspection.

7.2 THEORY OF ACCEPTANCE SAMPLING

Acceptance sampling is the practice of inspecting a small quantity of the parts in a lot for conformance to specification. A decision is then made whether to accept or reject the entire lot based on the findings of the sample. The justification for acceptance sampling is that it provides a cost savings over 100 percent inspection (see *7.5 Minimizing To-*

tal *Cost of Inspection and Repair* for Dr. W.E. Deming's proof to the contrary). The cost savings are the result of less time needed to inspect a sample, and requirements for fewer inspectors. There is the additional expense and overhead required to design and administer the sampling plans, however. Other reasons for sampling include:

- *Greater speed* — More lots can be evaluated faster and the "good" separated from the bad for Material Review Board (MRB) dispositioning.

- *Minimized handling* — By inspecting only a sample, fewer items are subjected to the possible damage which sometimes occurs in handling or measuring during the inspection process.

- *Greater accuracy* — The problem of inspector error due to monotony is minimized. With fewer items to inspect, more time can be used to ensure completeness and accuracy.

- *Faster corrective action* — Lot rejection due to sampling tends to dramatize quality deficiencies and to speed up corrective action over 100% sorting.

The disadvantages of acceptance sampling are administrative costs, sampling risks, and the fact that decisions have to be made with less information than is provided with 100% inspection or vendor supplied process control charts.

Acceptance sampling should be used when 100% inspection is causing errors due to monotony, for destructive testing, when the cost of inspection is high in relation to the cost resulting from passing a defective and when a particular parameter is a good indicator of a lot's overall quality.

Importance of a Random Sample

A random sample is one in which every item in the population has an equal chance of being chosen. The concept of sampling is based on the idea that a sufficient quantity of items is chosen in a random fashion. The sample must contain all the characteristics of the total population. One statistical tool developed to help in selecting a random sample is a *random number table* (see Appendix Table A-10 for an example). Before using the table it is first necessary to assign a number to each unit in the population. Simply enter any column or line and select the sequence of numbers as they occur. For example, if we require five samples, and choose to enter the table at line 21, we would get the numbers 26, 20, 46, 66, 36 (Table A-10). These are the corresponding numbers you would select from the population. Further information on random number tables can be found in various statistical textbooks.

Risks Associated With Sampling

In sampling, certain inherent risks are involved which must be addressed and understood. These risks can be broken down into the following two types:

- *The Producer's Risk* (also referred to as α Risk) can be understood easiest when thought of as the probability that a good lot will be rejected by the sampling plan. The quantified risk must be defined prior to adopting a given sampling plan. The risk is stated in conjunction with a numerical definition of good quality such as AQL (acceptable quality level). The sampling plan should have a Producer's Risk which is equal to or better than the AQL.
- *The Consumer's Risk* (also referred to as the β Risk) is the risk that a bad lot will be accepted by the sampling plan. The consumer's risk is generally stated in conjunction with a numerical definition of bad quality such as LTPD (lot tolerance percent defective).

Estimating Sampling Risks

The estimation of the *Producer's Risk* of a sampling plan includes the following steps:

- Plot an OC curve for the sampling plan in question. OC curves are described in the following section.
- Find the percent defective in the process when it is running at capability.
- It may be necessary to estimate it, but if greater accuracy is desired a process capability study should be done.
- Find the process capability percentage on the OC curve and follow it up to determine the probability of acceptance.

- Subtract this probability of acceptance from 1.0. This number is the producer's risk for the sampling plan.

To estimate the *Consumer's Risk* proceed as follows:

- Plot the OC curve for the sampling plan in question.
- Find the percent defective that the consumer wants to reject. This may be understood to be the worst case quality that the customer will accept.
- Find this value on the horizontal scale of the OC curve. Follow it up to determine the probability of acceptance. This number will be your risk of accepting bad quality material.

7.3   OC CURVES

Operating Characteristic Curves — The OC curve is a means of quantifying the producer's and consumer's risk. The OC curve for an attributes plan is a graph of the percent defective in a lot versus the probability that the sampling plan will accept the lot. The probability must be stated for all values of "P" (percent defective) since "P" is unknown. An assumption is made that an infinite number of lots will be produced.

It is characteristic of sampling plans that the probability of acceptance is high as long as product quality is good, but decreases as product quality becomes poorer. An example of an optimum OC curve is shown in Figure 7.3.1. Let us assume that we desire to accept all lots less than 2% defective and reject all lots greater than 2% defective. All lots less than 2% defective have a probability of acceptance of 1.0 (certainty) and all lots greater than 2% defective have a probability of acceptance of 0%. However, in reality, there are no sampling plans that are perfect. There will always be some chance that a good lot will be rejected or a bad lot will be accepted. The one major goal in developing a sampling plan should be to make the acceptance of good lots more likely than the acceptance of bad lots.

The shape of the OC curve can be affected greatly by the parameters of the sampling plan. Figure 7.3.2 illustrates this by showing the curve for perfect discrimination as well as the curves for the other sampling plans. As the sample size approaches the lot size and an appropriate accept number

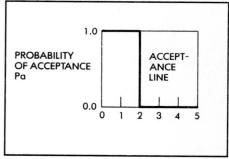

Fig. 7.3.1 *Optimum sampling plan performance.*

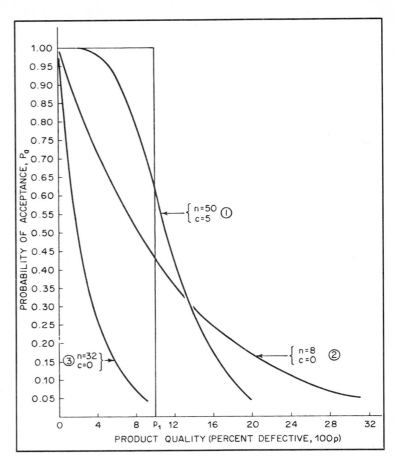

Fig. 7.3.2 How OC curves are affected by the parameters of a sampling plan. Reproduced with permission from J.M. Juran, "Quality Control Handbook," 3d ed., 1979, McGraw-Hill Book Company.

(c) is used, the OC curve approaches the perfect curve P1. With an accept number of zero, the resultant OC curve will be exponential in shape or concave upward as in curves 2 and 3. Increasing the acceptance number tends to push up the OC curve for low values of P (percent defective) as in curve 1. Increasing both the accept number and sample size at the same time (curve 1) gives a curve which most closely resembles the perfect discriminator curve of P1.

Constructing OC Curves

Construct an OC curve by determining the probability of acceptance for various values of P (percent defective in a lot). The probabilities of all values of P must be determined since P is unknown. There are three distributions which may be used to find the probability of acceptance. They are the *Poisson, hypergeometric,* and the *binomial*. Probably the easiest one to use is the Poisson distribution if all assumptions are met for its use.

Some assumptions which must be met are as follows:

- Sample size must be 16 or greater.
- Lot size must be at least 10 times greater than the sample size.
- Percent defective is less than 0.01.

Many tools exist which can aid in calculating and plotting the OC curves. One such tool is shown in Figure 7.3.3. The following steps are required to plot the OC curve for a single sampling plan.

- Set up a table like the one shown in Table 7.3.1 for various values of P (percent defective). Express "P" as a decimal. The "P" value range should cover both good and bad product.
- Complete the second column of the table by multiplying each of the "P" values in column 1 by "n" (sample size).
- Find the probability of acceptance (Pa) by using the table of curves in Figure 7.3.3.
- Plot the probability of acceptance (Pa) for each corresponding value of "P" as shown in Figure 7.3.2. When comparing a number of OC curves, make sure the same vertical and horizontal scales are used.

Sampling Plan $n = 200$, $c = 10$		
P	nP	Pa
.01	2	0.99999
.02	4	0.997
.03	6	0.96
.04	8	0.81
.05	10	0.69
.06	12	0.35
.07	14	0.28
.08	16	0.08
.09	18	0.02
.10	20	0.01

Table 7.3.1 Calculations for plotting an OC curve.

Inspection Lot Formation

Lot formation is one of the most important factors in acceptance sampling. It is imperative that we know the pertinent details about a lot (i.e. who, when, and what) before we can make intelligent decisions with the inspection data.

The following guidelines are given to ensure the validity of your inspection data. Others may exist, but are more related to individual processes.

- Do not mix products from different sources (process or machines, production shifts, raw materials, etc.) unless you can prove that variation is small enough to be ignored.
- Do not mix products from various time periods. Require suppliers to date code their products to allow you greater

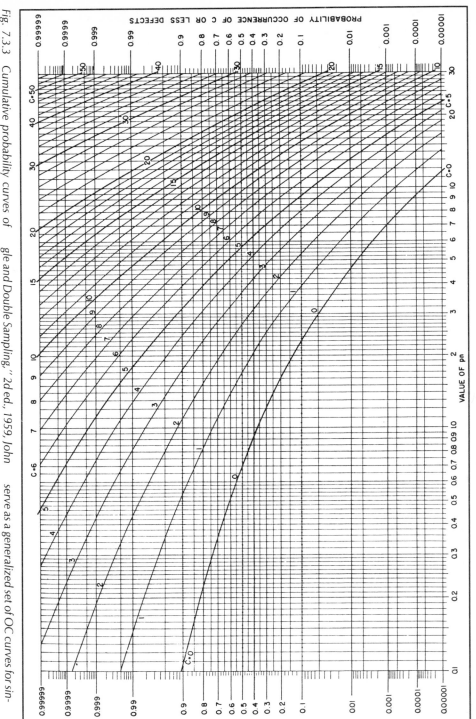

Fig. 7.3.3 Cumulative probability curves of the Poisson distribution. Reproduced with permission from Harold F. Dodge and Harry C. Romig, "Sampling Inspection Tables, Single and Double Sampling," 2d ed., 1959, John Wiley & Sons, Inc. (A modification of a chart given by F. Thorndike in "The Bell System Technical Journal," Oct., 1926.) These curves serve as a generalized set of OC curves for single sampling plans when the Poisson distribution is applicable.

flexibility in your lot information.

- Keep the lots as large as possible to take advantage of the fact that lot size has very little effect on the OC curve. Large lots may create some problems such as storage problems and production and delivery problems when rejected.
- Make use of additional information such as capability studies and prior inspection results in lot formation. This information can prove to be very helpful when the lots are few and far between.

Sampling Justification

Sampling may or may not be the most effective solution for a given situation. Each situation must be evaluated individually in deciding whether or not to sample. Three alternatives exist for a product. One can choose to do (1) 100% inspection, (2) sample, or (3) no inspection. A thorough cost analysis of all three alternatives should be completed prior to adopting any one method. Many excellent references exist which may aid in this process. One such reference is *Quality Planning and Analysis,* by Joseph Juran and Frank Gryna.

7.4   SAMPLING PLANS

A sampling plan basically consists of a sample size and the acceptance or rejection criteria. The required number of samples are taken from a lot or batch and the decision criteria for acceptance or rejection are applied to the results of the inspection. All samples must be taken randomly because they are supposed to be representative of the lot. As the lot size has essentially no effect on the probability of acceptance, many sampling plans do not include lot sizes.

Sampling plans are used to minimize the cost of inspection and should be carefully studied and chosen to adequately fill the needs of both the consumer and the producer. They should not be chosen merely for convenience. The plan should be easy to understand and administer, as overly complicated plans are often ignored, misinterpreted or result in poor information. No sampling plan ensures that only good lots will be accepted and that all bad lots will be rejected. The OC curve for the sampling plan will be helpful in recognizing inherent sampling risks.

Parameters Affecting Sampling Plans

Sampling plans are of two general types: lot-by-lot inspection plans, and continuous process plans. These two are the most often used types but there are other plans which can be used for special inspection problems.

Lot-by-lot sampling plans are used whenever product can be broken into distinct homogeneous lots. The lot size is the quantity of units in the lot. The sample size is a specified number of samples taken from the lot for purposes of inspection subject to acceptance or rejection. The sample plan will specify the criteria for acceptance or rejection. Many plans allow for single, double, or multiple sampling choices.

Continuous process plans involve product that is produced in a continuous stream and cannot easily be broken into separate lots. Initially, product is inspected 100% until some consecutive number of good units are found between successive bad units. When the required number of defect-free units is found, inspection proceeds on a sampling basis until a specified number of defects appear. When this happens, 100% inspection again goes into effect and the cycle begins again.

There are three types of continuous process plans. The plan described above is known as the CSP-1 plan. The CSP-2 plan is different in that it allows a single defect to be found and doesn't return to 100% inspection unless another defect is found in the next given number of samples. The CSP-3 plan differs in that an inspection period, i.e., a production shift, is incorporated into the plan.

The U.S. Department of Defense document H107 contains prerequisites for sampling plans for continuous production. The process must produce homogeneous products, be capable of 100% inspection and the inspection procedure must be fairly simple. The product has to move such that it can flow past an inspection station.

These continuous process inspection plans are ideal for controlling in-process subassembly operations. Encouraging your vendors to supply you with the results of these inspections could very well eliminate the need for incoming product inspection.

Classifying Sampling Plans According to: AQL, LTPD, and AOQL

Sampling plans are classified according to three quality indexes: AQL, LTPD and AOQL.

AQL Plans — Acceptance Quality Level is defined by MIL-STD-105D as "the maximum percent defective (or the maximum number of defects per hundred units) that for purposes of sampling inspection can be considered satisfactory as a process average," (*MIL-STD-105D,* 1963). AQL plans favor the producer as they give a high assurance of probable acceptance. They do not take into account the other side, which is the product that will be rejected, or the consumer's risk. Note that this is viewed as a process average. That does not mean that all lots will be at the AQL or better.

LTPD Plans — Lot Tolerance Percent Defective is defined in the Dodge-Romig tables as "an allowable percentage defective; a figure which may be considered as the borderline of distinction between a satisfactory lot and an unsatisfactory lot," (*Dodge, H.F., and Romig, H.G.,* 1959). When chosen, these plans tend to favor the consumer as they decrease the risk of accepting a lot equal to or below the lower quality limits. As AQL plans do not tell anything about the product that will be rejected, LTPD plans do not tell anything about the product that will be accepted. In order to obtain this information it is necessary to refer to the OC curve of the plan.

AOQL Plans — Average Outgoing Quality Limit is a sampling plan which is to be used only when product can be 100% inspected. The AOQL plan assumes that the average quality over many lots of outgoing product will not exceed the AOQL after rejected lots have been 100% inspected and all defects replaced with non-defective units. Thus some lots will be accepted based on the sample and rejected lots will be accepted only after they contain 100% good product. Beware of returning bad lots to the producer for 100% inspection as you have no control over whether they will actually do the inspection.

Types of Sampling Plans

Sampling plans are either attribute plans or variable plans. An attribute plan is one in which each sample is inspected and classified as defective or non-defective. The lot is accepted or rejected based on the number of defects found compared with the acceptance number from the plan. A variable plan takes a measurement of each sample and statistics (average, standard deviation, range) are calculated and compared with the acceptance limit of the plan, indicating whether the process is in or out of control. Attribute plans are often used for lot-by-lot production and variable plans for continuous process production. Chapters 2 and 3 describe ways of collecting and analyzing variable and attribute data.

In most cases, it is cost prohibitive to inspect all parameters or dimensions on a product, as each would require a pass/fail acceptance number. A limited number of critical parameters are often chosen to serve as measures of a product's acceptability.

MIL-STD-414 is an example of a variables plan and uses the AQL as its quality index. This plan requires that the distribution of the individual measurements be known and that the process is stable. It allows for the choice of three measures of variability: average range, known standard deviation, and estimated standard deviation. The plan provides five inspection levels with level IV considered normal. Levels I and II are used when small samples are required and higher risks can be tolerated. Level III is used when less discrimination is needed and level V when greater discrimination is needed. Two acceptance procedures are offered when working with single specification limits, Form 1 and Form 2, with Form 2 using an auxiliary table that provides an estimate of lot percent defective based on a quality index. Only the computational procedure is different between these two forms. The accept and reject results should be the same. A variables plan, such as MIL-STD-414, offers an advantage over an attributes plan in that the sample size is smaller.

One of the most widely used attribute plans is MIL-STD-105D, which again uses the AQL as its quality index. This plan protects the producer from rejection of lots which are at the AQL or better. If the quality history is unsatisfactory, more stringent acceptance criteria should be used to protect the consumer from accepting lots which are moderately worse than the AQL. This is where a tightened inspection comes in. The plan includes three general inspection levels: I, II, and III. Level II is the normal inspection level generally used. Level I is used when less discrimination is needed (reduced inspection). Level III is used when more discrimination is needed (tightened inspection). These levels are also used when switching rules are implemented. The switching rules are included in the front of the MIL-STD-105D and should be used to obtain maximum benefit from the plan. Another group of special inspection levels (S-1 through S-4) are included when destructive or extremely costly testing is required and higher risks can be tolerated. Different tables which list the number of samples to be inspected and accept/reject criteria are included for single, double, and multiple sampling plans.

Let's take a look at the MIL-STD-105D tables and how they are used.

Upon receipt of the incoming material, we must determine the size of the shipment and the specified AQL level to be used, which in our case will be 1.0. In our example, we have received 10,250 pieces. Refer to Table I in MIL-STD-105D (Table 7.4.1). Move down the "Lot or batch size" column until the lot size range of 10001 to 35000 is found. Move horizontally across the table until reaching the general inspection level II column. Note the letter "M" in our example. Continue on to Table II-A in MIL-STD-105D (Table 7.4.2). In the sample size code letter column, find the letter "M" from the previous table. The next column, "Sample size," lists how many samples to take from the lot. This is 315 pieces in our example. Remember, these samples must be pulled at random from the lot. We said previously that we would use an AQL of 1.0. Move horizontally across the table to the AQL column labelled 1.0. The value 7 is found in the accept (AC) area and 8 in the reject (RE) area. This means that if 7 or fewer defective samples are found, the lot is acceptable. If 8 or more defective samples are found, then the lot is rejected.

If we find an arrow in the AQL column for our sample size code letter, we must follow the arrow in the direction it is pointing. When the first set of accept/reject values are reached, we must move horizontally back to the sample size column and use that size for our inspection. For ex-

Lot or batch size			Special inspection levels				General inspection levels		
			S-1	S-2	S-3	S-4	I	II	III
2	to	8	A	A	A	A	A	A	B
9	to	15	A	A	A	A	A	B	C
16	to	25	A	A	B	B	B	C	D
26	to	50	A	B	B	C	C	D	E
51	to	90	B	B	C	C	C	E	F
91	to	150	B	B	C	D	D	F	G
151	to	280	B	C	D	E	E	G	H
281	to	500	B	C	D	E	F	H	J
501	to	1200	C	C	E	F	G	J	K
1201	to	3200	C	D	E	G	H	K	L
3201	to	10000	C	D	F	G	J	L	M
10001	to	35000	C	D	F	H	K	M	N
35001	to	150000	D	E	G	J	L	N	P
150001	to	500000	D	E	G	J	M	P	Q
500001	and	over	D	E	H	K	N	Q	R

Table 7.4.1   MIL-STD-105D Sample size code letters table.

ample, if we were using sample size code letter D and an AQL of 0.65, we run into an arrow pointing down. Following the arrow, we find we must use a sample size of 20 and accept/reject values of 0 and 1.

Further detailed information on how to use the other sampling tables and switching rules can be found in the MIL-STD-105D booklet. Grant and Leavenworth also have a great deal of information in Part 2 of *Statistical Quality Control,* 5th edition.

Single, Double, and Multiple Sampling

Many sampling plans offer a choice of single, double, or multiple sampling. In single sampling plans, a random sample of n items is drawn from the lot. If the number of defectives is less than or equal to the acceptance number, c, the lot is accepted. If not, it's rejected. In double sampling plans, a smaller initial sample is drawn and a decision to accept or reject is made if the number of defectives is either

Table 7.4.2 MIL-STD-105D Single sampling plans for normal inspection.

Sample size code letter	Sample size	Acceptable Quality Levels (normal inspection)															
		0.010	0.015	0.025	0.040	0.065	0.10	0.15	0.25	0.40	0.65	1.0	1.5	2.5	4.0	6.5	10
		Ac Re	Ac Re	Ac Re	Ac Re	Ac Re	Ac Re	Ac Re	Ac Re	Ac Re	Ac Re	Ac Re	Ac Re	Ac Re	Ac Re	Ac Re	Ac Re
A	2	↓	↓	↓	↓	↓	↓	↓	↓	↓	↓	↓	↓	↓	↓	0 1	↑
B	3	↓	↓	↓	↓	↓	↓	↓	↓	↓	↓	↓	↓	↓	0 1	↑	↓
C	5	↓	↓	↓	↓	↓	↓	↓	↓	↓	↓	↓	↓	0 1	↑	↓	1 2
D	8	↓	↓	↓	↓	↓	↓	↓	↓	↓	↓	↓	0 1	↑	↓	1 2	2 3
E	13	↓	↓	↓	↓	↓	↓	↓	↓	↓	↓	0 1	↑	↓	1 2	2 3	3 4
F	20	↓	↓	↓	↓	↓	↓	↓	↓	↓	0 1	↑	↓	1 2	2 3	3 4	5 6
G	32	↓	↓	↓	↓	↓	↓	↓	↓	0 1	↑	↓	1 2	2 3	3 4	5 6	7 8
H	50	↓	↓	↓	↓	↓	↓	↓	0 1	↑	↓	1 2	2 3	3 4	5 6	7 8	10 11
J	80	↓	↓	↓	↓	↓	↓	0 1	↑	↓	1 2	2 3	3 4	5 6	7 8	10 11	14 15
K	125	↓	↓	↓	↓	↓	0 1	↑	↓	1 2	2 3	3 4	5 6	7 8	10 11	14 15	21 22
L	200	↓	↓	↓	↓	0 1	↑	↓	1 2	2 3	3 4	5 6	7 8	10 11	14 15	21 22	↑
M	315	↓	↓	↓	0 1	↑	↓	1 2	2 3	3 4	5 6	7 8	10 11	14 15	21 22	↑	
N	500	↓	↓	0 1	↑	↓	1 2	2 3	3 4	5 6	7 8	10 11	14 15	21 22	↑		
P	800	↓	0 1	↑	↓	1 2	2 3	3 4	5 6	7 8	10 11	14 15	21 22	↑			
Q	1250	0 1	↑	↓	1 2	2 3	3 4	5 6	7 8	10 11	14 15	21 22	↑				
R	2000	↑		1 2	2 3	3 4	5 6	7 8	10 11	14 15	21 22	↑					

↓ = Use first sampling plan below arrow. If sample size equals, or exceeds, lot or batch size, do 100 percent inspection.
↑ = Use first sampling plan above arrow.
Ac = Acceptance number.
Re = Rejection number.

quite large or quite small. A second sample is taken if the first is inconclusive. Since it is only necessary to draw second samples in borderline cases, the average number of pieces inspected per lot is fewer than with single sampling. In multiple sampling, one or more still smaller samples are taken until a decision is finally reached. This process may result in fewer inspections but is more complex to administer. Figure 7.4.1 diagrams a multiple sample plan (Juran, 1979, p.24-5).

Double or single plans simply stop at levels A or B in the diagram respectively. In double or multiple sampling plans, the probability of acceptance is more difficult to calculate. It is also more difficult to calculate for a variables type plan. In the case of the most frequently used plans such as MIL-STD-105D, Dodge-Romig and Bowker and Goodes, OC curves are given selectively for single, double and multiple

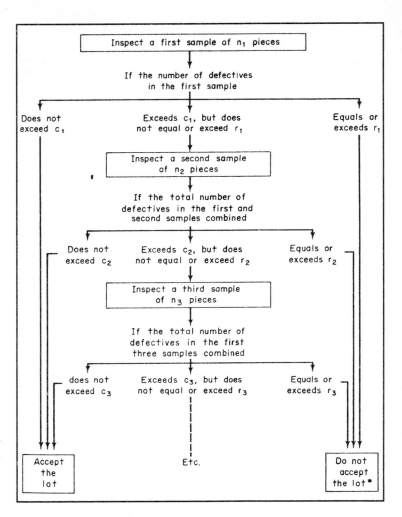

Fig. 7.4.1 Flow diagram for multiple sampling. The asterisk means that fully inspecting the lot may be necessary at this point. Reproduced with permission from J.M. Juran, "Quality Control Handbook," 3d ed., 1979, McGraw-Hill Book Company.

plans. Care should be used in drawing conclusions from published OC curves because they may not be plotted on comparable scales. However, in general, it is possible to derive single, double or multiple sampling schemes with OC curves. (*Western Electric*, 1956, p.242, *Juran*, 1979, p.24-1 and 24-24, and *Juran/Gryna*, 1970, p.418.)

Table 7.4.3 summarizes comparative advantages and disadvantages of single, double and multiple sampling. In cases where the cost of inspection per part is high, the reduction in number of pieces may justify multiple sampling despite higher administrative costs. On the other hand, single sampling is preferred if minimally trained operators are used or other costs are high.

Feature	Single sampling	Double Sampling	Multiple sampling
Acceptability to producer	Psychologically poor to give only one chance of passing the lot	Psychologically adequate	Psychologically open to criticism as being indecisive
Number of pieces inspected per lot	Generally greatest	Usually (but not always) 10 to 50% less than single sampling	Generally (but not always) less than doubling sampling by amounts of the order of 30%
Administration cost* in training, personnel, records, drawing and identify samples, etc.	Lowest	Greater than single sample	Greatest
Information about prevailing level of quality in each lot	Most	Less than single sample	Least

Characteristics of Effective Sampling Plans

Good acceptance sampling plans have several characteristics. They are listed below and elaborated in following subsections (*Juran/Gryna*, p.418).

- Selecting the sample must be thoroughly described to assure that parts are randomly chosen.
- The quality index (AQL, AOQL, etc.) is chosen to reflect real needs of the consumer and producer. In concept, the AQL should not call for a higher quality than actually required. It should represent a balance between the cost of achieving higher quality and the cost of permitting a lower level. In practice it is a compromise between vendor capability and buyer requirements.
- Sampling risks are realistically shared between producer and consumer according to quantitative terms. Sampling tables can be used to match required producer's risk with consumer's risk. If tables don't work, special sampling plans can easily be devised. Producer's risk is defined broadly as the probability or risk of a "normal" product being rejected by inspection. Consumer's risk is that of accepting product when the lot quality is relatively poor.

Table 7.4.3 Comparative advantages and disadvantages of single, double, and multiple sampling. Reproduced with permission from J.M. Juran, "Quality Control Handbook," 3d ed., 1979, McGraw-Hill Book Company.

(The *Statistical Quality Control Handbook*, by Western Electric Co. Inc., 1956, discusses these concepts in detail.)

- The total inspection costs are minimized. A complex set of costs can only be approximated because the primary and secondary costs depend on agreed upon procedures.
- Ancillary knowledge such as process capability, vendor data, etc., are built into the plan. Sampling data (either past or present) is only one source of information concerning probable acceptability. For example:

  1) Vendor's test data, operator's measurements, automatic machine records (these can be validated through the concept of decision audits and used to further understanding)
  2) Scientific and/or engineering knowledge pertinent to the process
  3) Separately acquired data on process or machine capability — for example, the standard deviation

There are no general procedures for defining how such data is used to alter published sampling tables. However, where specific knowledge exists, it may well improve or simplify the plan.

- The plan is sufficiently flexible to reflect pertinent changes. Flexibility is a definite asset. MIL-STD-105D has been noted as an example of a very flexible plan (*Juran/Gryna*, 1970, p.425).
- Measurements taken are exact and repeatable. Any quality measuring process loses credibility if it is not repeatable. Measuring tools must often be specially designed to ensure ease of use and accuracy of measurement. Automatic recording of data and conversion of data to computer format further reduces the likelihood of error and greatly speeds up the process.
- Measurement data becomes a data base for both short and long-run process control. Modern computer technology makes it relatively easy to catalog and store process and audit data. If collected data is converted to computer format and transmitted to a larger computer with appropriate software, archival record storage is not difficult.

- The plan is simple to explain and easy to administer. Some plans have gone some distance to achieve ease of understanding, but at the time of this writing, much remains to be done. Modern computer technology has much to offer in this regard.

Sampling Bias

When sample selection is left to human choice, human biases can inadvertently intrude. Here are some examples:

- Sampling from the same location in all containers
- Selecting only samples that appear to be defective or acceptable
- Only picking samples that are easily accessible

There is a classic story of an inspector who always picked samples from the four corners of each tray and the knowing production operator who carefully placed perfect product in each corner. Structured sampling plans assume randomness. To avoid distortions from biases, sampling must be planned. Once a plan is chosen, it must be policed to ensure that actual sampling occurs according to plan (*Juran,* 1979, p.24-7, and *Juran/Gryna,* 1970, p.434).

Sampling Plans Based on Prior Quality Data

Conventional sampling plans assume that the frequency distribution of sampling lots follow classical probabilities of occurrence. In short, they ignore any previous knowledge gained on the past quality sampling. If this knowledge could be rationally applied to future sampling processes, it should reduce sample sizes and thus inspection costs. This is called the "*Bayesian*" approach after the Bayes Theorem (*Juran,* 1979, p.24-33). A Bayesian plan generally requires a smaller sample than does a conventional plan with equivalent risk factors. But this is not always true because the assumption that prior distributions can be applied to present sampling practice is not always valid. Oliver and Springer have developed "*Bayesian Sampling Plans*" that provide sample size and acceptance criteria for single and double sampling plans, (*AIIE Technical Papers,* 1972, p.443). They do so by incorporating data on the quality of previous lots into the sampling tables. The methodology is straightforward:

- Collect quality data on previous lot sizes of N and sample n. Calculate the fraction defective p in each sample.
- Calculate the average fraction defective p, and the standard deviation of fraction defective using the basic formulas.
- Define the values for AQL, LTPD and the corresponding producer's and consumer's risks.
- Read the plan from the tables.

Juran and Gryna, in their book, *Quality Planning and Analysis*, page 435, illustrate that this technique does indeed result in smaller sample sizes. There is considerable literature available on this subject (see references at the end of this chapter). Hald provides an extensive analysis of sampling plans based upon prior product quality distributions (*Technometrics*, 1980, pp. 275-340). He also questions the assumption of transference of past data to present situations.

7.5 MINIMIZING TOTAL COST OF INSPECTION AND REPAIR

An entirely different perspective on incoming lot inspection is obtained if one views the goal to be minimizing the total cost of incoming parts inspection, plus the costs of repairing and testing assemblies that fail because of defective parts being incorporated into these assemblies. Dr. W.E. Deming offers some simple mathematical proofs to show that the least-cost method of production is either no incoming inspection or 100 percent inspection (*Deming, W.E.*, 1982, Chapter 13).

To illustrate this point some suppositions need to be made:

- Initially, a single incoming part will be considered. (The method can be extended readily to multiple parts.)
- Every assembly produced is functionally tested.
- If the incoming part is defective, and goes into an assembly, the assembly will fail its test. (If the part is not defective, by definition, the assembly will pass its test.)

To examine this premise mathematically we need a few definitions:

p — The average fraction defective in the incoming part receipts (say for one day).

k_1 — The cost to inspect one incoming part.

k_2 — The cost to fix and retest an assembly that had a defective part.

k_1/k_2 — The "breakeven" percentage point.

Now consider two cases:

Case 1 — The worst lot of incoming parts will have a fraction defective (p) less than k_1/k_2:

$$p < k_1/k_2$$

The least cost solution is zero inspection.

Case 2 — The best lot of incoming parts will have a fraction defective (p) greater than k_1/k_2:

$$p > k_1/k_2$$

The least cost solution is 100 percent inspection.

See Figure 7.5.1a and 7.51b for a representation of these cases. The proof for these rules is quite simple. Let:

i — A randomly selected part from an incoming lot.

k — Cost to test a part drawn from supply S and continue testing until a non-defective part is identified.

k_1 — The cost to inspect one incoming part.

k_2 — The cost to fix and retest an assembly that had a defective part.

x_i — 1 if defective and 0 if not defective.

If p = Average x_i, then the cost to inspect part $i(c_1)$ is:

$$c_1 = k_1 + kx_i$$

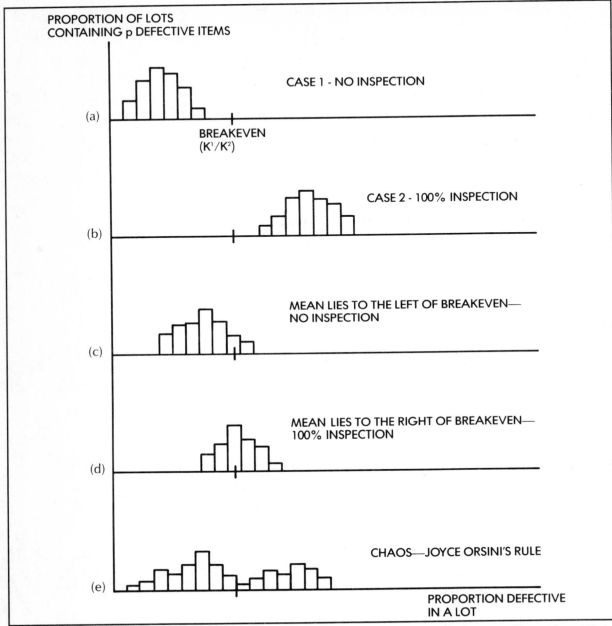

PROPORTION OF LOTS
CONTAINING p DEFECTIVE ITEMS

(a)   CASE 1 - NO INSPECTION

BREAKEVEN
(K^1/K^2)

(b)   CASE 2 - 100% INSPECTION

(c)   MEAN LIES TO THE LEFT OF BREAKEVEN—
NO INSPECTION

(d)   MEAN LIES TO THE RIGHT OF BREAKEVEN—
100% INSPECTION

(e)   CHAOS—JOYCE ORSINI'S RULE

PROPORTION DEFECTIVE
IN A LOT

Fig. 7.5.1   Illustration of how either 100% inspection or no inspection is the least cost
method.

The cost to repair the assembly (c_2) is:

$$c_2 = x_i(k_2 + k)$$

Note that c_1 and c_2 cannot both have values. If one is not zero, the other must be zero.

$$\text{The total cost } c = c_1 + c_2$$

Note that at the breakeven point where $p = k_1/k_2$, the total cost c is the same whether the part is inspected or not. Further, if $p < k_1/k_2$ no inspection will always give the lower total cost and if $p > k_1/k_2$, 100 percent inspection will always give the lesser total cost.
  Several conclusions can be drawn from this analysis:

- To treat a clear case 1 as a case 2 situation (test every part) will maximize the overall cost, as will the inverse.
- In cases where defective parts are binomially distributed around a mean p, the same case rules apply even though the distribution of defective parts straddles the breakeven point (Figure 7.5.1c and 7.5.1d).
- One must be sure, for a Case 1 decision, that p will be less than breakeven. This requires that the purchaser or the vendor keep the production process for the incoming part under statistical control — at a minimum, test small samples from every lot.
- If there is no clean grouping, the process is said to be in chaos (Figure 7.5.1e). In this case, and assuming that the breakeven point is greater than 0.002, Joyce Orsini's rule may be used (see the last reference in this chapter): Take a random sample of 200 parts from a lot. If there is no defective part in the sample, accept the remainder as is. If one or more defective parts are found, replace them and inspect the rest of the lot.
- In the event of work being done to a semi-finished part, analysis shows that the same case 1 and case 2 rules apply where k_2 is now the average loss from downgrading or scrapping finished product that fails.

Comparing the Deming method with standard acceptance plans such as the Dodge-Romig tables and Military Standard 105D leads to some surprising conclusions. For example, the theory behind the Dodge-Romig tables is said to be to minimize the cost of inspection to achieve a prescribed level of quality or percentage of acceptable parts rather than minimizing the total cost.

Anscombe says "It is time to realize what the problem really is, and solve that problem as well as we can instead of inventing a substitute problem that can be solved exactly but is irrelevant," (*Journal of the American Statistical Association,* 1958, pp. 702-719).

It is argued that, in many circumstances, prescribing an acceptance plan will cost considerably more in total than 100 percent inspection. It is argued further that if the process producing the parts were in good statistical control, tests of samples would provide no useful information. Therefore, the ideal situation is for the vendor to supply proof that his production process is in statistical control.

7.6 MIL-STD-2000 FOR MAKING SOLDERED ELECTRICAL AND ELECTRONIC CONNECTIONS

Solderability of printed wiring has been a consistent concern of the electronic industry in recent years. Modern electronics equipment assembly is increasingly becoming a solder application process. When this process is flawed by solderability problems of component leads, circuit boards or machine processes, solder joint repair is a significant part of the assembly operation cost. Until about six years ago, European industry was pursuing improved solder platings while the United States seemed to be more interested in using aggressive fluxes to solve the problem during mass soldering. As limits were approached in flux activity without really resolving repair action issues, some consultants and individuals began suggesting a look at basic solderability criteria of the components and circuit boards.

And recently the United States Department of Defense has come out with their own series of soldering standards, Mil-Std-2000. For more information on Mil-Std-2000 write to Commanding Officer, Naval Air Engineering Center,

Code 5321, Lakehurst, NJ 08733-5100.

Mil-Std-2000 is a reformatting of the DOD-Std-2000 series of soldering standards. Mil-Std-2000 requires 100% inspection except in cases where statistical sampling is done as part of a statistical process control program. All inspections must be documented internally and sent back to the government. The documentation includes individual article, summary and rework reports. All rework must be documented.

What follows is an example of how one aerospace contractor is using Mil-Std-2000 to improve the quality of their printed wiring assemblies. Mil-Std-2000 requires that 100% inspection be done on all soldered connections and assemblies. Any defects found must be documented and recorded for future reference. The manufacturer must calculate defect rates for the printed wiring assemblies and a daily rate of defects. This requires individual records for each printed wiring assembly as well as a summary of daily production.

The manufacturer divided the defects into three types: printed wiring board, component and part, and solder joint. Then the manufacturer divided the defective parts into two groups: rework and repair. From the different types of categories and two groups, the manufacturer had 77 total defects.

The manufacturer went from using a check sheet as shown in Fig. 7.6.1 to using the DataMyte 769 data collector. The bar code sheet that replaced the check sheet is shown in Fig. 7.6.2. Using the 769 data collector, the manufacturer is able to sort the data by defect type, by product type or by any other characteristic of the data collected. This allows the company to look at the data that affects their processes the most. And it also allows them to easily put together the reports that the government requires to satisfy Mil-Std-2000.

See Chapter 20 for more information about the DataMyte 769 attributes data collector that will help you inspect according to Mil-Std-2000.

Check Sheet for Department 33033

Inspector_____   Date_____

Assembly No._____

Assembly Norm. No._____

Solder Norm. No._____

Job No._____

Serial No._____

Defect																
A02.1																
A11.1																
A14.1																
A16.1																
B17.1																
B22.1																
B23.1																
B25.2																
B30.2																
B36.1																
B55.1																
C57.1																
C60.1																
C64.2																
C77.2																

Fig. 7.6.1   The check sheet the manufacturer used before they began using the DataMyte 769 data collector.

7-26

Inspection

DATA COLLECTION SHEETS DEPARTMENT 33033
DATE – 08/15/89

INSPECTOR

ASSEMBLY NO.

ASSEMBLY NORM NO

SOLDER NORM NO.

JOB NO.

SERIAL NO.

ATTRIBUTES

A02.1

A11.1

A14.1

A16.1

B17.1

B22.1

B23.1

B25.2

B30.2

B36.1

B55.1

C57.1

C60.1

C64.2

C77.2

Fig. 7.6.2   A bar code sheet printed from the
769 data collector used for data entry.

For Further Reference

Anscombe, Francis J., "Rectifying Inspection of a Continuous Output." *Journal of the American Statistical Association,* Vol 53, 1958: pp. 702-719.

Deming, W. Edwards *Quality, Productivity, and Competitive Position 1982,* Massachusetts Institute of Technology, Chapter 13.

Government Printing Office, *Sampling Procedures and Tables for Inspection by Attributes, MIL-STD-105D.* Washington, D.C.: Government Printing Office, 1963.

Grant, Eugene L., and Leavenworth, Richard S. *Statistical Quality Control,* 5th ed. New York: McGraw-Hill Book Company, 1980.

Hald, A. "The Compound Hypergeometric Distribution and A System of Single Sampling Inspection Plans Based on Prior Distributions and Costs." *Technometrics* (1960) 2: 275-340.

Juran, Joseph M., ed. *Quality Control Handbook,* 3d ed. New York: McGraw-Hill Book Company, 1979.

Juran, Joseph M., and Gryna, Frank M. *Quality Planning and Analysis.* New York: McGraw-Hill Book Company, 1970.

Oliver, Larry R., and Springer, Melvin D., "A General Set of Bayesian Attribute Acceptance Plans." *American Institute of Industrial Engineers. 1972 Technical Papers.*

Orsini, Joyce "Simple rule to reduce total cost of inspection and correction of product in state of chaos," dissertation for the doctorate, Graduate School of Business Administration, New York University, 1982. Obtainable for University Microfilms, Ann Arbor, 48106.

Schafer, R.E. "Bayesian Operating Characteristic Curves for Reliability and Quality Sampling Plans." *Industrial Quality Control* (1967) 14:118-122.

Schafer, R.E. "Bayes Single Sampling Plans for Attributes Based on the Posterior Risk." *Naval Research Logistical Quarterly* (1967) 14:81-88.

Western Electric Co., Inc. *Statistical Quality Control Handbook,* 2 ed. Easton: Mack Printing Company, 1956.

8. JUST-IN-TIME, TOTAL QUALITY CONTROL, AND SUPPLIER CERTIFICATION

8.1 INTRODUCTION

This chapter will help to acquaint the reader with manufacturing management practices used to improve quality and productivity. People who use SPC should become familiar with just-in-time (JIT) and total quality control (TQC) concepts because they provide a framework for addressing quality issues on a plant level. In addition, supplier certification programs can help address quality issues before they even reach the plant floor.

With SPC, the task of reducing variation in a process soon takes us beyond the process itself. Concerns about proper machine maintenance, adjustment and operation turn into other concerns like simpler methods, better raw material uniformity and more efficient parts handling. Further improvement in quality and productivity soon becomes an environmental issue; that is, SPC can document a need for improvement in just about any manufacturing environment, but there are some environments where the actual improvement can occur faster than others. This chapter is a look at those types of environments.

JIT, TQC and supplier certification practices can optimize the use of SPC. That JIT and TQC differ in methods has more to do with the broad nature of manufacturing itself and how people with different cultural backgrounds (the Japanese with JIT for instance) discover methods that work.

8.2 JUST-IN-TIME MANUFACTURING

There are many articles and books on the subject of JIT. The most notable source for the information found in this section is *Japanese Manufacturing Techniques, Nine Hidden Lessons in Simplicity,* by Richard J. Schonberger (Macmillan Publishing Co., Inc., 1982).

Just-in-time is a manufacturing strategy that is intended to increase profit and competitive position. It is not simply a way to reduce inventories, or force suppliers to deliver goods "just in time" for use in manufacturing. JIT is based on the idea that a company should buy or produce only what is needed and only when it is needed.

JIT aims at the timely delivery of materials and tools to each work station, and the many benefits that occur from working toward this goal. Some of the quality and productivity benefits are:

- Workers become responsible for making defect-free parts, resulting in less scrap, rework, material waste and wasted effort.
- There is more awareness of the sources of delay and error.
- There are higher levels of worker motivation.
- It creates a fertile environment for plant-wide quality improvement.
- There is greater productivity and lower cost — which fuels a continuous quality effort.

JIT was developed in Japan to improve companies that did repetitive manufacturing, which is capital and labor intensive in both Japan and the United States. JIT is a radical departure from traditional U.S.-style manufacturing with its shop-oriented plant layout, large inventories, elaborate material handling and computer controlled production management systems. Instead of focusing on factory automation and elaborate control systems, JIT aims at simplifying and streamlining the flow of goods and labor.

JIT and Quality

JIT manufacturing represents a commitment to quality. In a sense, a company using JIT bets its whole manufacturing process on the quality of its goods. If there are no defects the process runs smoothly; however, if there are defects the process grinds to a halt. Defects become diseases from which the whole body suffers and they are dealt with quickly and efficiently. This is unlike traditional batch-mode production methods, where a certain level of defects is assumed as unavoidable; in order to keep the process running, batches of parts are made so that there are always enough good parts around.

By reducing inventories and switching away from batch-mode production, producing defect-free parts becomes much more important. A part with a defect creates a break in the chain that can cause work stoppage all the way along the production process. With batch-mode production, a bad part simply gets thrown into a scrap pile or shipped to a customer, and production does not stop because a worker can simply pick another piece from the bin.

Not coincidentally, JIT practices apply wherever SPC can be used to improve quality. JIT requires each worker to be aware of the quality requirements of his work, and to assure that these requirements are met. Every worker must know what quality looks like at every work station, and feel responsible for it. The feeling of responsibility is not a dictated feeling either. It is a personal motivation continually reinforced by relationships with other production workers in the plant.

Although it is difficult to apply in principle, an ideal JIT practice is to have each worker physically hand his finished piece to the next worker in line. This tends to create a strong feeling of pride, social responsibility and teamwork. Each worker knows that the next worker depends on that piece being made well and on time in order to do his work. This motivation is lost in the case of a worker who simply puts the piece in a finished parts bin, and the bin moves by conveyor or lift-truck to some unknown destination on the other side of the plant.

JIT Practices

The JIT commitment to quality is part of a central goal of reducing waste and therefore the cost of manufacturing. JIT can be started in a plant by following these practices:

- Cutting lot sizes
- Cutting setup times
- Implementing total quality control
- Implementing a pull system
- Organizing the plant for continuous flow manufacturing
- Withdrawing buffer inventories
- Simplifying buying practices

Each of these practices will be described briefly. The biggest misunderstanding about JIT is that it should start with one's vendors — reducing carrying costs by forcing vendors to deliver just in time. Although JIT does take on that appearance, as a strategy it should start at the other end of the plant, in the final assembly area, and work backward.

Cutting Lot Sizes

Lot sizes are usually determined by the cost and time it takes to set up a machine and the carrying cost for the batch of parts. As machine tools become more performance oriented and more expensive, larger lot sizes occur. The negative ramifications of large lots are not examined as closely as machine setup and inventory carrying costs:

- Large lot sizes hide defects.
- They create waste.
- They create a "hurry up and wait" mentality where a fast machine produces a lot of parts that sit around and wait for the next manufacturing stage.
- They require elaborate shop floor control and material handling to schedule and move the parts to and from the machining center.

Many of the costs associated with large lots are assumed to be fixed. Cutting lot sizes may create more costs initially because it forces us to confront these fixed costs, such as setup times, production scheduling and material handling. But effort put into simplifying and streamlining the operation eventually can create many benefits, not the least of which is a more flexible manufacturing operation capable of responding faster to market demand.

With a lot size of one (an ideal with JIT), an order to ship is all the documentation necessary. Successive manufacturing stages are positioned right next to each other. The material is handled one piece at a time station to station. Setup time is reduced to seconds and work is apportioned so that each step takes an equal amount of time. Teamwork occurs because each worker knows that work will halt and quotas will not be met if there are defective parts or problems at any one station.

A lot size of one is impractical in most instances, but the goal of reducing lot sizes is not. Reducing lot sizes exposes quality problems and sources of delay. It is easier to trace a problem to the time it occurred and the circumstances behind it. It forces us to think of ways to create efficiencies between manufacturing steps rather than relying on the internal efficiencies of some high speed machine tool to make up the difference in waste and carrying costs.

Cutting Setup Times

Machine setup times must be cut in order to justify cutting lot sizes. Japanese companies expend a lot of effort in cutting the setup times of machinery to the point of making special fixtures and conveyors for dies and jigs, and drilling teams of workers from neighboring machines to join in and set up a particularly cumbersome machine when needed.

When setup times are no longer viewed as fixed, a whole new type of plant engineering can take place, that of developing manufacturing systems with minimal setup that are responsive to more instantaneous demand. A lot of the setup time caused by the machine being a general purpose machine tool is engineered out. The resulting more dedicated machinery can produce better quality in smaller lot sizes.

Implementing Total Quality Control

Total quality control will be elaborated on later in this chapter. Total quality control allows JIT manufacturing to function smoothly. JIT and TQC are self-perpetuating in that JIT exposes defects and TQC serves to eliminate them, which allows lotless production to continue.

Implementing a Pull System

Most manufacturing systems are push systems. Work piles up, or queues up in front of each machine with the idea that this guarantees continuous production. Work begins at the next station when there is a large enough quantity of unfinished parts to be worked on. To smooth over the whole production process, buffer inventories and safety stocks are put in front of machines that operate faster than others, so they won't be idle.

A pull system is production that is responsive to final assembly, which in turn is geared to customer orders. Rather than pushing materials through a factory, a part is made when that part is needed at the next stage in the process. Parts are thus delivered just in time to the next station. The Japanese Kanban (card) system is a shop order system where a station needing a quantity of parts places a card at the previous station to signal the production of those parts. Idle labor is not as big a concern with a pull system. Productivity eventually increases as management and workers divide up the work more equitably and the system smooths itself.

Organizing the Plant for Continuous Flow Manufacturing

Just-in-time manufacturing requires an abandonment of the job-shop layout of a plant in favor of a cellular layout. In a cellular layout, successive stages of a production cycle are physically situated right next to each other — a punch press next to a grinding/deburring station next to a drill press and so on. The idea is that it is much more important to have workers on the same part work next to each other and understand successive steps in the making of the part than to group like machinery in separate shops. A cellular layout removes much of the need for material handling apparatus, such as fork lifts going between stations. With reduced buffer inventories between stations, the floor space occupied by the layout also decreases.

Withdrawing Buffer Inventories

To expose sources of problems, Japanese managers use an offensive strategy of actually removing buffer stock between machinery. This stimulates activity to remove sources of delay, remove potential quality problems, and continue perfecting the process. Once the process smooths out in operation, the managers remove a bit more buffer again to expose the workers to other problems.

Simplifying Buying Practices

Buying practices for JIT are aimed at dock-to-assembly line movement of parts rather than dock-to-inventory. Arrangements are made for more frequent deliveries of less parts, without a lot of the formal paperwork. The vendor is made more responsive to the needs of the plant and coached to develop his own resources so that he can play the role profitably. Long-term relationships and quality are stressed rather than competing for a low bid.

8.3   TOTAL QUALITY CONTROL

Total quality control was definitively described in the book *Total Quality Control* by Armand V. Feigenbaum

(McGraw-Hill Inc., 1961). TQC has also become a catchall phrase for quality practices in Japan that are part of JIT manufacturing. TQC is an expansive subject, and only a few definitions and concepts will be dealt with here.

Total quality control is a management system for an entire organization, not just the manufacturing area, according to Feigenbaum. It is a system of integrated controls which ensure customer quality satisfaction and economical costs of quality. It includes engineering, purchasing, financial administration, marketing and manufacturing. Fundamental to the concept is that quality must be designed and built into a product. The idea that quality can be inspected into a product is rejected. Defect prevention is emphasized rather than defect detection.

TQC is a systems approach that recognizes both the advantages and inherent flaws of the division of effort required in an organization. It seeks to complement this with quality improvement through an integration of effort. The characteristics of TQC are that it has a point of view of continuous improvement and it requires the thorough identification and documentation of tasks and responsibilities. As a result, TQC becomes a foundation for ongoing organizational quality control and provides for the systematic engineering of order of magnitude improvements in quality.

Cost of Quality

TQC requires an understanding of the costs of quality. Cost of quality has been elevated in a total quality control system to a financial control on a par with labor costs and material costs. The cost of quality is more than just the cost of scrap. It includes the costs of control and the costs of failure.

The costs of failure are divided into internal failures and external failures. Internal failure costs include:

- Scrap
- Rework
- Engineering changes
- Idle time

External failure costs include:

- Market research

- Technical support
- Warranty repairs
- Lost business

There is debate as to whether some expenses, such as technical support and marketing, are a cost of quality. Technical support is aimed at increasing customer satisfaction and, thus, lowering external failure costs. Marketing is aimed at establishing market expectations for a product. Market research helps to identify the specifications a product needs to be successful. Both marketing functions tend to reduce product failures if successful. At the same time, product failures, and their resulting quality costs, occur despite good engineering and manufacturing if the product specifications did not match the needs and expectations of the customers.

The costs of control can be divided into prevention costs and appraisal costs. Prevention costs are the costs of preventing failures from happening in the first place and include:

- Design reviews
- Specifications reviews
- Quality training
- Preventive maintenance

Appraisal costs are the costs of determining whether specifications are being met and include:

- Prototype process development
- Data collection
- Inspection
- Quality control
- Test and inspection equipment

Total quality costs can be graphed as shown in Figure 8.3.1. For each company there is an optimum cost of quality, where failure cost is least but control costs are not so excessive as to make quality control unprofitable.

An analysis of cost of quality can be quite illuminating to a company. The cost of quality can amount to as much as 40 percent of production costs and 20 percent of sales. If a

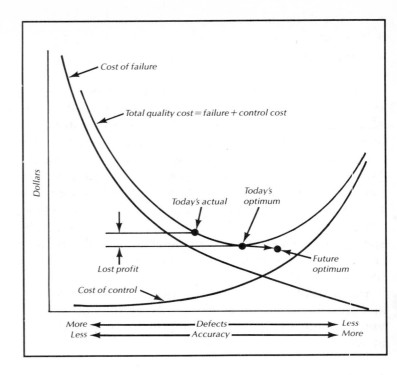

Fig. 8.3.1 Cost of quality curve.

$50 million a year company has a cost of quality of 15 percent of sales, that's $7.5 million a year being spent. If that cost of quality could be reduced to 3 percent, or $1.5 million, that's a $6 million savings every year.

More important than how much is spent is where it is spent. A company that is spending most on failure costs could certainly afford more control costs. A company that spends an inordinate amount on engineering specification reviews and changes may be able to justify an investment in a CAD/CAM (computer-aided design/computer-aided manufacturing) system to help design more quality into its products.

Japanese Total Quality Control

The Japanese took Feigenbaum's statement "The burden of quality proof rests . . . with the makers of the part," to heart. Their efforts to simplify production systems tend to decrease the amount of systems documentation, staffing specialization and controls for quality, but increase the dependence on the worker. The following are some TQC practices from Japan (Schonberger, 1982):

Responsibility for quality is placed in the hands of the production department — Workers are required to know what quality looks like, and to ensure quality piece by piece. Quality problems are addressed at the lowest level, with quality control staff serving in an advisory role.

Statistical process control — Workers use SPC to discover ways to reduce variability.

A habit of continuous improvement — TQC is seen as a dynamic rather than a static system. There is no such thing as "acceptable quality levels," rather, there is instilled in each worker the desire to think of ways to improve quality and productivity and to help implement them.

Measurable standards for quality — Visible indicators of quality are placed at each process and throughout the plant. Charts are kept and displayed and gaging and inspection equipment are made highly conspicuous.

Vendor inspections — Quality inspections by customers of their vendors take on the form of a military inspection, with a list of demerits and a timetable.

Insistence on compliance — Management must continually make everyone aware of the quality expectations and how they can be achieved. Compliance must be strict in that no product can be shipped that does not pass all criteria.

Line stop — Workers are given the authority to stop a production line in order to solve a quality problem.

Correcting one's own errors — Rather than have a separate rework line, rework must be done by the same worker who produced it.

Project-by-project improvement — Workers meet and develop a list of projects and activities for quality and productivity improvement. These promote teamwork and lead to improvements faster than what one worker or an isolated staff could accomplish alone.

8.4 START SMALL, MAKE A SUCCESS OUT OF SPC

Most people need to be convinced on their home turf that statistical process control, or SPC, is worth it. The skeptical ones might say "Isn't SPC another of those Japanese business practices, like lifetime employment, morning calisthenics, company songs and quality circles? Must we try all of them?"

Some attitudes can be well justified and dead wrong at the same time. It's true, over the last five years we've sent plenty of people to Japan to study their business practices. After a few weeks, sometimes months, they come back with all sorts of theories. It's enough to make you a little wary. The problem with this viewpoint is that it just isn't true about SPC. First of all, its pioneers are a pair of Americans, W.A. Shewhart and W.E. Deming. Although it seemed discredited in the 1950s and 1960s, when American industry had little competition for its products, SPC is *the* management tool to achieve and maintain product quality. William Conway, former chairman of Nashua Corporation of Massachusetts called Dr. Deming's methods the third wave of the industrial revolution. The first wave was the mechanization of English textile plants in the 18th century. The second wave was the introduction of mass production by Henry Ford.

Strong testimonies notwithstanding, there are some definite problems with SPC and they have to do with that dark fearsome beast called implementation. When you read about Dr. Deming's management philosophies, and study SPC as it is currently practiced in Japan, you find out how rigorous a practice it can be. Some of the conditions deemed necessary for SPC to work are:

- Educating management on the philosophy of SPC.

- Training management and hourly workers in simple statistics and using control charts. Sometimes basic mathematics must be taught first.

- Retooling fixtures and inspection systems to measure variability. Reducing or eliminating final inspection is an eventual goal.

- Dedicating workers, process engineers, and management to the task of process improvement based on statistical techniques. This can involve restructuring jobs, work standards, and even wage contingencies.

- Accepting the results — which can affect product deliveries, inventory, tooling, machinery purchase, and the whole manufacturing process.

As they say, strong medicine can sometimes kill the patient. What perhaps ought to be whispered in back of the call to get started is: "You can start small." Most of the examples that are mentioned here involve American companies that have really just started. They didn't send all their personnel to statistics boot camp. They assigned responsibilities to a small group, usually the quality or process engineering group. For fear of choking, they didn't swallow the whole retooling argument either. What they did do is pinpoint one or two processes that could stand improvement, and they gave SPC a try.

Edgewood Tool and Manufacturing Co. of Taylor, Michigan, is one example. They had a problem with misformed parts on hood hinges for Ford light trucks. The problem was traced back to the blanking stage. One critical dimension — the distance from the edge of a pierced hole to the edge of the part — was monitored using a control chart. They found that parts variation increased whenever the operator loaded a new coil onto the machine. The solution was an inexpensive gaging block which made loading and positioning the coil a more precise operation.

The significance of Edgewood Tool's use of statistical process control is not so much that they were able to solve a problem, and decrease scrap, rework and inspection, but that they did it by monitoring only one characteristic. They needed only a few trained personnel to achieve results.

A bumper supplier for Ford Motor Company had a similar experience. The supplier used SPC to monitor a plating bath. By determining the capability of the plating bath, and working on one characteristic — bath temperature — they realized a 25 percent energy savings on that part of the process. Once again, a limited application paid off in both real terms, and in an increased awareness of the power of SPC.

Ford, in its efforts to encourage their suppliers to implement SPC, offers these suggestions:

- Start with a pilot program.
- Select just a few characteristics.
- Train the group involved in statistics.
- Adopt gaging for variability.
- Document your results.
- Create management awareness.

One more suggestion is to look into purchasing a data collector/quality control computer. It will offer a lot of help in collecting data and making statistical sense out of it — areas in which newcomers to SPC can have quite a few problems. But more on the hardware later. First, let's expand on Ford's suggested methodology.

Starting a Pilot Program

The chief criteria for a pilot program should be its visibility and its potential for success. There is an advantage in selecting a new operation for these reasons. A new operation may also have newer machinery, and people more inclined to do things differently. If an older process is used, select one that needs improvement, and where improvement can be measured. As many variables as possible should be isolated, so that the results of your work will be dramatic.

Selecting a Few Characteristics

There is a tendency for newcomers to try to monitor as many characteristics as possible. ITT Hancock, a supplier of front door hinges for Ford, had this problem. Their quality control manager and supervisor attended a five-day seminar on statistics conducted by the American Supplier Institute (formerly Ford Supplier Institute), and came away determined to implement a program. Their initial approach was to try to control 28 characteristics that contributed to a hinge torque problem. They found that they could make little headway. After three months, they cut the number of characteristics to 12 and then to five for ongoing control. One of ITT Hancock's conclusions was, " . . . an SPC program will generate the earliest results and enthusiastic local support when only few characteristics are selected for a first application." Their tangible gains were a 10 percent reduction of machine downtime, a 15 percent reduction of labor for rework, a 10 percent scrap reduction, and improved relations with Ford. But they also learned a bit about implementing SPC by the experiences of their own operation.

Training in Statistics

Training should include some of the prime decision makers involved in your pilot program. That would normally include the QC manager, production manager and supervisors, and plant manager. Since, according to Dr. Deming, management is responsible for 85 percent of quality problems, upper management training is needed to provide the rationale for change when change is required.

Those directly responsible for the pilot program, including the production engineer, QC auditor, and machine operators, should receive some training also. The training would include how to make and keep control charts, and how to correctly interpret results. This will help eliminate the common pitfalls of making corrective adjustments too soon in a process.

If you are an automotive supplier, excellent training can be obtained through the American Supplier Institute seminars. Unfortunately, most other industries have not developed extensive training resources. Hiring a statistical consultant to help with the program is one solution.

Adopting Gaging for Variability

Part of the selection process for the pilot program must involve choosing characteristics that can be measured easily, and to which you can apply statistical techniques. The aim of SPC is to reduce variability, regardless of engineering specifications. This means that gaging and fixturing must be capable of measuring any significant deviation from the nominal. GO/NO-GO gages and other pass-fail types of fixtures will not do.

It is in the accuracy of measurement devices where the power of statistics in identifying problems lives or dies. It is here where you should take advantage of some of the latest measurement technology. Since it is necessary to collect data systematically and render it meaningful with the least amount of error, the best system would be one that ties measuring devices directly to data collection equipment. Gages with readouts still require you to copy down readings. They invite transcription errors and a lot of intermediate paperwork. The ability to handle the data collection dictates to a large extent the number of characteristics you can control. You can imagine how bogged down in paperwork an ambitious SPC program can get.

Documenting Your Results

The primary document used is the control chart. It takes some training to read them properly; however, once understood, they are what the people who run the process use to get the process in control. Once in control, histograms and capability studies can be used to predict whether design tolerances can be met.

Histograms and capability studies are documents that managers will find valuable. With them you can estimate the percentage of defective parts. How much scrap or rework is expected will determine your inspection load. You can also use the histograms to minimize the cost of production. Presuming that scrap is more costly than rework, cost can be minimized by shifting the mean of your parts distribution away from the scrap end, and accepting just a bit more rework.

Creating Management Awareness

The final suggestion is the most important. The selling aspect of SPC pilot programs cannot be overemphasized. You must create management awareness by making a success story. Most companies have plenty of experience with the problem-ridden periods of a process startup. One strong point right off the bat is that SPC is a morale builder. Workers feel that they gain a controlling influence over the process. Most companies using SPC find that the workers become the best advocates of it. That's why the publicity about the program should flow upward to management and to other areas of your company that could use SPC.

A well documented SPC program provides the kind of statistics and charts that managers can understand, and can translate into dollars. Pontiac Division of General Motors, for example, boasts that their implementation of SPC in one plant cut the cost of their engine production by 30 percent in 18 months.

Choosing a System

There are a number of computer-aided SPC systems on the market today. Most require a desktop computer and software to perform analysis. They depend on the manual input of data, which usually produces results slower than you would want. Automated data collection systems solve that problem.

What automated data collection systems do is quickly collect and process data, providing control charts, capability studies, and histograms simply by connecting to a computer or printer. No other equipment is required to generate graphs and reports. One QC auditor can perform both the data collecting and reports generation.

Such a system can manage your data collecting and do the statistics for you, leaving you free to concentrate on analysis and problem solving. You can build a bit more potential for success into your pilot program. It also provides a systematic approach to data gathering that you can literally carry over to other manufacturing processes.

Conclusions

To repeat, American companies have so far achieved quite a bit of success using SPC. This is in spite of most of them being in their infancy — limited to one or two processes and a small group of people. In many ways, success is assured by this approach, because any negative ramifications will never be more than that involved with the trials of improving a single process.

Using a pilot program makes the learning process easier, and it cuts through the hierarchy of job responsibilities, making production of a part the main goal, and process control a unified endeavor. Using current technology right at the start, and by promoting the success to others in your organization, a pilot program can spearhead the greater usage of SPC, and the revitalization of quality control in your company.

8.5 SUPPLIER CERTIFICATION

As an important part of JIT, many companies have implemented a *supplier certification program*. The goal of this type of program is to develop a win/win relationship that ties your company and your supplier together for mutual success. The reduced lead times and frequent deliveries of JIT cannot be asked of the supplier until he has made some progress with JIT principles. Until the supplier has demonstrated his ability to produce zero defects using SPC and has reduced his setup time, the risk of having deliveries made on a JIT schedule is too great.

Certification Benefits

Both the customer and the supplier benefit with a supplier certification program. The customer's benefits can be seen from incoming product through end customer satisfaction. Here are some of these benefits:

- The customer receives zero defect product.
- There is mutual cooperation and trust.
- Teamwork is established between the customer and supplier.
- There are fewer buyer contacts.
- It is a joint improvement program.
- There are stable and repetitive deliveries to the customer.
- There is better accountability which makes it easier to trace the source of a problem.
- There is less paperwork.
- The customer can eliminate safety stock as a buffer against quality and delivery problems.
- The customer can eliminate incoming inspection and count verification.
- There is direct delivery to the production line.
- The customer can minimize parts handling.
- The program reduces inventories.
- The customer reduces lead times.
- There is a higher visibility of problems with specifications and drawings.
- The customer can determine if the supplier is capable of making parts to specification.
- It lowers total cost.

Because supplier certification programs are only successful if they are cooperative efforts, the supplier also benefits.

These benefits can be seen from contractual setups through deliveries and continuing business. Here are a few of these benefits:

- The supplier can expect 100% of the customer's business when meeting requirements.
- There is improved quality with a focus on process controls.
- The supplier has less rework and scrap with SPC.
- It is a joint improvement program.
- The program improves schedule visibility and stability.
- There is more security regarding survival.
- The supplier has improved competitive ability.
- Specifications are clear.
- It eliminates administrative formalities.
- The supplier can reduce setup time.
- The supplier has a known process capability with SPC.
- The program eliminates inspection with SPC.
- There is reduced inventory when the program is carried to the supplier's supplier.
- The supplier gets first consideration for new business.
- There is mutual cooperation and trust.
- There are closer functional ties and improved teamwork.
- The supplier can lower total cost.
- There is a sharing of cost savings.

Most of these benefits can be achieved by setting up mutual goals in the areas of:

- Total quality control and SPC
- Total cost reduction (not price reduction)
- Smaller lot sizes and more frequent deliveries

With less rework, less scrap, fewer specification and drawing errors, fewer receiving and invoicing errors, the total cost of the product can be reduced. With the guarantee of zero defects, inventories and safety stock can be reduced, and inspection can be eliminated. Smaller lot sizes and more frequent deliveries can be accomplished with reduced setup and streamlining of administrative paperwork.

Supplier Certification Criteria

The use of SPC should be a requirement of any supplier certification program. This will assure you that your vendor's process is capable of producing the part to your specification. The use of SPC on the process will then guarantee that all parts produced are defect-free and can be delivered to your production line, bypassing inspection by the supplier and the customer. In addition to requiring that the supplier use SPC, you should consider these areas when evaluating suppliers' abilities and performances:

- Management commitment to partnership
- Technical capability and support
- Quality history
- Delivery performance — on time and quantity
- Total cost competitive
- Preventive maintenance
- Problem solving
- Financial stability
- Business base — customer and industry
- Purchasing support
- Flexibility
- Location
- Lead time
- Waste elimination
- Research and development
- Education and training
- Labor conditions, morale
- Employee turn over and management stability
- Capacity
- Environmental concerns
- Safety
- Ethics
- Sales support

These criteria should be defined in measurable terms whenever possible to minimize the subjectivity of your rating. On-site surveys are necessary as are frequent visits to the supplier's location to cement the relationship aspect of this program.

Internal Supplier Certification

After developing your criteria and measurements, perform an audit on your own company. How can you ask your suppliers to do something you aren't doing yourself? You don't have to have perfect performance in all areas before starting your certification program, but you should have commitments and goals in place for improvement. After all, we are all suppliers to our customers internally and externally. It will be much easier to carry this program to your external suppliers when you have experience and results to present to them. Your team will be more credible and enthused when it has some success of its own to talk about.

Certification Team

Team members should always include persons from the purchasing, engineering, and quality functions. At varying times, you will also want to include production and financial people. It is important that these people establish a relationship with their counterparts at your supplier's facility to enhance communication and problem solving.

Conclusion

The certification process is very long and time consuming. It often takes a year or more before you will be able to classify a supplier, or a part, as being certified. Change takes time and what you are doing in a supplier certification program is changing the culture of your own organization as well as your supplier's organization. This is where management commitment comes in. Without it there will be little chance of success. But the benefits of the program are so great, that any enlightened management team should find the program easy to support and encourage.

For Further Reference

Crosby, Philip B., *Quality is Free*. New York: Mentor, 1979.

Deming, W. Edwards, *Quality, Productivity, and Competitive Position*. Massachusetts: MIT, Center for Advanced Engineering Study, 1982.

Feigenbaum, Armand V., *Total Quality Control,* 3rd ed. New York: McGraw-Hill Book Company, 1983.

Grieco, Peter L. Jr., Gozzo, Michael W., and Claunch, Jerry W., *Supplier Certification: Achieving Excellence*. Plantsville, CT: PT Publications, Inc., 1988.

Kume, Hitoshi, "Business Management and Quality Cost: The Japanese View." *Quality Progress,* May 1985.

Schonberger, Richard J., *Japanese Manufacturing Techniques, Nine Hidden Lessons in Simplicity*. New York: The Free Press, 1982.

Taguchi, Genichi, *Introduction to Off-Line Quality Control*. Japan: Central Japan Quality Control Association, 1979.

PART II Applications

9. AEROSPACE AND DEFENSE INDUSTRIES

CASE 9-1 MISSILE MANUFACTURER

Automated Data Collection Allows Government Supplier to Have Quick and Easy Access to Data for Audits by Government Groups

A major missile manufacturer is required to provide documentation and proof of quality to its government customers. This company needed a way to make their data easily accessible when they are audited by government groups.

Problem

As part of their data collection program they needed to be able to automate the collection of dimensional and visual inspections as well as a variety of tracking information required for traceability and data retrieval purposes. As a government supplier they are subject to frequent and comprehensive audits by various government groups. Consequently, their data collection system had to allow for quick and easy retrieval of part and process information. They needed to collect information such as work order numbers, part revision numbers, serial numbers, operator identification and step number as well as other process related information such as machine number, cause, and corrective action information. This tracking information needed to be tied to the dimensional and visual inspections done on each individual part and transferred to their computer system for storage and retrieval purposes. In addition the process data needed to be available for immediate process analysis at the workstation.

Solution

The company automated their data collection system using the DataMyte 900 series data collector. The 900 series data collector allows for quick and easy part retrieval of information. It allows them to track the data by visual and dimensional inspection, and they can look at the data immediately or store it on their computer.

See Chapter 18 for the 900 series data collector and the OVERVIEW system.

CASE 9-2 HELICOPTER MANUFACTURER

Data Collector Allows Inspection to be Done Quickly and Accurately and in Real-Time

A major helicopter manufacturer needed to implement attribute data collection in their printed wire board area and harness assembly area. They wanted to collect and analyze the data quickly and easily at the workstation, transfer it into a data base program for further analysis, and then move it to their mainframe computer for long term storage and archiving.

Problem

In the printed wire board area, they are inspecting for over 100 defects as identified in a government weapon specification. In this area, they need to be able to break down the board inspection into sub-areas on the circuit board and be able to cross reference occurrences of defects to specific locations on the board. They also need to maintain documentation to show when inspections were done, what defects were found, and all other information required for traceability and archiving purposes.

Fig. 9.2.1 Rework Report

```
Unit ID:  Helicopter Company              Printed: 06/07/89,13:51
769 REWORK REPORT



RECORD #     2
                    06/07/89,13:35     Harness I.D.      1986453-A
Inspector Number  57643               Grid Location     1A
Cable I.D.        S3                  Connector I.D.    U2
ATTRIBUTES        1
Mis Wired         1



RECORD #     3
                    06/07/89,13:35     Harness I.D.      1986453-A
Inspector Number  57643               Grid Location     1A
Cable I.D.        J5                  Connector I.D.    C3
ATTRIBUTES        1
Wrong Connect     1
```

In the harness assembly area, 25 defects are identified in the weapon specification. Due to the large size of the harness assembly, they need to be able to inspect different locations on the harness independently and cross reference defects to locations. They also need the harness inspection to include information such as serial number, locations inspected, and defects found.

Solution

See p. 20-4 for the 769 data collector.

The solution included a DataMyte 769 data collector and a personal computer (PC) based data base program. Using the data collector their inspectors are able to collect and analyze the data quickly and easily at the workstation. The data collector allows for data entry with a bar code reader so the inspection can be done quickly and accurately. In addition, the data collector's data base capabilities allow the data to be analyzed at the workstation so the results can be fed back into the process real-time. The data collector also allows them to print out rework tickets (Fig. 9.2.1), data reports and charts (Fig. 9.2.2) at the workstation, and download the raw data into the PC based data base program.

Fig. 9.2.2 Pareto Chart

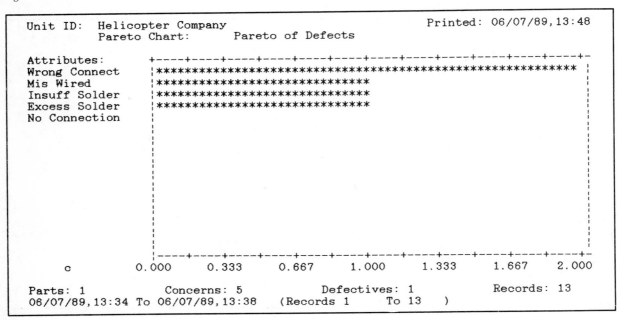

CASE 9-3 AEROSPACE SUBCONTRACTOR

Automated Data Collection Drastically Reduces Inspection Time

A major subcontractor to military defense manufacturers needed a way to automate their data collection system. Due to the multiple precision grinding and milling processes their products require, their manual method soon became time consuming, cumbersome, and inaccurate.

Problem

This company produces guidance and tracking systems for military missiles. Due to the sophistication of the technology used in the design of this part, each step in the process is critical. During the production process there are over 1000 critical dimensional checks that need to be completed and a final inspection that takes one person forty hours to complete. With the manual data collection system, all of the data needed to be keypunched into a computer before any analysis and reporting could be done on the process.

Fig. 9.3.1 Capability Report

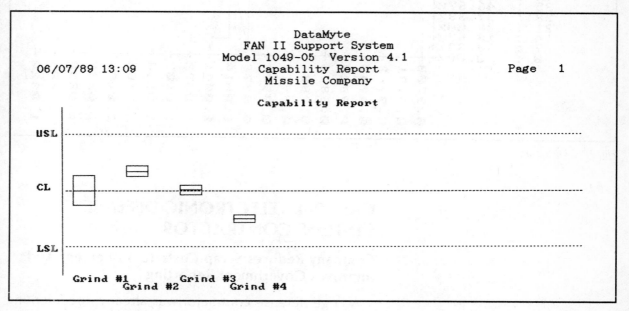

Solution

See p. 19-4 for the 862 data collector.

The solution was to implement an automated data collection system using the DataMyte 862 data collector, DataMyte FAN software and electronic measurement devices. Using a data collector at each step in the production process allows all the in-process data collection to be automated and data analysis to be done at the workstation without a personal computer. See Figures 9.3.1 and 9.3.2. After the data has been collected, it is downloaded into the desktop computer for long term storage, analysis, and report generation. The benefits of this system include eliminating days of keypunch time previously required by the manual system, immediate feedback of process statistics to the operator at the workstation and less time required for the actual data collection.

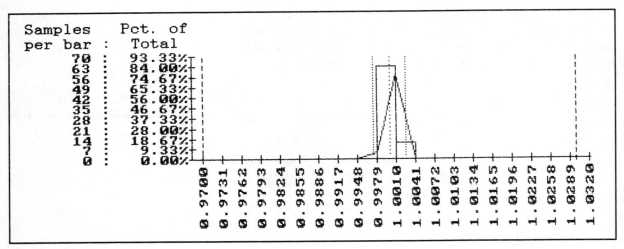

Fig. 9.3.2 Histogram

CASE 9-4   ELECTRONIC DEFENSE SYSTEMS CONTRACTOR

Company Reduces Scrap Costs 10-15 Percent And Improves Government Reporting

A large defense contractor was already documenting their inspection when it saw a need to implement SPC. The company believed that SPC could reduce costs and make them more profitable in the face of increasing cost control pressure. The result of their efforts was a million dollars saved in the past year.

Problem

Although they were collecting the data, their methods were too slow to respond to process changes. The result was an unacceptably high scrap rate in their PC board assembly operation.

When they introduced control charting in the assembly area, it proved ineffective for two reasons:

• The analysis was too slow to do any good.

• The workers didn't keep them current.

The workers obviously had not been "sold" on their value. They found them time-consuming and annoying.

The assembly area had low production volumes of any different types of parts. All of the parts were tracked by part number. The manufacturer had to have some way to keep the jobs separate.

Solution

The quality improvement plan had several key goals:

• Streamline the data collection effort.

• Get analysis in time to prevent scrap.

• Maintain flexibility for short-run low volume production.

The task force did not feel that re-educating workers in SPC would be as effective as simply making it less time-consuming, so that a few auditors could do as much data collection as all the assembly workers.

For their incoming inspection area, where they measured small parts such as metal plates and rivets, they installed a DataMyte 762 fixed station data collector. Using micrometers and calipers, inspectors could record readings at the touch of a button. All of the data was traceable by time, date and part number.

See p. 19-6 for the 762 data collector.

For the assembly area they used handheld data collectors, which could keep track of all the part numbers and instantly produce charts. Data from the data collectors was also archived on an IBM PC running FAN software. Not only was the data available in a single format for government reporting, but the data could be transferred back to the data collectors so the operators could compare the current run to a previous run.

In-process inspection of the PC boards was also streamlined using a DataMyte 769 attributes data collector. Using bar code entry, visual inspection could be made and Pareto charts and p-charts produced on the video display terminal.

Overall, the company estimates it is saving 25% in labor costs using the DataMyte system. Scrap rate was reduced by 10 to 15% because of the shorter time between collecting and analyzing data.

CASE 9-5 AEROSPACE CONTRACTOR

Attribute Data Collection to Meet Military Specifications

A large manufacturer of instruments for aerospace and process industry applications uses precision fabricated and machined metal components in their products. The manufacturer buys some components, but does its own precision machining on proprietary components. The manufacturer builds its electronic assemblies to military workmanship requirements. Most new military programs require the parts to be built to specification Mil-Std-2000. The specification includes extensive data collection on all electrical and electronic assemblies involving soldering.

Problem

The manufacturer had to conduct 100% visual inspection of all soldered connections and assemblies. Any defects found had to be documented and recorded for future reference. The manufacturer had to calculate defect rates for the printed wiring assemblies (PWA), a percentage of defects for each PWA and a daily rate of defects. This required individual records for each PWA as well as a summary of daily production.

DATA COLLECTION SHEETS FOR Mil-Std-2000 DATA
DATE — 01/14/87

Fig. 9.5.1 Bar code sheet used for data entry.

The manufacturer had divided the defects into three types: printed wiring board (PWB), component and part, and solder joint. The manufacturer divided the defective parts into two groups: rework and repair. From the different types of categories and two groups, the manufacturer had 77 total defects.

Solution

See p. 20-4 for the 769 data collector.

The company purchased DataMyte 769s to help collect data. The 769 data collector has a bar code wand for bar code entry, which makes data entry faster than typing it into a computer. Figure 9.5.1 is an example of bar code sheet used for data entry. The manufacturer liked the ability of the data collector to sort data by defect type, by product type, by date or by any other characteristic of the data collected. This allowed the company to look at the data that affected its processes the most.

The data collector displays its charts on a video monitor, which gives operators immediate feedback on the quality of parts they are building. Transfer of data to a computer is easily done with the DataMyte 769 Software program. The data can then be presented to government inspectors, to satisfy the requirements of Mil-Std-2000. From the computer, the data can easily be used in a spreadsheet or data base program.

Because of the quantity of data the company had to collect, the data collector easily paid for itself within a few months.

CASE 9-6   MISSILE PARTS MANUFACTURER

Company Gets 20 Percent Productivity Improvement While Controlling Scrap

A missile parts manufacturer automated the tracking of trends on a low volume job and provided more time for their operators to do machine maintenance and adjustment.

Problem

The Department of Defense required the company to document 25 critical dimensions on each missile part. Although each machine operator produced only one or two parts per hour, there was still not enough time to take measurements and properly analyze the process.

Fifteen of the less critical dimensions were checked with go/no go gages. The ten most critical dimensions were measured by hand and the operators plotted control charts. The documentation effort overall was costing them 30 percent in reduced production. In addition, each scrapped part cost $900 - 1200.

Solution

Because of the small sample size, $\bar{x}$ & R charts would not be sensitive to process changes. The plant elected to use moving average and range charts.

Fig. 9.6.1 Moving Average & Moving Range Report

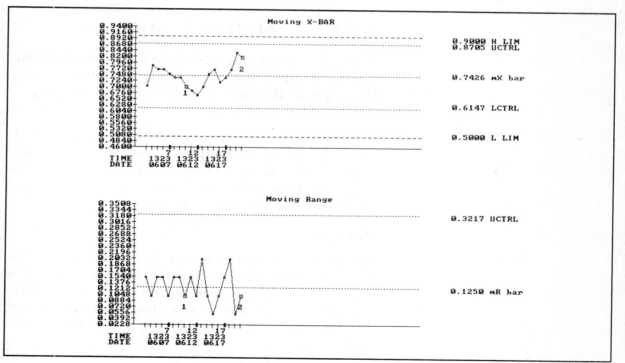

Aerospace and Defense Industries

See p. 19-4 for the 862 data collector.

DataMyte 862 data collectors were set up for moving average and range charts, and programmed with the following trend rules:

- 1 point out of 3 sigma
- 2 consecutive points between 2 & 3 sigma
- 7 points up or down
- 7 points above or below the mean

If an operation was out of control for any of these reasons, the data collector would beep and display a warning. The operator would then have to enter a code into the data collector describing the change he made to the process.

Operators now had time to decide if a new tool or machine adjustment was required. The operator also knew when he had to involve an area supervisor.

Overall, the use of this system reduced documentation time from 30% of each hour in production down to 10%. The system also helped predict tool life, and reduced the incidence of scrap.

CASE 9-7   AEROSPACE SUPPLIER

Company Uses Automated Data Collection to Prove Their Quality to Their Customers and Improve the Accuracy of Their Documentation While Keeping Paperwork to a Minimum

A supplier to aerospace needed to prove their quality to their customers. The company wanted their quality program to be visible on the factory floor. The workers are divided into teams, and each team is responsible for the quality of the part that they manufacture.

Problem

The engineers were not listening to the operators suggestions on how to improve the production process. Each time they began machining a helicopter manifold they had to go through the check sheets before they could run the part. One of the teams was producing parts with sporadic dimensions.

Solution

The company implemented DataMyte data collection

Fig. 9.7.1 A DataMyte 862 data collector and a micrometer.

systems for those teams that wanted to use them. Since they began using the DataMyte systems, they can take an issue to an engineer and recommend a change. They have prove to back up their recommendations. They keep the data in the DataMyte system for the helicopter manifold, so that from the last time they ran the job they can see any problems, correct them and get consistency between runs. Since they began using the data collection systems they have cut the charting task in half for jobs such as machining a helicopter manifold. The DataMyte system improves the accuracy of their documentation and keeps paperwork to a minimum.

See p. 19-4 for the 862 data collector.

CASE 9-8   MAJOR GOVERNMENT SUBCONTRACTOR

Inspection Department Joins Apollo Work Station Network, Improves Response Time

A major defense subcontractor selected the DataMyte system to control manufacturing processes. By linking the DataMyte system to their Apollo multi-tasking personal work stations, they dramatically improved response time and reduced scrap rework.

Problem

This company was looking towards the factory of the future in manufacturing information systems. The goal was to produce a true paper-free environment shared by quality control, manufacturing, engineering, and administration.

The company selected the Apollo DOMAIN System network as its principal architecture. See Figure 9.8.1. Each Domain workstation is a node on a ring, serving as gateways to an IBM mainframe running an MRP program, mechanical drafting, computer aided design, a scanner for supplier documentation, and other manufacturing data bases.

The biggest problem with the new system was that quality control engineers were still required to manually record data which had to be keypunched later into the Apollo network.

Solution

As part of a joint effort with Apollo, DataMyte provided a system to automate the quality control inspection process

Fig. 9.8.1 Apollo network configuration.

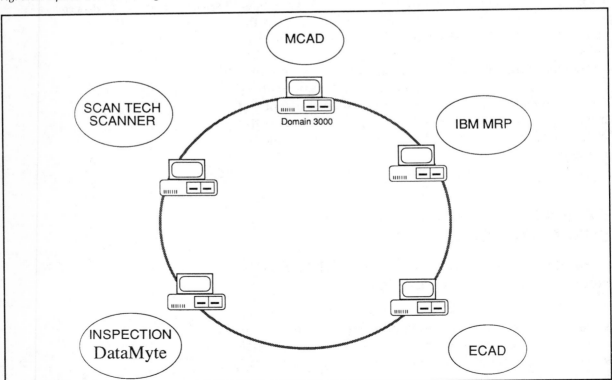

and provide an interface to Apollo work station (Figure 9.8.1).

DataMyte FAN software was installed on an Apollo DO-MAIN 3000 work station, which had an IBM PC compatible co-processor option. Data from individual DataMyte data collectors could then be uploaded into FAN software and made available to anyone on the network.

Although the initial objective was to automate the data collection task, it was quickly discovered that DataMytes used for on-line operator-based process monitoring allowed rapid analysis of the data and reduced the response time for corrective action.

Scrap and rework was reduced substantially. In fact, payback calculations presented to upper management revealed a ten month return on a $150,000 investment.

CASE 9-9   MUNITIONS COMPANY

Company Uses Data Collection Systems for Highly Secretive Manufacturing Documentation

A privately run operation for the Department of Energy manufactures cruise missile fuses. It has been required by new military specifications to provide SPC data ensuring the quality of these high precision component parts. Due to the defense nature of the product, the highest possible quality must be maintained to ensure workability in the war environment.

Problem

The major problem being experienced by this company was in the method of data collection, and in the error rate experienced in their data collection. The company is a highly secretive operation, in which outsiders are not allowed anywhere near the manufacturing areas.

The major reason for their interest in automated data collection was in the reduction of error rate, and the reduction of time consumed in SPC data collection. The high precision parts that they manufacture run in very small dimensional checks. These checks require special gaging and use of such things as column gages to allow precise measurements on all of their close tolerances.

The customer has been taking these readings manually and utilizing tally sheets to control the information flow. However they found that the error rate in a manual data collection system was in excess of 18%, which is way above their goals for quality product. Therefore, the data was erroneous enough to be almost ineffective in their uses. The data also would not at that point, match the military specifications they had for quality.

Solution

See p. 19-6 for the 762 data collector.

The customer evaluated types of data collectors for several years. They chose the DataMyte 762, 2003, and 769 data collectors for applications within their operations due to the ease of use and capabilities of the units. DataMyte helped the company design specialized systems with special gages for each of the applications according to drawings that they supplied.

The majority of the fixed station applications used DataMyte 762 data collectors with multiple types of gaging interfaces, and DataMyte 769 data collectors for attribute fixed station data collection. In addition, 2003 units were utilized as auditing features for the floor inspectors to ensure that data received was within specifications.

One of the major problems with military auditing procedures is that the military will not check any data for long periods of time, and then will require data that may be six or seven months out of date. One of the major software features for the company was the fact that you can specify starting and ending dates for data retrieval, which allows them to provide the charting immediately for analysis by GAO inspectors.

CASE 9-10   MUNITIONS MANUFACTURER

Company Automates Data Collection to Meet U.S. Government SPC Contractual Requirements Included With MIL-Q 9858A

A munitions manufacturer needed to monitor eleven critical dimensions involved in the production of grenades.

Problem

The contract required SPC data on the total production run of over 25,000,000 pieces. Without finding an economical approach to this challenge the company could possibly lose the contract.

Solution

Special analog gaging was used to send data automatically to a DataMyte 761 data collector. Gage stations were developed and placed in three strategic locations on the factory floor. This enabled one inspector per shift to harvest the parts on a periodic basis and collect the needed data.

See p. 19-10 for the 761 data collector.

Each critical dimension could be individually monitored on the video display terminal. The graphs generated allowed for rapid interpretation of trend analysis information so potential problems could be anticipated before the occurrence.

The data was then transferred to an IBM PC for archiving and easy retrieval. The additional feature of the date and time stamping of the data collected especially impressed the auditing group. Due to tremendous numbers involved it was obvious this challenge could never have been met economically and competitively without automated data collection.

As a result, this contract was satisfied and the government is offering additional business to this supplier.

CASE 9-11 HELICOPTER COMPANY

Company Uses SPS Sensor I Torque Wrench to Solve Graphite Composite Material Assembly Problem

On graphite composite material components used for helicopter blades, a major aircraft manufacturer was experiencing assembly problems where these parts were bolted together.

Problem

Graphite composite parts were being crushed and their reliability being put into dispute when tightening bolts through them under conventional torque control.

This was leading to a high rejection level of very expensive parts.

Solution

The SPS Sensor I Wrench was used to tighten the bolts to torque while in the JCS mode. If a yield of the joint was detected before the required torque value was reached the operator stopped, knowing if he went further he would destroy the components. This resulted in a considerable component cost saving. By tying the wrench to the DataMyte 2003 data collector, the company stored the torque data along with serial number, time and date, making documentation easier.

See p. 20-8 for the 2003 data collector.

10. AUTOMOTIVE SUPPLIERS

CASE 10-1 CAR ENGINE MANUFACTURER

Hot Test Area Uses the DataMyte 769 to Record Inspection Results

A car engine manufacturer needed to monitor the defects that they were finding in the hot test area. They choose the DataMyte 769 attribute data collector because it allowed them to display pareto charts instantly.

Problem

As the hot test inspectors found defects on the engines they would log in what was wrong with the engine such as a busted hose or a broken fly wheel. The problem was that it was difficult to track the engines by serial number. There was too much paperwork and margin for error.

Solution

The company decided to use the DataMyte 769 attributes data collector since they were already using many

See p. 20-4 for the 769 data collector.

Fig. 10.1.1 The conformance auditor talks with the hot test inspector about a pareto chart produced by the 769 data collector.

DataMyte variables data collectors. The hot test inspector now logs in the engines with defects using the 769 data collector. That way management can see pareto charts and other reports sorted by serial number.

CASE 10-2 SMALL ENGINE MANUFACTURER

Corporate-Wide Quality Commitment Spurs Many Applications For Automated SPC

A large manufacturer, primarily known for its small engines, has improved its ability to use SPC by installing new systems for measurement and data analysis. Although the company is family-owned and thought of as rather conservative, more innovative methods were being advanced for the following reasons:

- Their existing SPC program depended on handwritten methods for data gathering and record keeping. These methods could not cope with the volume of data.
- Several of their customers were demanding better documentation of their SPC efforts.
- Searching for improved methods was always part of their corporate commitment to quality.

Problem

The company had metalworking applications in their foundry, key and fastener plant, engine assembly plant and die casting plants. They were interested in a system for SPC that had these characteristics:

- The system would collect and process data faster and more accurately than handwritten methods.
- At the user level, it must be adaptable to a variety of gaging and fixturing needs.
- At the system level, it must have a great deal of uniformity and upward compatibility.

The company realized that there may not be a single supplier that could help implement all the SPC applications.

Many applications required in-line gages and automatic triggering of the data collection system. Other applications required full operator involvement in measurement and data analysis.

Solution

Because of the volume of engines and parts being produced by the company, and the massive amount of data required to perform SPC, the company purchased a DataMyte system, knowing the system could pay for itself in less than six months in labor savings alone. Being completely modular, the DataMyte system could be used in a great variety of individual applications (see *Operation*), and be expanded to handle a virtually unlimited amount of data. The system also provided complete analytical capability at the operator level, where day to day quality problems were expected to be solved.

The DataMyte system consisted of DataMyte 762 data collectors at machining and assembly stations, a DataTruck® to harvest accumulated data, DataMyte 2003 data collectors for roving auditors checking both variables and attributes data, and software for their IBM PC computers.

See p. 19-6 for the 762 data collector.

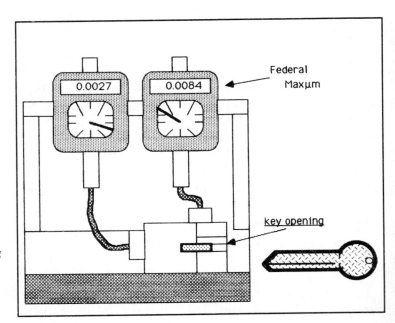

Fig. 10.2.1 DataMyte used with key checking fixture.

Operation

What follows are descriptions of several gaging applications used at the company to collect data for SPC.

Foundry — DataMyte 762s are used with digital indicators to measure critical dimensions on flywheels. Six dimensions, including overall thickness, and height from the vein down certain key position slots, were all previously checked by hand. Using a 762 and a multiplexer for Fowler Ultra-Digit indicators, all six dimensions can be measured and recorded in one 762. Control charts for each characteristic are displayed on a monitor with the press of a button.

Key and fastener plant — The plant that produces raw keys (without individual grooves) for automobiles needed a faster method of maintaining process control. Two thousand keys were produced each hour. A fixture which resembled a steering wheel lock is used to measure overall length, thickness and fit. Two Federal Maxum dial indicators attached to the fixture were used to display the measurements, but the gages had to be read and the data recorded by hand. See Figure 10.2.1.

A DataMyte model 762 was interfaced directly to the Maxum gages, so that readings on the gage are recorded with the press of a button. The data collector compares each reading to specifications and beeps audibly if any reading is out of specification. Control charts are displayed on a monitor, so an immediate indication of the statistical state of the process is available. Readings are now easier to record, and therefore the operators are able to maintain a sampling rate more in keeping with the speed of the process.

The tap-ream-grinding operation at the plant required a completely automated approach to data collection. Ten unmanned machining centers each had a carousel-type part handling system. See Figure 10.2.2. The carousel would bring a part to each of the eight stations, where it would be drilled, shaped, redrilled, a keyway cut, and finally put into a test fixture. The test fixture had several LVDT indicators, which would come down on the part at the same time and display readings on ten column gages. A DataMyte model 761 uses its auto-scan feature to input each of the ten column gage readings in sequence. The process can then be monitored by displaying histograms and control charts for each dimension on the CRT monitor. Accumulated readings are unloaded from the 761 into a DataTruck and then transmitted to an IBM PC for archiving and long term analysis.

For incoming inspection, DataMyte 2003 data collectors are used with calipers and micrometers. Auditors check bins of small parts, such as fasteners, springs and bar stock. Individual parts are compared to specs, and a histogram is printed out to judge the acceptability of a lot. The DataMyte 2003 is also used by a roving auditor to check parts coming off low volume processes.

Engine plant — DataMyte 2003s and 753s are used to audit fastener torque at the engine assembly plant. A DataMyte 2003 is used by a roving auditor to check breakaway torque on the assembly line, using a handheld wrench. A DataMyte 753 is tied in-line with a nut runner. Each nut runner has a torque transducer mounted in-line with its spindle. The transducer is then tied to a 753 which uses its skip counting feature to record every tenth reading while fasteners are being tightened down.

DataMyte 761s are used in the machining area of the assembly plant to check variability in milling and boring operations in their crankshaft, piston rod and camshaft lines. Both two-cycle and four-cycle engines are manufactured and assembled at the plant, and the data collectors allow instant recognition of problems on the continuous high volume processes.

Die casting plants — The company's two die casting plants are using DataMyte 2003s for rough overall measurements. The data collectors allow them to measure dimensions as well as process variables such as pressure and temperature. A temperature probe is used to check mold cavity temperature, and they are realizing a tremendous savings in scrap by controlling just this aspect of the process.

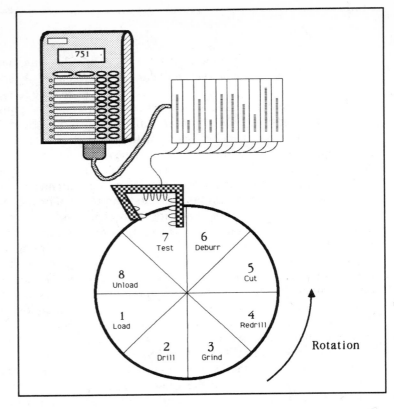

CASE 10-3   PISTON MANUFACTURER

New SPC Charting Methods Create Time Savings and Operator Enthusiasm

A large piston manufacturer had each operator running two different machine tools, and a need for statistical process control. To keep productivity up and to ensure quality at the same time, the company adopted automated gaging and control charting.

Problem

Handwritten methods of control charting were not keeping up with production demands or the need to recognize process changes and take corrective action. Each operator had to measure three or four characteristics on pistons at regular intervals. With two machines to maintain, the operators could not effectively do SPC.

Solution

The company investigated the possibility of automating SPC by installing a single DataMyte 861 at a work station. The data collector interfaced with a cradle gage through Marposs Series E column gages. The operator did their measurement checks, only now the data was logged automatically. A control chart was displayed on the video display terminal at the press of a button.

See p. 19-8 for the 861 data collector.

The operator response was immediately positive. They got the impression that management believed enough in SPC to supply a system for them to do it more effectively. Management also found that a quick look at the video display kept them in touch with what was going on.

The company has since implemented the system at 25 gaging stations. The result has been better control of the processes, more quality characteristics could be monitored, and documentation that they could supply to their customers.

CASE 10-4   METAL FABRICATING SHOP

Data Collection System Saves 15 - 18 Minutes Per Hour At Each Machine

A medium size metal fabricating shop specializing in stamped parts for the automotive industry wished to find a rapid way to implement SPC at the machine level.

The company wanted operator involvement to the extent that operators must be able to view and interpret control charts so trends could be spotted. Being able to spot individual in or out of control points was not sufficient. Control charts could also identify poor setup procedures and excessive tool wear.

Problem

Hand calculated control charts were drastically lowering productivity. Operators were becoming discouraged with the time and arithmetic involved. A method for automating data collection and control charting had to be implemented.

A system had to meet certain plant criteria:

- Factory hardened to withstand mist of oil
- Multiple gage inputs
- Easy to use
- Low cost
- Able to interface with either a personal computer or mainframe

Solution

See p. 19-4 for the 862 data collector.

A DataMyte 862 data collection system was purchased and placed at all presses. Operators could now pick up a part and record three to five dimensions automatically by pressing a button on a caliper, micrometer or dial indicator. Data was recorded in one minute rather than two and a half minutes the old way. Overall, the DataMyte system saved 15 to 18 minutes per hour at each station, counting the time saved for recording five characteristics and entering the data into a calculator to get summaries. Operators knew immediately after measuring how each dimension was set up and whether a change was necessary.

The DataTruck was used to harvest data periodically from each data collector in the shop. The data was then transmitted to a computer for long term storage. This eliminated the need for cabinets full of two and three year old control charts, since their customers requested they retain full documentation.

After transmitting the previous day's data, each data collector could then be re-programmed by the computer or DataTruck for a new job or part number. Workers were very pleased with the system and felt it had enhanced their jobs while giving them time to control their own process. After six months, scrap rates fell and many dollars were saved.

CASE 10-5   TIRE COMPANY

Tire Tread Width Monitored

A tire manufacturer needed to control the width of tread in a cutting application. They needed to automate the entire operation, including the actual cutting, measurement of the tread width, collection of data, and report generation.

Problem

Handwritten data collection and computation did not work. By the time operators got the results, they already produced a lot of bad products which had to be scrapped. Operators were over-adjusting the cutting operation because valid statistical evidence was not available in time.

Solution

The solution was to automatically measure tread width by mounting a laser micrometer behind the slicing blades. This accurately measured the product on the fly. They then interfaced a DataMyte 762 to the RS-232C port on the laser micrometer which sends a dimensional reading to the data collector every few seconds. The data is captured and stored by the data collector in subgroups. By using a CRT connected to the data collector, the operator can view a continuously updated $\bar{x}$ & R chart, and make adjustments to the process as required. If the chart did not indicate a problem, the process was left to run.

See p. 19-6 for the 762 data collector.

All the data from the laser micrometer was stored in the data collector memory. The quality manager could then come to the machine, connect a DataTruck to the data collector, transfer all of the new information that was collected, and take the DataTruck back to his IBM PC. He could then transfer the data to the PC, merge it with existing data files, and generate ongoing management reports.

CASE 10-6   SPARK PLUG MANUFACTURER

Use of Resistance Transducer and DataMyte Reduces Variability

A world leader in the manufacturing of spark plugs had determined that traditional methods for checking resistance during the manufacture of resistor spark plugs were not effective. Spark plugs are an integral part of any gasoline engine. If a plug does not fire, the cylinder in question goes through 720 degrees of wasted movement. Much depended on the high quality of this company's product.

Problem Definition

In the past, spark plugs produced by this manufacturer had exhibited a relatively high level of variability in resistance. In fact, it was estimated that 1% of production was being rejected as unsuitable due to resistance values that were out of specification. The problem was due in large part to the process used to determine the resistance of a resistor spark plug; this process involved the formation of a resistor within the plug. During the process, a special powder was carefully dropped into the ceramic shell that is the insulator for the plug. This powder was then fused under pressure and at high temperature to form the resistor. For the process to be effective, it was necessary for the powder to be dropped, tamped (compacted), and heated. Furthermore, all steps of the process required a high degree of accuracy and repeatability.

For several years the process was monitored by traditional methods; that is, by measuring the resistance of the completed plug with a multimeter. Essentially, all that was being done was a simple "GO" or "NO GO" evaluation. (Either the plug was good or it wasn't.) The data accumulated during this process did little to indicate the capability of the process or if it was being operated in a consistent manner.

To compound the problem, it was known that if the resistance couldn't be maintained within a range of from 2,500 to 12,000 ohms, radio and TV interference as well as ignition problems could occur when the plugs were used in an engine.

Solution

The manufacturer had defined the known problems and areas of concern, and began to investigate the best way to implement improvements. Several solutions were considered, ranging from acquiring new production equipment to performing 100% inspection. Cost for these choices were prohibitive, however. They determined that prerequisites for the overall program must include:

- A method that did not require the need for new production equipment or additional manpower.
- Equipment that was portable.

- Equipment that could interface with existing computing equipment and peripherals such as printers.
- Equipment that would allow all necessary x̄ & R charts and capability reports to be generated on an "as needed" basis.

Based on their analysis, they decided to use a DataMyte 2003 handheld data collector connected to a specially developed resistance transducer. This equipment arrangement provided a data collection and analysis capability that was portable, user friendly, and could be used to produce all necessary charts.

Once the equipment was installed, the manufacturer's quality control personnel used the special transducer to measure and record the resistances of the production spark plugs. Data was fed directly into the DataMyte 2003 for analysis. From the DataMyte, the data could in turn be fed into the computing system for further analysis or dumped to a printer. When the data was plotted on x̄ & R charts, the magnitude and source of any variation in resistance was identified.

See p. 20-8 for the 2003 data collector.

The first control charts generated indicated that the resistance powder dropping process was out of statistical control. Analysis of the process indicated that the mechanism controlling the flow of resistance powder was operating in

Operation

The information that follows provides a brief description of how the DataMyte 2003 and transducer were used by the spark plug manufacturer. Early on, because of the relatively simple operation of the data collector, it was felt that the best way to implement an SQC program in the firing furnace area was to involve the operators themselves. They had the ability to measure and record the resistance data and they were in the best position to be able to take action on their results. To facilitate the collection of data, the furnace operators took production samples to a centrally located SQC area, where a resistance checking fixture was located. Each operator followed a simple procedure for recording the resistance measurements.

The fusing furnace's line number was used as the file number in the data collector. During production, the plugs were set in furnace racks of 60 plugs (122 rows by five deep). The file was therefore set up in a 12 item by five item sample file. The operators generally took a tray of 60 plugs once or twice each hour over to the SQC area, where they used the data collector and transducer to measure and record the resistance of the samples. Several times a day, the data collector was connected to a printer and the x̄ & R charts for each resistance station were plotted.

Within several weeks of the start of the program, all the operators were taking sample measurements and recording the resistance of their production. x̄ & R charts became routinely used.

an inconsistent manner. Further, it was found to have a great deal of variability in the volume of powder deposited over a relatively short interval of time. Once the problem had been isolated using the DataMyte, changes were implemented in the dropping mechanism to reduce the variability.

CASE 10-7   GLASS MANUFACTURER

Current Automobile Designs Demand Tighter Tolerances

A large manufacturer of glass used in automobiles needed to maintain closer tolerances. Tolerances became a concern when the automobile industry, using more sophisticated design methods, began to measure the drag coefficient of cars. A significant increase in scrap and rework had resulted in their attempts to produce higher quality windshields and by using conventional quality control equipment.

Problem

The problem had to do specifically with holding tolerance with off-form measurement of curved glass. A 0.200 tolerance was no longer sufficient, and the auto maker had eliminated the large windshield moldings to reduce drag. Their current procedures could not maintain the tolerance because of the variables involved in the glass bending process. The glass forming process required precise controls over humidity, line speed, and quench time (first stage of cooling). The bending bar can remain in contact with the glass for only a certain amount of time. Vacuum can only be applied for a certain time period also. If a piece of glass explodes during quenching, the next several pieces could become scrap because of broken glass on the conveyor belt system.

Lack of control was compounded by the fact that little statistical analysis was performed on raw data being collected. This was simply a result of the limitation of the inspection equipment being used and the fact that not enough personnel were available to manually collect and analyze data.

Solution

A group of management personnel was organized to evaluate the needs of the process control system. Several conclusions were reached about data collection needs:

- Little or no variability due to measurement devices, or to differences in methods from inspector to inspector, could be tolerated.
- The data collection system must be easy to use. Training must be kept to a minimum.
- Statistical information must be available immediately to the inspectors.

The solution seemed to lie in a combination of automatic data collection, precise gaging and fixturing. They chose a DataMyte 1556 handheld data collector because it could interface with a gap gage for data collection, and then be connected to their computer to produce capability graphs and control charts. The 516 gap gage was selected because it was precise enough for their measurement needs, and could be modified easily to work with a new fixture design.

See p. 20-12 for the 1556 data collector.

The modification of the gap gage was the first step. A local machine shop was contracted to modify the gage fingers with a positive locating pin that would fit in a glass checking fixture. Figure 10.7.1 shows the positive pin locaters and the SQC analysis system.

The checking fixtures were then designed with slots at the measurement points for the gage locating pin. Operator variability was virtually eliminated with this scheme because the gage mated positively with the fixture. As long as the glass was fixed in position in the fixture with no variation in positioning, the gage measurements would reflect the variability of the glass alone.

The data collector could also be programmed with prompt messages, indicating the check point location, and specification limits to instantly alert the operator to bad pieces. This further increased the validity of the data.

By using automatic data collection and well designed fixturing, statistically meaningful data could be obtained. The positive locating pins had reduced operator variability and increased the ease of measurement. This eliminated the need for an extensive operator variability study, and al-

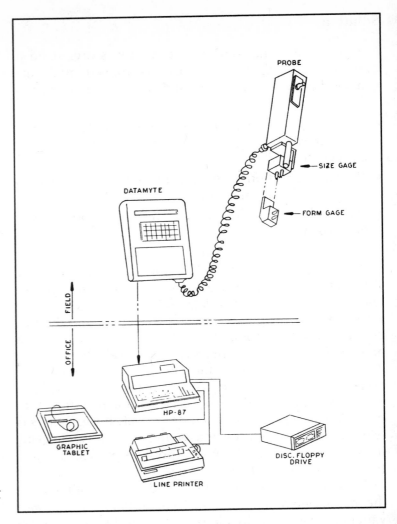

Fig. 10.7.1 Positive pin locaters and SQC system.

lowed the inspectors to concentrate fully on data gathering and process control.

CASE 10-8   ENGINE BEARINGS MANUFACTURER

Use of DataMyte Reduces Scrap From 5% to Less Than 0.5%

A large manufacturer of engine bearings needed to reduce its scrap costs and to produce parts that were more consistently acceptable to its customers. This company produces thousands of engine bearings a day and is a major supplier to the automotive industry.

The cost of quality in this company has been exceedingly large, and they faced stiff pricing competition in the marketplace. Maintaining precise, capable dimensions would allow them to satisfy their demanding customers and provide a significant cost savings due to reduced scrap.

Problem

One particular dimension in the production of these engine bearings that was a tremendous cause for concern was "split-line height." See Figure 10.8.1. This dimension had a critical tolerance of 15 tenths, but in production it would range as high as 30 tenths. Such variations caused this dimension alone to account for over five percent of all engine bearing defects. Auditors were attempting to control the variations in split-line height, but the time required to process the data manually was so great that many bad parts were being produced.

Solution

The solution to this problem was to have the auditors use a DataMyte 2003 to collect and analyze the data. Each machine was assigned a file in the data collector's memory.

See p. 20-8 for the 2003 data collector.

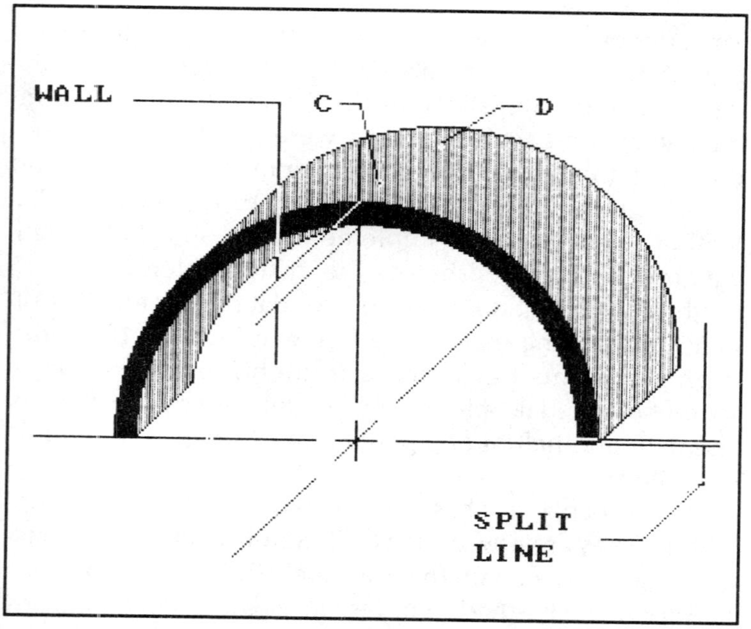

Fig. 10.8.1 Critical dimensions on the bearing cross section.

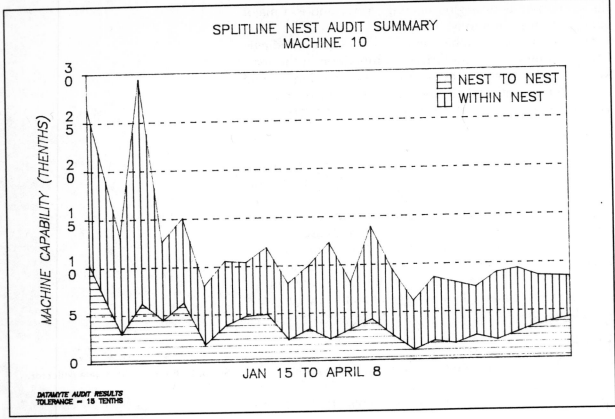

Fig. 10.8.2 Chart showing reduction in split-line height variability.

The auditors measured split-line height using their existing gage fixtures and manually entered the readings into the data collector. Statistical information was calculated immediately and capability charts were provided to the operators so that they could adjust their machines. The results of the DataMyte program were dramatic. See Figure 10.8.2. Within a few weeks of implementation, split-line height was reduced to well below the critical tolerance of 15 tenths. With continued use of the data collector, this dimension is being maintained below a level of 10 tenths. Along with this has come a tremendous reduction in scrapped material. Split-line height defects now account for less than one-half of one percent of all scrap — a ten-fold reduction.

The data collector has been readily accepted by the operators. They are now asking the auditors to check their machines to make certain that they are maintaining capability. The company has had to order more data collectors to satisfy the demands of the auditors and operators who are not willing to allow anyone to borrow the data collector for even a short period of time.

In addition to solving real-time factory data collection needs, the DataMyte 2003 has also proved to be a time-saver for management. Information in the data collector is transferred directly into the supervisor's IBM PC-AT. Using the utility software program supplied by DataMyte, the supervisor is able to take the data and put it into a data interchange format (DIF) on his hard disk. This DIF file is then read directly by his spreadsheet program for instant analysis.

CASE 10-9 TRUCK ENGINE MANUFACTURER

Torque Data Collection Time Decreased Three Fold

A large manufacturer of truck engines wished to increase the productivity of their torque auditors and obtain information faster for analyzing fastener clamp load. Auditors were using electronic torque wrenches, but were forced to write all information down, walk over to a personal computer, keypunch the information, and wait for analysis.

Problem

The company wanted a way to eliminate the double and triple handling of data and the associated delays. Company management felt that the time delay made problems hard to track down. However, they wanted to retain some of the features of the system, namely, immediate identification of a bad fastener, and archiving of the data onto a computer system. Also, a range of torque wrenches, from 50 to 350 lb. ft. had to be accommodated.

Solution

The company found that simply by upgrading some of the equipment they could retain all of the features of the present system and enhance performance as well. The new equipment consisted of a DataMyte 2003 data collector, printer and cables for their existing electronic torque wrenches.

The data collector provided these features:

See p. 20-8 for the 2003 data collector.

- Connection to a variety of wrenches, including in-line torque transducers.
- Direct recording of torque values. Auditors would no longer have to record them by hand.
- Immediate identification of bad fasteners. When an auditor took a reading the data collector would compare it to preset specifications and emit an audible signal if it was not within spec. The data collector also compared subgroups of readings to preset control limits and would alert the auditor if the last subgroup was out of control.
- Graphic analysis, in the form of capability studies and control charts, obtained by connecting the data collector to a printer.
- Full communications with their existing computer system, for the archiving of data.

The company now uses one data collector to monitor 140 different fasteners. Breakaway torque is monitored by checking fasteners with handheld wrenches. Peak torque is monitored by connecting the data collector to an in-line transducer mounted on a nut runner.

The company notes that the speed of data collection has improved threefold. Data analysis was reduced from four hours to thirty minutes. The performance in improvement allows rapid changes of tight or loose bolts, reducing the possibility of failures in the field.

Data is transmitted to the office personal computer and transmitted over a phone modem to the company headquarters 500 miles away, where it is stored by the mainframe computer.

CASE 10-10   BENCH SEAT MANUFACTURER

Company Monitors Torque Tools to Meet Federal Guidelines and JIT Schedule

A bench seat manufacturer that produces for a truck and bus final assembly plant needed to improve their fastener torque monitoring to meet federal guidelines. By using a DataMyte 2003 handheld data collector with in-line torque transducers they established an ongoing program for maintaining proper torque.

Problem

Federal guidelines required a certain torque on bolts used to fasten bench seats, because some of these seats would be on school buses and buses used for public transit. The company had recently failed to meet these guidelines.

It seems that the powered torque tools used to fasten the seat bolts either over or undertightened the bolts. Overtightening could stress the joints and cause fatigue, and undertightening did not achieve the required clamping force.

In addition, the truck and bus assembly plant was adopting a just-in-time (JIT) production system that required on-time delivery of defect free parts. The company was strategically located about thirty miles away from the assembly plant, and felt confident with a better SPC program for torque they could meet both delivery and reliability requirements.

Solution

The company was using Ingersol Rand torque drivers. The key to ensuring consistent torque on the fasteners was to be able to read the torque as the tool was running down the bolts on the seats. To read the torque they installed in-line torque transducers between the Ingersol Rand drivers and the sockets. The transducers provided a signal which could be interpreted by a DataMyte 2003 handheld data collector.

The DataMyte 2003 was then used to continuously audit the tools. The auditor would go from station to station, connect the 2003 to the torque tool, and do a capability study. The study would compare a sample of readings against the specifications.

See p. 20-8 for the 2003 data collector.

The DataMyte 2003 displayed the study results on its LCD (see Figure 10.10.1). Once capability was verified, the auditor could proceed to the next station. A tool could be adjusted immediately if the results indicated it was not meeting specifications.

They also did visual inspection of defects on finished bench seats with the DataMyte 2003. They would record a defect, its location, bench seat type and other notes. The data collector would display a Pareto chart showing the fre-

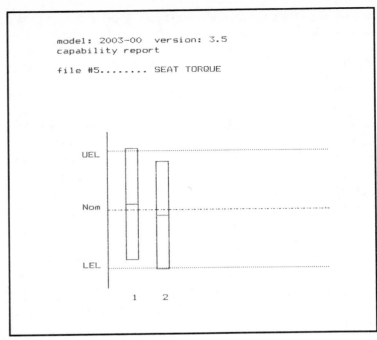

Fig. 10.10.1 Example of a capability graph used to check the torque tools.

Fig. 10.10.2 Example of a Pareto chart used to analyze the results of visual inspection.

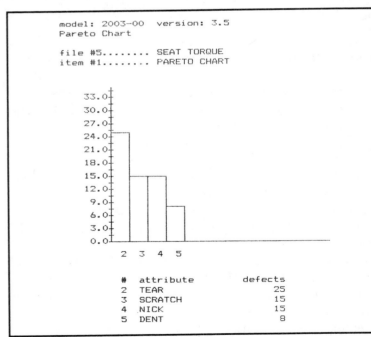

quency of defects in rank order by type. See Figure 10.10.2. They would then construct p-charts and continuously monitor the most frequent defects.

The establishment of this program led to improved torque capability, their meeting federal guidelines, and on-time delivery to their customer.

Case 10-11   AUTOMOTIVE BATTERY SUPPLIER

Real-Time Math Capability Helps Company Monitor Battery Performance

A leading manufacturer of automotive batteries not only needed to collect the data but apply some formulas to obtain the final analysis. Without the use of a DataMyte Model 762 the task could not be accomplished in a timely manner.

Problem Definition

To meet the specified power requirements for the batteries it is necessary to maintain control of how much electrolytic material is being applied to the individual plates which make up the cells. This task is complicated by the fact that the material is applied wet.

At the time this was strictly a laboratory procedure. The results were provided at the convenience of the lab not manufacturing.

Solution

They found that the mathematical capabilities of the DataMyte 762 data collector could help calculate and chart electrolyte. The data collector was preprogrammed with the necessary formulas and placed on the factory floor. The operators then periodically removed an individual plate already filled with wet paste. This plate was weighed and the results automatically transmitted to the data collector. The data collector stored the results in item 1. It then proceeded to subtract off from this reading a constant value equal to the weight of the plate. Now stored in item 2 was a net value equal to the paste only. The operator then took the plate with paste and placed it in a moisture balance. The amount of moisture was established and this value was automatically sent to the data collector and stored as item 3.

See p. 19-6 for the 762 data collector.

10-21

The data collector then subtracted this value from 100% and stored the value in item 4 identified as percent solids. Then without any direction the Model 762 multiplied item 2 by item 4 and arrived at the needed value of "dry electrolyte" applied to the individual plate.

Control charts were generated by the data collector and put up on the video display terminal for immediate review. They could now predict the expected performance that the battery would provide.

In addition, the raw data was available so problems could be traced back. When the lab provided results there was always confusion regarding whether the paste was being applied incorrectly or was of the wrong consistency.

As a result, production levels are better than ever since the rework has been reduced so dramatically. The operators also have benefited in that they now control their own destiny.

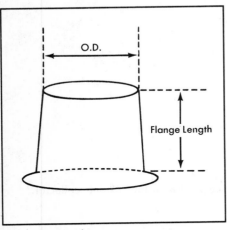

Fig. 10.12.1 Bushing measurements.

CASE 10-12   BUSHINGS MANUFACTURER

DataMyte Data Collectors Reduce Inspection Time

A manufacturer of metal and rubber bushings and sleeves for the automotive market needed to reduce the amount of time they spend doing inspection. They choose to automate their data collection by using a DataMyte 862.

Problem

The company needed to measure the O.D. and length under the flange to the end of the outer shell of a suspension bushing. See Figure 10.12.1. They needed a way to cut down inspection time without sacrificing quality or their contract with an automotive manufacturer.

Solution

The company found that by having a DataMyte 862 at each press, inspection time was cut drastically. The company will not need to add an additional inspector and they will be able to supply the necessary reports to keep their contract with the automotive manufacturer.

CASE 10-13   WINDSHIELD MANUFACTURER

Data Collection System Pays for Itself Within Nine Months

A manufacturer of windshields was collecting data manually on 30 lines. Data collection was too time consuming and costly.

Problem

The company needed a way to speed up data collection. With 30 lines to collect data on, much of their day was spent collecting and analyzing data.

Solution

The company found that a DataMyte 762 on each line dramatically reduced the time spent collecting data. They use their data collectors to measure the depth of bend for their windshields. See Figure 10.13.1. The company calculates that the data collection system will pay for itself within nine months.

Fig. 10.13.1 Quality characteristics on a windshield.

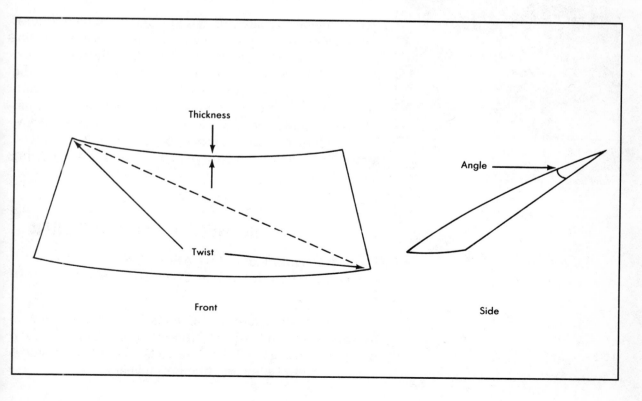

CASE 10-14   CAR ENGINE MANUFACTURER

Engine Supplier Gets Inspection Results More Quickly Using DataMyte 769 Data Collection Systems

A major manufacturer of car engines needed to get inspection results more quickly from their final inspection process. It took a salaried person a couple of hours each day to tally the reject numbers and compute the results.

Problem

The engine supplier was not able to get the inspection results right away. Therefore, they could not focus on their quality problems very quickly. They also needed to document where the engines with problems were coming from. They needed to sort the engines by assembly line or repair area. They look for defects such as data omitted, engine damaged, and build up wrong.

Solution

The engine supplier purchased DataMyte 769 attributes data collection systems for the final inspection area. The inspectors enter the defects into the data collector by serial number. Because the data collector automatically tallys the reject numbers and computes the results, they save a couple of hours each day. And the data collection system has allowed them to focus on quality problems more quickly, because they can look at a pareto chart from the data collector immediately and see which defects are occurring most frequently.

Fig. 10.14.1 An inspector uses the 769 data collector to display a pareto chart of final inspection.

CASE 10-15   BUMPER MANUFACTURER

DataMyte Data Collectors Allow Operators to Check Between 30 and 64 Characteristics Per Bumper

A large manufacturer of bumpers for Ford and General Motors is able to inspect more characteristics on each bumper by using the DataMyte 2003. This plant produces approximately 1.8 million bumpers a year.

With production of 1.8 million bumpers each year collecting data on all of those pieces was very time consuming and error prone. The company knew that they needed to automate their data collection method. The data collection system they chose must satisfy two main requirements:

- The data collector must interface to their J & S gap and flushness gages.
- And, the data collection system must be easy to use.

Solution

The company chose DataMyte data collection systems because they met their requirements. The DataMyte 2003 interfaced to their J & S gap and flushness gages and the inspectors found it easy to use. Two bumpers per hour are brought to an inspection room, where they are placed in a checking fixture. An inspector measures the deviation from nominal for many checkpoints on the bumper using a J & S gap and flushness gage. He records the measurement in a DataMyte 2003. See Figure 10.15.1.

The SPC coordinator uses DataMyte FAN software to print capability charts of the data collected by the inspectors. When starting up a run of a bumper type that has been produced before, the production line starts with five "tryout" parts. The SPC coordinator will not approve full-scale production unless measurements on four of the five bumpers are within ± 1 sigma of the last time the bumpers were run.

Fig. 10.15.1. A DataMyte 2003 and J&S gap and flushness gage.

See p. 20-8 for the 2003 data collector.

CASE 10-16   STEERING COMPONENTS MANUFACTURER

SPC Helps Manufacturer Maintain Standard of Quality Required to Pass Annual Inspection by Ford and GM

A large manufacturer of steering components for Ford, Chrysler, GM and John Deere needed to automate data collection. Their customers require proof of quality with the product.

Problem

The problem this manufacturer faced was how to put together the reports that their customers required without disrupting their production process. Their machine operators were busy machining parts. They did not have the time to produce control charts by hand.

Solution

The manufacturer looked at several data collection systems and chose DataMyte 862 data collection systems. See Figure 10.16.1. The 862 data collector interfaces directly to the gages that they are using such as Mitutoyo micrometers, and the data collector allows the machine operators to produce control charts in just minutes.

Fig. 10.16.1 A Mitutoyo micrometer and 862 data collector are used to measure and record the pitch diameter on threaded socket assemblies.

11. ELECTRONICS AND COMPUTER INDUSTRIES

Case 11-1 SEMI-CONDUCTOR MANUFACTURER

Better Control Leads to 15 Percent Scrap Reduction in Chip Cutting Operation

A consistently uniform, high quality silicon wafer is critical to a top quality finished product as well as keeping production costs to a minimum. By paying close attention to the early stages of production of a semiconductor this manufacturer has been able to remain a leader in this field.

Problem

The company was using SPC and was searching for a more efficient way of collecting data for controlling seven very similar processes.

The process involved the cutting of silicon wafers used in the production of semi-conductors. In order to avoid problems downline and reduce internal failures it was necessary to control wafer thickness to very tight specifications (± 7.5 mils). Additionally, silicon bar stock being very expensive, scrap reduction was also a goal.

The two operators working the seven cutting saws were having difficulty keeping accurate charts on all saws simultaneously.

Solution

The solution was to automate control charting with a system that allowed the two operators to quickly record data and review the results for all seven saws, randomly as the processes dictate.

See Chapter 19 for the 762 data collector.

The DataMyte 762 data collector with data grouping capability was the system chosen to solve this problem. Because thickness was the only dimension tracked at each saw and all data was collected at one inspection area the 762 provided the capacity and flexibility required.

The grouping feature was of particular importance to the working of this system. It permitted each operator to enter data for a given saw in any sequence. This way if one saw was shut down for a blade change, or any reason, data collection could continue without disrupting the collection routine. Likewise, it was not necessary to have one operator wait for the other to complete a subgroup before moving to the next saw.

By giving the operators the flexibility of random data entry and immediate feedback of variability of the cutting process significant efficiencies were gained. Freed from the task of manual recording and charting of data operators could concentrate on solving problems, adjusting the process and predicting blade change intervals. This translated into a 15% decrease in failure and scrap.

The DataMyte 762 was set up for seven items each corresponding to a saw. Specification and control limits were set up for each item and edited as needed for product changes. Each item was assigned as a separate group so operators could easily move from item to item as the collection tasks required. After collecting the thickness data with a hand micrometer operators review $\bar{x}$ and R charts and make necessary adjustments. Capability studies are also run for each saw.

Case 11-2   METAL CIRCUIT PACKAGE MANUFACTURER

Variable and Attribute Data Collection System Tied Together by a LAN 1 Network Used to Increase Productivity and Yield

A manufacturer who assembles metal circuit packages used in microwave, fiber optics, and power hybrid applications needed to increase the quality of the finished product yet reduce inspection time. With a manual SPC program already in place it was apparent that collecting data itself would not solve their quality problems. This data had to be put in a form the inspectors could analyze. The main issue is to find and control specific problem areas related to the assembly procedure in a timely fashion.

The base of the metal circuit package consists of the body or can. Wire pins, heat sinks, mounting frames and/or connection tubes are then fused to the can using either glass or copper depending on the function of that pin in the electronic application. The final product must be hermetic. Infrequent use of data collected manually and time consumed in collecting and analyzing meant inefficient process control.

Problem

This company had to get a handle on the inconsistent quality of the final products and also reduce the inspection time. Implementing an automated data collection system would give the inspectors real-time feedback of their process so they could effectively analyze the process.

The critical characteristics on a part are both variable and attribute in nature. The pins are inserted through the holes of the can. Each hole has an eyelet of either copper or glass between the can and pin. Next the can is placed in a fixture which holds the pins in the correct position. The fixture is then placed in a furnace to hermetically fuse the pins to the can. Due to fixture wear, pins tend to fall from their correct position and the process deteriorates. Positioning of heat sinks, mounting frames, and connection tubes are also critical to the final product. The availability of real-time process control and process capability charts are essential to determine when the fixture should be replaced.

The actual condition of the fuse is also very important in tracking down and pinpointing problems. There are two major attribute categories: refire or rework. A defect which can be corrected by putting the package through the furnace again is a refire condition. A package that must be chemically stripped and reassembled is a rework condition. Again availability of real-time pareto and control charts are essential in pinpointing why these problems are occurring.

There are approximately 50 major package types with each type having many different variations. Because of this product mix, process changeover is frequent and quick. In order to maintain effective and efficient control on each process much time was spent collecting and sorting data manually.

Solution

An automated data collection system complete with 862 variable and 769 attribute workstations tied together with an Allen-Bradley LAN 1 Network gave this manufacturer a total system solution. Because the inspection process was being done at several different locations, stand alone

See p. 20-4 for the 769 data collector.

DataMyte workstations that provided online, real-time feed back was a key feature.

The DataMyte 862 provides an electronic interface to their gages and has multiple file capability to address the frequent process change over. The inspectors can use the x̄ & R charts to determine a trend developing thus monitor their fixture wear.

See p. 19-4 for the 862 data collector.

With the DataMyte 769 inspectors are able to sort and filter the data right at their work cell. This provides them with a tool to target frequently occurring defects immediately. A LAN 1 Network was installed to automate the file transfer process to and from the computer and data collectors. On line monitoring from a remote sight was also accomplished via the network.

Inspectors now spend more time analyzing the control charts provided by the DataMyte data collectors and spend less time collecting/manipulating the raw data. Since this market is driven by delivery, quality, and price, of which everything hinges on quality, it was easy for this manufacturer to justify an automated data collection system.

The benefits already realized by this manufacturer include the following:

- They have increased awareness of quality issues.
- They have reduced man hours for data collection.
- They have online, real-time feedback.
- They have the ability to keep up with the change over process.

Operation

The information that follows provides a description of how the DataMyte 862 was used by this manufacturer to monitor their assembly process. The critical dimensions are pin height and pin length. Using electronic gages interfaced to the 862, the inspectors would take 3 finished cans and take the pin height on all 3 cans. Then the operator would measure the pin length on all 3 cans. At the end of each subgroup the 862 displays a x̄ & R chart so the inspectors can analyze the fixture wear. Histograms are used to show the capability of each process. Separate files are used to monitor different furnaces, can models, and fixtures.

The 769 is used to detect the most commonly occurring defects related to their fusing operation. Attributes such as bubbles, hyperbolic, voids, piling, off center pin, pin height, partial melt, and misloads are recorded.

The ultimate goal is to reduce man hours associated with data collection by 10 -15 %, eliminate intermediate inspection routines, and increase productivity.

CASE 11-3 ELECTRONIC CONNECTOR MANUFACTURER

Single Format System Improves Pin Dimension Checking

A large manufacturer of electronic connectors, including gold-plated pins, recognized the need for initiating a statistical program for monitoring their manufacturing processes. Areas critical to this manufacturer were the size of the pins themselves and the plating thickness on the pins. To remain competitive in their particular market, this manufacturer needed to implement an SQC program as soon as possible.

Problem Definition

Current measurement techniques used by this manufacture involved "eyeballing" measurements with mechanical gages; this included the length, diameter, and head diameter of the pins and measurement of the plating thickness by using an x-ray fluorescent device that transmitted data directly to an Apple II computer. Because this manufacturer was using traditional methods for measurements, no statistical analysis was being performed on the data. This was also due in part to the fact that the Apple II computer was only dedicated to retrieving and storing data generated by the x-ray device.

The manufacturer had several plants, and each was experiencing failure problems with their connectors. While they knew that a serious problem existed, they were not able to pinpoint the causes for the problem. This was due in large part to the vast amount of data that required manual processing.

Solution

The manufacturer determined that prerequisites for the overall program must include:

- The ability to perform statistical calculations on the data coming from the x-ray fluorescent machine.
- Eliminating the need for 100% inspections of the sizing of the pins, yet obtaining a significant increase in the quality of their products.
- Eliminating manual charting of the statistical data.

- Monitoring several phases of the production process, not just the "finished product."

Based on their analysis, they chose a DataMyte 862 hand-held data collector along with an electronic caliper. The DataMyte would provide the data collection and analysis capability and could also be used to generate the necessary charts and capability reports. Further, the DataMyte would be able to retrieve the x-ray data from the Apple II computer and then perform statistical calculations on that data.

By using a DataMyte as part of the overall SPC program, the manufacturer was able to eliminate manual charting and reduce data turnaround time. Areas of the process that were to be monitored included pin length, pin diameter, and head diameter.

See **Chapters 19 and 20** for data collectors and **Chapter 21** for calipers.

Operation

The information that follows provides an overall description of how the DataMyte 862 was used by the manufacturer. The critical area of concern with respect to plating thickness was on shaft "B", as this is where the connection to other devices takes place once the pins have been imbedded in the connector. It was determined that two files would be set up to monitor the measurements statistically.

The first file would be dedicated to the pin measurements. This included four items: A = length; B = head diameter; C = "A" shaft diameter; and D = "B" shaft diameter. (See Figure 11.3.1.) The data would be used to keep track of x̄ & R chart values for each item. The first file was set up as follows:

ITEMS = 4 (one for each area to be monitored)
SAMPLES = 10

The associated prompts were set up as follows:

ITEM 1 = LENGTH
ITEM 2 = HD DIAM (head diameter)
ITEM 3 = A DIAM (A shaft diameter)
ITEM 4 = B DIAM (B shaft diameter)

Limits were established as follows:

ITEM 1 (Lo = 20.0 mm, Hi = 20.6 mm)
ITEM 2 (Lo = 3.25 mm, Hi = 3.75 mm)
ITEM 3 (Lo = 2.25 mm, Hi = 2.75 mm)
ITEM 4 (Lo = 1.25 mm, Hi = 1.75 mm)

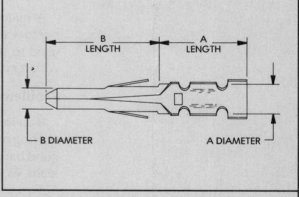

Fig.11.3.1 Pin dimensions

CTRL LIMITS = NO (No control limits were entered at first. When sufficient data was collected, the DataMyte would calculate the limits itself.)

Once the DataMyte had accumulated sufficient x̄ & R history, charts could be generated daily and process control could be maintained accordingly.

The second file was set up to handle the data received from the Apple II computer, which received its data from the x-ray fluorescent device. The Apple II was configured to communicate with the DataMyte via an RS232-C serial interface card. The DataMyte, was connected to the Apple II computer with a cable.

The second file consisted of one item and 10

samples. Once the data collection process was completed, data resident in the Apple II computer was downloaded to the DataMyte. Upon receiving the specified amount of data, the DataMyte exited Remote Control mode. After sufficient history had been created, the DataMyte was connected to an electronic printer where the $\bar{x}$ & R charts of the x-ray data were dumped.

With the SPC program in place as described in the previous paragraphs, the connector pin manufacturing process could be statistically monitored, with all of the critical measurement areas covered. All that was required at this time was to periodically clear the data in the files and change the date in the Data Collect mode.

As can be seen from this operational process, the DataMyte was able to accommodate automatic data capture while eliminating mechanical, "eye-balling" gages. It was also able to statistically summarize and analyze data from extremely sophisticated electronic testing equipment.

CASE 11-4   DISK DRIVE MANUFACTURER

Control Charts Help Predict Tool Maintenance

A major disk drive manufacturer attaches terminal connectors to wire leads in motors and brakes of machinery used to produce disk drives. The connectors are attached with a crimping tool after the wire insulation is stripped.

Problem

If an inspector could determine when a crimping tool was starting to go bad, instead of finding out after it had already gone bad, he could have the tool rebuilt before useless wire leads were produced. The QC engineer was using SPC to determine when the crimping tool needed to be rebuilt (see Figure 11.4.1). He would take crimp height measurements with a micrometer and plot them on an $\bar{x}$ & R chart.

After the tool was rebuilt, the engineer noticed that the crimp heights were still out of control, so he constructed a histogram (Figure 11.4.2). The histogram showed that the distribution was skewed to the high side. This was already suspected because the points on the $\bar{x}$ & R charts were moving up and down together. He concluded that the tool only allowed the variation in crimp height to be biased one way and the natural distribution would be skewed.

Closer examination of the crimp height measurements (Figure 11.4.1) showed that the measurements did not have enough precision. The engineer was trying to make decisions on data calculated to 0.0001 inch and the readings were only accurate to 0.001 inch.

```
*********************                          FORM NO. 2462D
*  CAPABILITY STUDY  *
*********************
```

PROCESS CONTROL CHART (X & R)

ASSEMBLY _CRIMP CONTACT PIN_ PART NO. _51905809_ REV _1_

MFG. MGR. _DICK WELLER_ MFG. ENG. _LARRY HANKE_ Q.C. ENG. _KEN FASTNER_

CHARACTERISTIC _CRIMP HEIGHT_   SPECIFICATION _.052±.002_

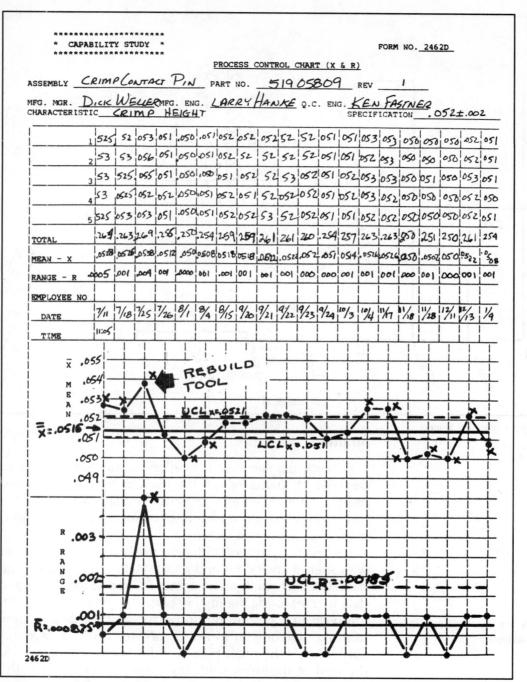

Fig. 11.4.1 Crimp height control chart

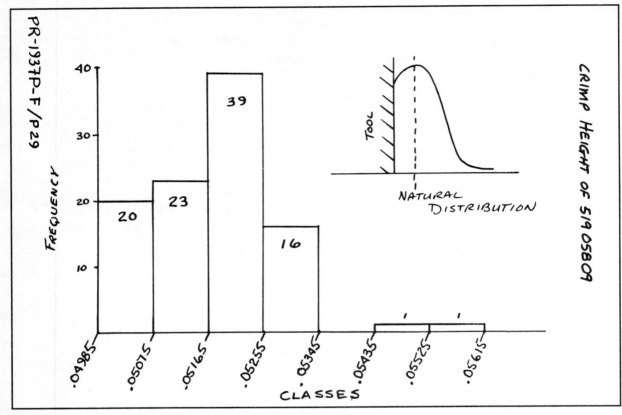

Fig. 11.4.2 Histogram of crimp heights.

Solution

Because the specification called for the crimp height to be 0.052 ± 0.002 inches, the micrometer had to be read more carefully. The engineer could either train his people how to take more accurate measurements or use an automatic data collection system.

The manufacturer decided to use a digital micrometer with special anvil ends and a DataMyte 762 to collect the data. On each shift, an inspector trained in SPC collected data by measuring the first 50 parts for crimp height. With the push of a button, the inspector sent a digital reading from the micrometer to the DataMyte. The DataMyte recorded all of the digits necessary for the control charts. At the end of the third shift, the inspector connected the DataMyte to an IBM PC to transfer and store the data on a floppy disk. After the data was transferred, the inspector cleared the data from the DataMyte so it was ready for the next day's data collection.

See Chapter 19 for data collectors and Chapter 21 for calipers.

The third-shift inspector generated the x̄ & R control charts for analysis. With the charts, the inspectors discovered that a downward trend in the x̄ chart indicates possible tool wear. Even though the readings are within the control limits, the inspectors had the crimping tool rebuilt, to prevent bad parts from being produced.

The cost savings associated with the DataMyte system justified the purchase in only 2-1/2 months.

CASE 11-5 DISK DRIVE MANUFACTURER

Both Variables and Attributes Data Collection Used to Verify Disk Drives.

A world leader in the manufacturing, design, and implementation of large computer disk drives needed to reduce scrap and rework. ANSI standards had to be met. However, the cost of 100 percent inspection was just too expensive.

The base material for the magnetic storage media was developed into round disks, with very critical inside and outside diameter dimensions. Any discrepancy outside of engineering or ANSI specs caused the disks to fail and damage the drive system in which they were placed.

Problem

Implementing statistical quality control on the critical dimensions would help minimize the scrap and rework. There still had to be a method, however, when performing first article inspection, to screen rejects until the process is "in control." During the initial inspection, all disks would be monitored so that no rejects (those that exceeded limits) could pass through the system. Once the process was in control, then sampling could be used. The parts to be inspected are shown in Figure 11.5.1.

Three outside diameter measurements were taken to identify any out-of-roundness conditions. The ID was also measured at three separate locations for the same reason. The data would then be fed to the mainframe computer system for processing and report generation. The inspection procedure shown in Figure 11.5.2 was developed for data collection. All measurements were taken with a 6 inch caliper that was mounted on a special extension arm fixture. See Figure 11.5.3.

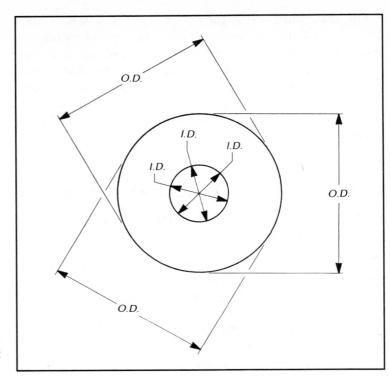

Fig. 11.5.1 Three sets of inner and outer diameters were measured on a disk to check out of roundness.

Fig. 11.5.2 Sample inspection form used.

Batch # _____   Quantity _____

Outside Diameter Measurement 1: _____

Outside Diameter Measurement 2: _____

Outside Diameter Measurement 3: _____

$\overline{X}$ = _____   R = _____

Inside Diameter Measurement 1: _____

Inside Diameter Measurement 2: _____

Inside Diameter Measurement 3: _____

$\overline{X}$ = _____   R = _____

Pass ☐   Fail ☐

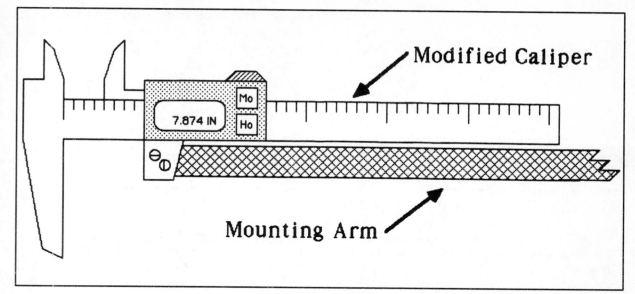

Fig. 11.5.3 Caliper modified with extension arm to measure disk diameters.

Along with variables data on the ID and OD measurements, attribute data collection was also important. Any excessive scratches, dirt or nicks in the raw material could cause failure of the disk when the magnetic storage medium was placed upon it. At that point in the process, visual inspection of the disks was needed. Without some way to collect and analyze the data the inspection process would provide no basis for eliminating common problems.

Solution

Two data collectors were adopted for use, the DataMyte 862 for variables data collection and the DataMyte 769 for attributes data collection. The 862 was used with an electronic caliper to take diameter measurements. The operator was warned instantly of any out-of-limits condition.

By using the DataMyte 862, manual data collection errors were eliminated. What previously took an inspector an average of five to six hours of inspection per day was now condensed to one hour. This time included inspection of the disks and manual calculation of the $\bar{x}$ & R values, which the data collector performed "on the fly." This also allowed them to get the process in control much more quickly.

Using the 769 for attribute data collection reduced the turnaround of inspection reports from three days to im-

See Chapter 19 and 20 for a description of data collectors.

mediate real-time reports. This not only led to faster inspection of disks, but allowed criteria for visual examinations, such as what is a scratch or what is a nick, to be more easily defined and maintained.

Operation

What follows is a brief explanation of how a DataMyte 862 and a caliper were used to implement variable data collection, and how the 769 was used for attributes.

The 862 file was set up as follows:

Items = 2 (ID measurement, OD measurement)

Samples = 3 per Subgroup (each subgroup would represent the average and range for each dimension on a disk)

Advance by sample = One disk inspected at a time (3 ID measurements, 3 OD measurements)

Eng limits = Yes (they were to be used)

The nice feature about the data collector in this application was its ability to warn the operator for readings outside the engineering limits as well as an out-of-control condition, thereby making it very easy to screen rejects. The information for engineering limits was set as previously indicated. Since multiple files can be stored by the data collector, separate files were used for the different models they were inspecting. Because the special fixture was used in conjunction with the caliper, a math formula was used to set an offset value to the specified amount, depending on the model of the disk being inspected. One value was set for the larger models and a different value was set for the smaller models.

As for attribute data collection, the DataMyte 769 was set up to track the following identifiers and attributes:

Single Area Data Entry Mode

Date (Each record is automatically time and date stamped by the 769)   Operator - Choice Field - Ditto On   Batch No. - Text Field - Ditto Off   Qty Inspected - Quantity Field - Ditto Off   Scratches - Attribute Field - Ditto Off   Dents - Attribute Field - Ditto Off   Grooves - Attribute Field - Ditto Off   Dirt - Attribute Field - Ditto Off

Note that the Data Entry Mode was set to Single Area, allowing the operator to enter multiple defects per disk. Also note that the operator needed to be entered only once for a data collection cycle because the ditto feature was on. The prompts for this routine show up as listed above (excluding Date).

After the last defect has been entered, the next prompt shown would be "batch no."

The DataMyte 769 generates bar codes (for use with the wand) and/or keycodes (for use with the keypad) for the data collection process. An inspector would just enter them accordingly. The 769 then sorted the information in the form of a pareto or control chart and indicated problem areas to the engineers and inspection team. Management reports were also generated to indicate defects per pre-determined quantity of disks inspected (usually defects per 100). It used to take three days for manual inspection of the disks, the accumulation of all the data, inputting into the computer, and waiting for the reports to be generated. Now it takes six hours. Attribute data collection is now efficient and practical.

CASE 11-6   COMPUTER MANUFACTURER

Torque Auditing Helps Monitor Incoming Inspection

A large manufacturer of desk-top computers was experiencing problems related to the quality of products being supplied by some of their subcontractors. It seemed that certain subcontractors were failing to adequately test their parts and assemblies prior to shipment to the manufacturer. Costs associated with incoming inspections, rework, and customer relations were soaring.

As a leader in the area of desk-top computers, this manufacturer knew that problems related to untested parts and assemblies were hurting their reputation. To regain control of the situation, the manufacturer decided to implement a program that included the use of DataMyte 2003 data collectors to handle the inspection of various incoming parts and assemblies.

See p. 20-8 for the 2003 data collector.

Problem Definition

Many of the major assemblies that make up a desktop computer consist of printed circuit boards that have critical fasteners attached to them. It was found that if these fasteners were not properly torqued prior to shipment, the printed circuit boards could easily crack.

Furthermore, many of the subassemblies had dimensions that were considered critical to the overall computer design. If the dimensions for these subassemblies were allowed to vary significantly, supply voltages could be shorted. This in turn could lead to a complete power failure in the computer.

Solution

A method was needed to monitor the torque on fasteners as they arrived from suppliers and monitor the spacing between these assemblies as they are assembled into the computer.

Hard copy results were needed immediately after data collection. They considered manually taking the measurements, charting the readings, and submitting data to their data processing department for reports and copies. They also considered automating the process with an intelligent data collector. It was determined that the best solution would be to connect DataMyte 2003 data collectors to either a torque wrench or a gap measurement device. Using these arrangements, the manufacturer was able to verify fastener torque and critical dimensions at the incoming inspection level.

In addition to being able to measure torque and check critical dimensions, other benefits realized by using the DataMyte 2003 data collector included the following:

- The ability to have the data collector generate capability reports and charts on demand.
- A significant cost reduction due to a decrease in rework.
- A significant reduction in data acquisition time.
- The manufacturer did not have to purchase any additional computing or test equipment since the data collector was capable of being interfaced to the existing hardware.

Operation

The information that follows provides a description of how the DataMyte 2003 was used by the manufacturer to meet their requirement for being able to perform torque and critical dimension measurements.

One file in the data collector was established to deal with the problem of fastener torque. This file was set up to measure six of the fasteners. See Figure 11.6.1. Since suppliers usually submitted these assemblies in lots of 50 the sample size was set to 5. Overall, this file was set up as follows:

ITEMS = 6 (one for each of the six fasteners)
SAMPLE SIZE = 5
SUBGROUPS = 50

Item IDs were established as follows:
ITEM 1 = LTF (left top fastener)
ITEM 2 = LBF (left bottom fastener)
ITEM 3 = LCF (left center fastener)
ITEM 4 = TCF (top center fastener)
ITEM 5 = RCF (right center fastener)
ITEM 6 = RSF (right side fastener)

Engineering limits were established as follows for each of the items:
LO = 30.0 in/lbs
HI = 38.5 in/lbs

For this file, the same transducer was used for each item. To keep it simple, gage file 16 for the torque wrench was used. After the setup information was entered, the DataMyte's data collect mode was entered. Footnotes were set up with the supplier identification code. In this case, three digit codes were assigned as follows:
Supplier A = 123
Supplier B = 456
Supplier C = 789

Once data was collected on all the suppliers, the 2003 was taken to the computer and down loaded. In FAN II software, reports were processed by the supplier code.

To accommodate testing of the critical gap measurements, a second file was set up as follows:
ITEMS = 3 (one for each area of the gap to be tested)
SAMPLE SIZE = 1
SUBGROUPS = 10

Item IDs were established as follows:
ITEM 1 = TOP
ITEM 2 = MIDDLE
ITEM 3 = BOTTOM

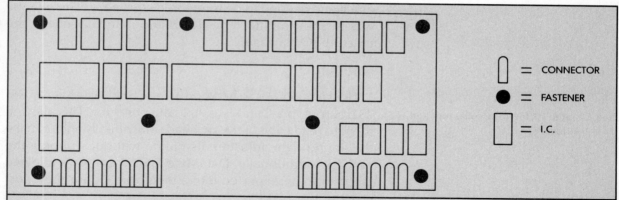

Fig. 11.6.1 PC board torque checks.

Limits were established as follows for each of the items:

ENG LO = 7.0
ENG HI = 9.5

For this file, the same transducer was used for each measurement. The gage file used was file 30 for gap gage.

When entering the Data Collect mode, the file name was set up to reflect the area of measurement on the final product. Footnotes were entered to identify the particular shift when improper installation took place.

CASE 11-7 PROCESS CONTROLLER MANUFACTURER

Company Reduces Variability on Circuit Board Pick and Place Machine

A manufacturer of process controllers and valves for the petrochemical industry needed to reduce PC board failures. By automating data collection at their circuit pick and place operation, they found they could control variability and reduce defects.

Problem

The company's concern was with their automated pick-and-place circuit board assembler and solderer. It has six control heads loaded with belts of circuits and resistors to be automatically inserted in the appropriate spot on a circuit board. The problem they had was with critical pressure points when the head, loaded with the resistor or circuit, comes down and inserts the appropriate component into the circuit board. The critical variable was the pressure from the head on the circuit board. The pressure fluctuated too

11-17

much and needed to be monitored constantly. The old way was to shut the machine down and manually check the pressure with a gage.

Solution

The solution was to hook up an electronic gage via an RS-232 output to a DataMyte 762 data collector. This allowed them to read each of the six heads individually and to make the appropriate adjustments online without stopping the process. Additionally, DataMyte FAN II software allowed the data to be exported into a data base file for their post process analysis.

See Chapter 19 for data collectors and for FAN II software.

12. FOOD, COSMETICS, AND HEALTH CARE INDUSTRIES

CASE 12-1 PROCESSED MEAT COMPANY

Packaged Meats Company Improves Margins With Better SPC Program

A major food processing company is the largest in the United States specializing in sliced meats and processed foods, turkey products and seasonings. The company has its own seasoning plant and it also has its own plastics facilities for making the wrapping and packages. By automating its data collection process and by using SPC they reduced scrap by 5 to 12 percent and reduced labor by even more.

Problem

The company had been using manual SPC methods to keep track of the weights, thicknesses and temperatures of some of their processed meats and bologna. The food processor needed real-time, or near real-time data analysis in order to reduce scrap and waste. Charting by hand took too much time and the charts ended up almost useless because they were too messy and dirty.

The data also had to be keypunched into a computer so the company could archive the data. The archived data was for their own use as well as for meeting government regulations and presenting to inspectors.

Solution

The employees collected data from an electronic weigh scale, which had RS-232C output, a caliper and a temperature probe, which also had RS-232C output. Using a DataMyte 762 data collector they could record the readings automatically. After collecting the data, the employees transferred the data to an IBM PC with the FAN II software program. Current data could readily be presented to government inspectors. The FAN II software program also allowed the employees to bring data back from the computer and load it into the DataMyte 762 the next time the product is run. This gives an instant comparison to previous runs.

The company's time lag between collecting data and

See p. 19-6 for the 762 data collector.

seeing control charts was practically eliminated. This helped reduce their scrap and waste by about five to ten percent—helpful in a low margin industry, such as the food industry. Manpower requirements were reduced by about 20% because one day per five day work week had been devoted to recording, keypunching and calculating the charts.

CASE 12-2   FROZEN DINNER COMPANY

Statistical Sampling Reduces Variability in Frozen Breakfast Portions

A well-known company in the food industry was introducing a new line of frozen breakfast entrees. The automated cooking and packaging line could produce several varieties of meals, including eggs, sausages and hash browns, or pancakes and sausages, at the rate of two meals per second.

Problem

An automatic sensor on the packaging line kicked out any meals that were underweight before they were boxed and frozen. The trouble was, there was no good way to automatically check for overweight meals. The company was concerned about not putting too much of any one item in each meal, but checking the gross weight of the meal did not supply criteria for adjusting the egg, sausage and hash brown portions.

Company managers decided that statistical sampling of the weights of portions going into the meal would provide data to control the process. The speed of the packaging line required an efficient method of gathering and analyzing data, however. Since the line was automated to a fairly high degree, the company felt that a labor-intensive SPC activity would not be suitable.

Solution

The SPC system that would be most effective needed to record and analyze weights automatically. The results of statistical sampling must be made visible to workers, who could then respond with solutions such as adjusting the process. The company selected a digital weigh scale with

See p. 19-6 for the 762 data collector.

RS-232 output, a DataMyte 762 and a CRT monitor. The DataMyte would record weights and plot $\bar{x}$ & R charts on the CRT. The chart could be updated for each subgroup, allowing instantaneous verification of the process.

It was soon discovered that of the three items being plotted, the egg making process had a lot of variability and contributed most to the problem of overweight meals. The eggs were supplied already cracked and in barrels. They were fed into a machine that scrambled them and then dumped them into a bin. The bin metered the eggs into the dinner trays. A control chart of the egg weights found that the amount metered into each meal varied according to how many eggs were in the bin. The bin metering mechanism was rebuilt to reduce the influence.

SPC resulted in a savings on the usage of eggs in this line. As the variability in the process was reduced, the frequency of sampling could be reduced also. This further justified their assumption that dedicating extra labor to the SPC task was unwise.

The company has now purchased additional DataMyte 762s and monitors for each shift. Each shift has a separate line setup, and a different type of frozen meal is produced. Having an easy to use, standardized system for SPC allows workers from each shift and managers to talk about the breakfast making process in the same terms.

CASE 12-3 COSMETICS COMPANY

Company Automates SPC in Plastic Molding Operation

A manufacturer of baby care products trained their employees on the use of SPC, but they still needed the tools to do the job. The quality assurance manager became convinced that SPC does not work without automated data collection.

Problem

The company was already practicing SPC manually, and they needed a way to automate the process. They found that manual SPC was time consuming, error prone, and after the fact.

Fig. 12.3.1 An operator displays a pareto chart using the DataMyte 769 data collector and monitor.

Solution

The company is using several DataMyte 762s for variables data collection and a couple of DataMyte 769s for attribute data collection. They are monitoring the quality of their machines with $\bar{x}$, R, sigma charts and histograms. The 769 attribute data collector is used to inspect the finished bottles before they are boxed for the powder filling area. See Figure 12.3.1. The company experienced an 80% reduction in the overall in-process scrap rate for out of spec variable dimensions on powder bottles in the plastic molding area.

See p. 20-4 for the 769 data collector.

CASE 12-4   PERSONAL CARE PRODUCTS COMPANY

Company Uses Digitizing Tablet and Data Collector to Audit Diaper Product

A large consumer products company, manufacturing many types of foods and toiletry articles, began a corporate wide SPC program to help cut costs and improve quality. The company retained Dr. W. Edwards Deming as a consultant to develop a top-down management commitment.

Problem

The company was looking for better ways to automate data collection, and also looking for specialty gages to fit their needs.

In one division, the company was interested in measuring the critical dimensions on a new adult diaper. Because of the nature of the product, they needed to hold close tolerances to ensure the comfort and wear of the product.

A major problem with the diapers is that they are not designed to lay flat on a surface and be measured. In the past all measurements were done with a ruler and written down. Control charts were done by hand. This proved to be extremely tedious, time consuming and inaccurate.

Solution

Company engineers thought that a digitizing tablet could simplify the measurement process. A digitizing tablet is a flat table on which the process engineer could tape the diaper down onto the surface. By using a light pen he could mark an xy coordinate reference point. Each time the light pen is moved to a different portion of the diaper the tablet automatically references back to the original xy position, and takes its measurements in relation to the original coordinate.

See Chapter 20 for a description of the 2003 data collector.

The digital signals from the tablet were fed directly into a DataMyte 2003 data collector. The engineer set up a file in the DataMyte for each of the dimensional characteristics. The DataMyte made the subtraction and calculation steps for each coordinate in relation to the reference point, and calculated the between coordinate dimensions.

The company found a 75 percent labor savings by using the digitizing tablet and DataMyte 2003 compared to the previous data collection method.

CASE 12-5   PERSONAL CARE PRODUCTS MANUFACTURER

Manufacturer Streamlines Incoming Inspection, Fill Weight and Torque Monitoring

The quality department of a major manufacturer of cosmetics and other personal care products faced the task of efficiently monitoring data being collected in different areas of the plant. Monitoring was needed to insure that increased production schedules and product quality could be maintained and improved.

Problem

In incoming inspection the company faced the need to monitor numerous vendor product containers. The characteristics to be monitored were:

- Cap and bottle neck diameters
- Container (internal and external) dimensions
- Part thicknesses

The containers are made from various materials including glass, plastic and metal. The monitoring of these and other dimensions were critical to product quality. Lack of adherence to proper dimensional requirements caused such problems as:

- Improper package sealing
- Package unattractiveness
- Proper application of the product became difficult or impossible.

The company previously collected data using various dimensional gages. Samples were collected from the incoming lots and data recorded by a staff of quality technicians. Later the data would be manually calculated, and the results would determine if the lot was to be accepted or rejected. The technician would complete a chart recording the past history of the particular part. What was needed was a means of streamlining this data collection process while eliminating the volume of paper which was being produced by the manual system.

Streamlining the collection processing of data would result in the following benefits:

- Shorten the time between receiving the product at the shipping dock and releasing it to the plant floor.
- Reduce the time and expense of incoming inspection.

- Provide a comprehensive source of archived data to aid in analyzing problems in vendor parts.

Solution

The company streamlined the incoming inspection area, and achieved the above mentioned goals by implementing a DataMyte data collection system. The system was composed of DataMyte 862s, several different types of electronic gaging equipment (including electronic calipers, digital indicators and a horizontal axis measuring device), DataMyte FAN™ software and Lotus 1-2-3 software.

Once a particular lot was received in the shipping area, samples were gathered and taken to the measurement lab for evaluation. At the same time, a DataMyte 862 would be uploaded with the various part numbers and part specifications required for the technician to gather the data. The technician would then take the data collector and measure the parts with electronic gages.

During the measuring process, the data collector would notify the technician of out of spec conditions, at which time the technician could further review the situation using the data collector screen. Once the data had been collected, he would then return the DataMyte 862 to the PC, where the data would be processed using the FAN software and Lotus software. Lotus software allowed the Quality Department to customize its reports and charts to meet its particular incoming inspection needs. The application software streamlined the data collection process, but also enabled them to produce customized reports that not only aided in the accept/reject decisions for incoming parts, but also allowed a complete data base history to be maintained in an efficient manner.

See p. 19-4 for the 862 data collector.

Fill Weight and Removal Torque Monitoring

Once the vendor's containers were released to the plant floor, a new challenge of monitoring a dozen different fill lines was faced.

The product containers being filled ranged from 1 oz. to 12 oz. and required removal torque between 12 to 25 in.lb., depending upon the particular package being monitored. Removal torque is the force required to remove the lid from a jar. The task of monitoring these two characteristics required the operator to sample each product line every half-hour. Manual sampling methods included:

- Establishment of a standard tare value for the amount of containers being used.
- Setting of the scale for the given tare value
- The weighing and manually recording of the weight of each sample, and
- Calculating the $\bar{x}$ & R values and manually plotting the charts

A similar process was then repeated for the monitoring of removal torque on each production line.

Although the manufacturer felt the implementation of SPC had been quite successful in monitoring both the fill weight and removal torque, it became quite apparent that the time the operator needed to produce the charts was overly burdensome. There was also no easy way for management to archive and review the past data.

Solution

The adoption of a full scale DataMyte system provided solutions for these applications also. Each product line was equipped with a 762 and monitor interfaced to an electronic weigh scale and a removal torque test unit with RS-232 output. The DataTruck was then used to set up part files on each of the product lines and to harvest data on a daily basis. The data was sent to an IBM PC equipped with FAN support software. FAN software allowed the quality department to establish computerized files on each of the various parts being manufactured. When a part is scheduled, the file is loaded to the DataTruck and transported to the 762 located on the line. The operator then monitors the production run with half-hour samplings. If adjustments or changes are needed, the operator indicates the nature of the adjustment or change by coding the $\bar{x}$ or R control point with a special assignable cause code. This code will then appear on all future $\bar{x}$ & R charts that the operator generates. The data is then harvested from the DataMyte and loaded into the PC for archival and report generation. The installed system created numerous new benefits which include:

- Decreased time the operator needed to spend in col-

See p. 19-6 for the 762 data collector.

lecting and recording the x̄ & R charts,

- Allowed the operator to easily review and analyze the process charting,
- Eliminated errors which were encountered in the manual system, and
- Provided management with an organized data base for use in evaluating product quality and machine productivity.

The information became useful in maintaining and improving product quality, and also became a valuable tool for the machine maintenance department in evaluating various machine repairs and for planning preventive maintenance.

CASE 12-6   PHARMACEUTICAL COMPANY

Company Improves Monitoring of Bottle Cap Opening Force Using the DataMyte 762 and 769

A large manufacturer of pharmaceutical products was faced with reducing its work force and needed to make its existing workers more productive. The company uses SPC to monitor the production of their plastic containers as well as various aspects of their packaging line.

Problem

See p. 19-6 for the 762 data collector.

The company had been using a manual SPC system for the molding process of plastic containers. Their SPC problem solving activities were too slow to effectively help their molding process. They needed a faster response time so the workers could adjust their machines when necessary.

The company was also monitoring the fill weights of certain products on their packaging line. Other data they needed to keep track of included the amount of force necessary to "pop" the caps off bottles and the crush strength of the container.

The company wanted to keep track of attribute data in the final packaging end area to make sure the products were labeled and packaged correctly. All in all, the company had several requirements that it wanted for its SPC program, including ease of use.

Solution

The company purchased a DataMyte 762 data collector to use with various hand tools in both the packaging and container-making areas. The tools the company used for monitoring the container molding process included calipers, micrometers and optical comparators. Workers used force gages to test the force needed to uncap filled containers. They also used electronic weigh scales to monitor various fill weights of the containers. See Figure 12.6.1. Using the DataMyte FAN II software program and a Data-Truck, the workers stored the data on an IBM PC. This gave them easy access to historical data to track improvement and to provide reports to government inspectors.

The DataMyte 769 attributes data collector suited the company's need for attribute data collection. The workers found the 769 easy to use, with its bar code wand for data entry, and its ability to print its own bar codes.

The company arranged for DataMyte to train its workers to get the most from their DataMyte data collection systems. After purchasing and using the data collection system, the company has decidedly increased the productivity of a shrinking work force.

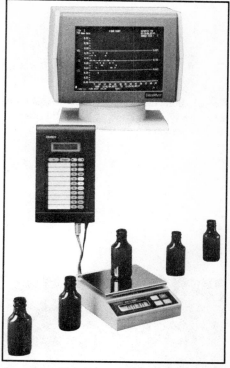

Fig. 12.6.1 Checking weights of pharmaceutical vials.

CASE 12-7   HEALTH CARE PRODUCTS MANUFACTURER

Company Expands SPC Operations

A major manufacturer of health care products successfully implemented a SPC program to monitor a medical gauze pad product.

Problem

With SPC successfully implemented, the manufacturer's interest in expanding operator duties not only included monitoring, but boxing and palletizing of the product as well. These new demands on the operator's time quickly showed a need for more efficient means of maintaining $\bar{x}$ & R charts.

Solution

The x̄ & R charts were produced on 2-hour increments with five samples for each subgroup. The weight and length of the gauze pad are critical for the product to retain proper absorbency. The solution was to install a DataMyte data collection system consisting of DataMyte 762s with monitors and a DataTruck to harvest and archive the data using FAN Support Software on an IBM PC.

The 762, interfaced to a scale and caliper, quickly increased the speed in which the operators were able to collect and produce x̄ & R charts. When a point on a chart required action, the operator would note the change on the x̄ & R charts using a set of established assignable cause codes to illustrate the action which was taken to correct the processes at the particular point and time. This code will appear on all future x̄ & R charts the operator generates.

The manufacturer was able to recoup the system cost with the operator time savings. The time required for machine shut down to perform the manual SPC chart was greatly reduced. This became a very important issue to the operator acceptance of the DataMyte system, since their compensation package was based on an incentive system. Management was also able to take advantage of being able to archive data collected by the DataTruck and FAN Software.

See Chapter 19 for data collectors, Chapter 21 for weigh scales and calipers and Chapter 19 for software.

Fig. 12.8.1 DataMyte 862 data collector and a toolmaker's microscope.

CASE 12-8   BIOMEDICAL MANUFACTURER

Final Inspection Using the DataMyte 769 Helps a Biomedical Manufacturer Increase Productivity, Decrease Costs and Provide Traceability

A producer of biomedical components including parts for pacemakers, needed to increase productivity, decrease costs and provide traceability for all parts requiring conformance to government regulations.

Problem

The company was having trouble with their suppliers. One supplier was shipping out of spec parts for approximately 12 months. The people in receiving inspection were spending a lot of time doing incoming inspection.

The final inspection group was finding dimensional problems — approximately 60% of the parts were rejected by final inspection.

Solution

The company decided that automated data collection would help them monitor their suppliers and decrease the dimensional problems that the final inspection group was finding. They use the DataMyte 862 to check the dimensions (Figure 12.8.1) and the DataMyte 769 to chart supplier histories.

In the receiving inspection department, they have improved productivity by 200 to 300% by being able to inspect a higher number of lots received in a given time and by reducing the number of samples per lot. Receiving inspection has also started a supplier certification program. The data collectors are being used in receiving inspection, in-process inspection and final inspection. The DataMyte 769 is used to plot all suppliers on a pareto chart. This lets the group focus on suppliers that need the most help with their SPC program.

See p. 20-4 for the 769 data collector.

CASE 12-9   DESSERT TOPPING COMPANY

Automated SPC Program Takes The Subjectivity Out of Dessert Topping Line

The company runs three shifts of dessert topping. They have been practicing SPC for several years since they decided that they no longer wanted to control a process within specifications — they wanted to control it within its own capability.

Problem

This company has four process operators, and each of them was adjusting their machines a little differently. They needed to take the subjectivity out of the line. They use a checkweigher to reject overweight product, but needed something to control the average weight.

Solution

The company decided to automate data collection and set standard operating rules to take the subjectivity out of the process. They use the DataMyte 862 data collection system. See Figure 12.9.1. The company was adjusting the machine based on the average of the subgroup of five. They were not plotting it on a control chart. With the DataMyte 862, the operator displays a control chart after every subgroup of five, and adjusts the machine based on the control chart.

CASE 12-10 ORTHOPEDIC IMPLANTS COMPANY

Data Collectors Used to Ensure Quality of Hip and Joint Replacements by Recording Inspection Results in the DataMyte 769

A company making orthopedic implants needed a more comprehensive inspection system to meet federal guidelines. They found that the use of data collectors for both dimensional and defect inspection provided the feedback they needed for manufacturing and the documentation to meet federal guidelines.

Problem

Federal guidelines are very stringent on these types of products. The company was required to document the dimensional characteristics, such as length, diameter and circumference, and also attributes. The attributes included the presence of scratches, nicks, blemishes and other defects. The reason for defects inspection is because they can affect the outcome of the implant surgery if they harbored bacteria that could infect the joint.

Solution

The company decided that a standardized data collection system would help alert the manufacturing processes making the parts. The real-time analysis would allow them to make adjustments before an unacceptable amount of defects occurred.

DataMyte 762 data collection systems connected to indicators and calipers were used for recording and analyzing

Fig. 12.9.1 An operator uses the DataMyte 862 data collector and monitor to display an x-bar chart.

dimensional data. DataMyte 769 data collection systems were used for recording and analyzing defects. Operators at each inspection station were responsible for both dimensional and defects inspection. Both systems provided a standard documentation scheme as well as instant feedback on a video display terminal.

The results of their system implementation has been an increased vigilance by their operators on quality improvement as well as automatic documentation to meet federal guidelines.

See Chapter 19 for a description of the 762 and Chapter 20 for the 769 data collector.

CASE 12-11   PHARMACEUTICAL COMPANY

Company Automates Data Collection in Aseptic Packaging Line

A major pharmaceutical company wanted to automate data being collected with regard to the filling of plastic drug delivery pouches.

Problem

This is an aseptic process performed in a clean room environment. The process involves two sets of four filling machines. Each set of four is tied to a common conveyor that brings the pouches to a packaging operation. Because of the sterility requirements, the company wanted to minimize the amount of time personnel need be present in the room and also minimize the amount of equipment present in the room. The measurements being tracked in this process are fill weight and the thickness of the seam on the plastic pouch.

Solution

The company chose a system consisting of two DataMyte 762 data collectors with monitors as well as scales and micrometers to collect the measurements.

The DataMyte 762 is able to handle a group of items (weight and thickness) for each of four different filling operations. This allows them to monitor each operation separately. An additional variable was the need to track production by lot number.

This system decreased the time needed to collect and analyze the data, eliminated errors which occur in the manual system, and provided real time feedback for controlling the filling process.

CASE 12-12 PHARMACEUTICAL COMPANY

Automated Raw Materials and Packaging Lab System Increases Productivity Five-Fold

A major drug company was very concerned with productivity and felt that a tremendous profit could be obtained by improving their plant operations. By automating SPC data collection of variables and attributes data in their packaging and receiving areas, they improved productivity five-fold and paid for the system in about one month.

Problem

The raw materials and packaging department had to inspect 13 samples of every lot of approximately 70 different parts per month. Each sample had to be checked for weight, length, width, volume capacity, neck length, neck OD, neck ID, bottle OD, cap opening/closing, and torque testing. The time it took to measure 13 samples per lot was 45 minutes to 1 hour 15 minutes. All of the statistics were kept by hand and documented on paper. They used a computer to summarize the data, and then the part was either released to production, rejected, or put on hold. If a part was rejected, a raw materials card was generated and sent to purchasing. Purchasing, in turn, would contact the vendor for appropriate action to be taken.

The primary problem in the raw materials and packaging lab was the incredible amount of time needed to generate reports on incoming samples. When a shipment of raw materials or incoming production material arrived, it took 10-14 days before the material was released to the production line. In conjunction with their just-in-time program, just starting to ramp up, if they could reduce the time the raw material sat prior to release to production, they could save money on payments to their vendors and speed up the packaging of their products. This was a just-in-time production goal.

Inspectors spend most of their time logging and documenting the necessary statistics for their incoming inspection, which allowed them to do only about four samples a day. The typical attributes report would classify the defects according to priority. They would reject a lot depending on the priority of the defect.

Solution

The company investigated ways of automating the inspection process. As a part of the productivity improvement they began using DataMyte data collectors a DataMyte 862 data collector was used to check 10 critical dimensions. The data collector could interface with their caliper and weigh scale. They entered measurements from a ruler by using the keypad. A DataMyte 769 data collector allowed them to input all the possible visual defects on all their products, using bar code entry.

See Chapter 19 for data collectors, Chapter 21 for calipers and weigh scales and Chapter 19 for software

The use of the data collectors cut data collection time per lot from a maximum of 1 hour 15 minutes to about 10 to 15 minutes. The time savings was quite substantial, but more importantly was the ability to store the data on DataMyte FAN II software. This allowed them to generate reports and approve incoming lots substantially faster. Additionally, when they tried to give performance reports to their suppliers of raw materials, it was extremely time consuming and many times did not get done. They had not been getting reports out on time the last three quarters.

The payback on the data collectors for the raw materials and packaging lab took one month.

CASE 12-13 FROZEN VEGETABLE AND COMBINATION MICROWAVE DISHES COMPANY

Automated Data Collection Is Used To Tighten The Upper And Lower Control limits For Product Weights

A well know manufacturer of frozen vegetable and combination microwave dishes finds it important to keep data about their process handy and in understandable form because of government requirements and checks.

Each state is responsible for protecting the consumer by regulating underfill in products. However the regulations indicate that discrepancies (overfill or underfill) should be distributed equally above or below the target. Extremes on either side of the target can potentially cost a company hundreds or thousands of dollars in fines or waste.

Problem

The company needs to keep accurate records on their process in case they are checked by the government. The company has long term storage needs for government checks. And they need to keep their process in control from day to day.

Solution

The company chose to automate their SPC system using DataMyte data collectors for variables data. The company has used automated data collection to tighten the upper and lower control limits for product weights. See Figure 12.13.1. DataMyte data collectors have reduced the data collection time per day by approximately 1-1/2 hours. DataMyte systems have also eliminated the possibility of human error in writing down numbers and reduced the

Fig. 12.13.1 Checking the weight of packaged green beans with a DataMyte and a weigh scale.

chance of inconsistencies in defining "out of control" from shift to shift. The company saves more than $25,000 annually because of DataMyte data collectors.

CASE 12-14   CANNED VEGETABLES COMPANY

Inspection of Canned Vegetables Done Quickly and Easily With the DataMyte 769

A major producer of canned vegetables found their data collection process was time consuming and not conducive to later analysis. A system was needed to grade products quickly and to transmit the data to headquarters for weekly staff meetings.

The company needed a better method to track the quantity of each product grade that was produced to increase market share and customer satisfaction. Also, a better system of traceability was needed to follow up on customer comments and complaints by using product batch codes.

Problem

This company has many facilities all canning vegetables during the summer months. Samples are pulled regularly and visually examined for defects prior to canning. Based on the number of defects, a grade is assigned to the vegetables. C-charts were done manually on each grade of product and notations made as to the types of defects found. This manual charting was time consuming and difficult to use in management reporting and later analysis. A system was needed to standardize the data collection between facilities and have the data available in a format which allowed computer analysis at each facility and the company headquarters. This standardization of the data would also allow customer service to trace customer complaints and comments more quickly and efficiently.

Solution

The company needed to track the date and shift of production, canning line, inspector, grade of product and defects found in each sample. This information allows product traceability once it reaches the market. Also, in many cases,

See p. 20-4 for the 769 data collector.

a grower number needs to be recorded. This allows feedback to the grower concerning possible changes in herbicide programs.

A DataMyte 769 data collector and bar code entry sheet at the inspection station are used to enter the grade and defect information. A sample is pulled off each line every 15 minutes and visually inspected for defects. A grade is then assigned to the product and the information entered into the 769 data collector. The data is analyzed by displaying c-charts and pareto charts on the video monitor at the inspection station. This immediately allows the inspectors to see the amount of each grade being canned and what defects were found in the samples.

The data from the 769 data collector is then downloaded to a personal computer using the 769 Support Software. Then the data is used in a spreadsheet program to produce the customized reporting required by corporate management. Based on a total poundage figure input by the inspector, the spreadsheet program calculates the pounds of each grade of product canned and supplies a summary report. The report takes three minutes to produce.

Using the DataMyte data collection systems in the growers' fields during harvesting is also under evaluation. This would facilitate control of the harvesting process and possibly lead to lower defect rates (higher grades) during canning.

13. FURNITURE AND APPLIANCE INDUSTRIES

Case 13-1 OFFICE FURNITURE SYSTEMS MANUFACTURER

Data Collection Decreases Variability in Attribute Inspection

A leading manufacturer of modular furniture systems for offices established an employee participation program for quality control in 1981. The program made the employees responsible for the quality of their own work. At the time, 4% of the work force participated in the program. The program has grown so that now the company has 65% of the work force involved. Last year, the company conducted over 25,000 employee audits of finished products. In the past six years, productivity has risen about 70%, while the cost of quality has fallen 80%. DataMyte data collectors have contributed to this boost in the company's efficiency.

Problem

When the company started the quality control program, it was primarily collecting attribute data, looking at the fit and finish of the finished products. However, because of the way the data was collected, not everyone collected data the same way. What one person might call a good product, someone else might not. The company needed a way to look at some of its processes with variables data, so that everyone was inspecting from the same perspective.

For example, one product that this affects is a shelf called a "flippy" in the trade. This is an enclosed shelf that mounts on a wall-like panel. The shelf has a front cover that flips up and slides over the top of the shelf. The shelf is made of laminations of particle board and plastic laminate. An edging material is applied to the edge of the shelf. The edging material is a fixed, consistent width. One of the defects the company inspected for was the fit of the edge to the shelf. Because the thickness of the shelf could vary, the edging material may or may not fit properly.

Solution

The company found that by tracking the thickness of the shelf during production, a more reliable product could be obtained. This involved using Mitutoyo calipers and micrometers to monitor each stage of the shelf lamination process. Employees had been using DataMyte 752s and 762s to measure dimensions of their products. The company has been introducing the DataMyte 862, to take advantage of the 862s ability to store data from many different parts at once.

See p. 19-4 for the 862 data collector.

By using numerical data from these data collectors, the company is able to do several things. It can monitor the performance of each operator and workstation. Adjustments can be made to ensure a consistent thickness for the product. The company can also monitor the process capability over time. This becomes important because of the variation in seasonal conditions: summers are more humid than winters, and the humidity affects the thickness of the pieces as they are assembled.

The DataMyte 769 is used in a final inspection area to collect attribute data with a bar code wand. The inspectors pull different pieces of the furniture system at random and assemble them. They then inspect the fit and finish of the assembled product.

CASE 13-2   RETAIL APPLIANCE COMPANY

Column Gage — DataMyte Linkup Provides Real-Time Operator Involvement

A leading retail appliance manufacturer needed a data collection and analysis system that was capable of interfacing directly with two different types of column gages. Up to ten of these column gages were tied into one fixture. In addition, the system had to be small enough that it would not crowd the operator's work station and still give the operator real time feedback for SPC.

The measurements were extremely precise. The system had to provide enough data to help produce a more consistently dimensioned part as well as improve overall plant productivity. The company was also interested in maintaining all raw data to do further analysis, but was not interested in having filing cabinets full of paper. Paper records were considered essentially "worthless," since the information would never really be used by management in a timely manner.

Problem

The difficulty of managing data generated from 25 production lines and interfacing with (on an average) six column gages at one time seemed monumental. The company was doing the data collection in a very traditional manner and they were forced to:

- Manually collect data from their existing column gages.
- Write the data down on forms for each process that they were monitoring.
- Manually calculate the mean of the data.
- Manually plot the $\bar{x}$ value on a control chart at the process.
- Later attempt to accumulate the raw data and calculate manually the control limits for the process.

Needless to say, the problems in this system began to grow because of the amount of data that had to be collected on just a few production lines. The time that it took to evaluate the data was also considered prohibitive. In addition, the company had begun to install new electronic column gages that were being viewed as being more difficult to work with by the operators. This added to the problems of overall data collection and analysis.

Solution

See p. 19-10 for the 761 data collector.

It was decided that a DataMyte model 761 data collector, with a ten column gage interface and junction box, would be the best system for data collection. This particular system solved some cumbersome problems:

- Up to ten column gages could be interfaced to each data collector — the operator simply had to put a part in a fixture and hit a foot switch and all of the measurements from the columns were immediately recorded in the data collector — this eliminated the operator's difficulty in reading the often difficult column gages. The operator no longer had to write down all of the information and do the calculations. The data collector immediately alerted the operator if a trend was beginning to appear or if parts were out of specification.

- When the operator completed his last sample in the subgroup, the data collector told him if he was in or out of control. The appliance manufacturer went a step further and tied each data collector into a video display at the process. By doing so, the operator simply had to push a button and his most recent 30 $\bar{x}$ points were displayed on a graph on the screen. At the same time, the operator also had the ability to look at a histogram of each characteristic being measured.

As a result of their success with the DataMyte system, real-time process control was established at each of the manufacturing processes in the plant. Because the DataMyte 761 maintains all the raw data collected at each process, the data could be harvested with a DataTruck twice each shift. Data was then transmitted to an IBM PC with FAN Software. The software package is used to manage all of the manufacturing setups in the facility as well as generating hard copy $\bar{x}$ & R charts. Capability reports and histograms furnished by FAN software are used on a real-time basis by management to further reduce the variability of their processes.

See Chapter 19 for a description of FAN software.

CASE 13-3   ELECTRIC METER MANUFACTURER

Interactive SPC Data System Boosts Machine Shop Production

A leading manufacturer of kilowatt-hour meters wanted to completely computerize their screw machine facility. Their goal was to develop a totally interactive system allowing operators to down load new setups directly from the factory floor and send the reports back to the operations office. At the same time operators needed to generate their own graphic reports at the work station for statistical process control.

Problem

One of the problems was how to economically provide the hardware to support eleven operators and eighty screw machines. Each operator had to have complete access to the system to match the changing hourly and daily requirements. The system had to be easy to work with.

Solution

They decided the best approach would be to supply each operator with his own workstation data acquisition system. Available for easy access would be a DataMyte 762 equipped with three gage ports and a monitor for immediate display of control charts.

In addition, each location was connected to the Allen-Bradley Vista LAN/1 Network. See Figure 13.3.1 The operations department could down load the desired part numbers from the 762 to the computer in the operations department in a matter of minutes.

The implementation of the Allen-Bradley Vista LAN/1 Network has provided this company with one of the most advanced SPC facilities in the world today. They are enjoying the benefits of better quality, reduced defects, timely information, and haven't added one bit to the work load of the operators.

See p. 19-6 for the 762 data collector.

Fig. 13.3.1 Flowchart of communications network.

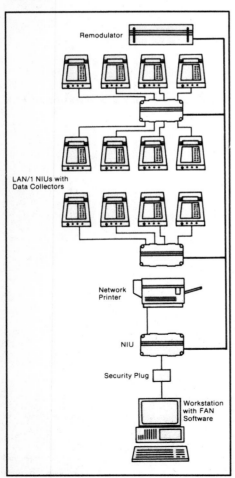

CASE 13-4   FLOOR SWEEPERS MANUFACTURER

Coordinate Measuring Machine — DataMyte Linkup Provides SQC Report Consistency

A world leader in the manufacturing of industrial floor sweepers needed a data collection and analysis device that was capable of analyzing the outputs from a coordinate measuring machine. They also wanted all statistical reports generated to be consistent in format. Since this manufacturer was already a user of DataMyte handheld data collectors, it was their desire to have data generated by their coordinate measuring device fed into the data collector for subsequent analysis and reporting. By using the data collector in this manner, the manufacturer was able to detect areas in the manufacturing process where machining costs be reduced. The manufacturer was also able to achieve the desired consistency in report format.

Problem

The process of analyzing data generated by the coordinate measuring device and generating reports was time consuming. It also did not fully meet their requirements for statistical process control (SPC). Before the data collector was implemented to handle the data analysis and reporting functions, the manufacturer was forced to:

1) Manually accumulate the measurements.
2) Load the data collected into the main computer memory.
3) Load in other applications programs that would in turn process and format the appropriate data.
4) Output the raw data collected to a printer.
5) Perform manual analysis of the raw data.

Because data analysis and reporting was so cumbersome, inspectors were spending far too much time collecting and analyzing data, and not enough time looking for ways to further their use of SQC.

Fig. 13.4.1 DataMyte data collector and coordinate measuring machine.

Solution

It was decided that by simply feeding the output from the coordinate measuring device into the data collector, they could virtually eliminate steps 3 through 5 of their current data processing procedure. See Figure 13.4.1.

Since the computing system in use had a serial RS-232-C interface, inputting data to the data collector was easy; only one cable was required to connect the two devices. Once connected, X values were transferred to the data collector, followed by the Y and Z values from the computer's memory. Once sufficient data was transferred, the data collector was used to generate the required capability studies, charts and graphs.

CASE 13-5 GARAGE DOOR SYSTEMS MANUFACTURER

Manufacturer Saves Fourteen Hours Each Week in Documenting Their SPC

A manufacturer of transmitters, receivers and motor controls for garage door opening systems had a very labor intensive final assembly operation that made defect auditing difficult. Seventy to eighty percent of the parts for a remote control were inserted at the final assembly stage. Sometimes the wrong parts were used or the parts were not cut properly. The manufacturer had to monitor these and other defects in the assembly.

Sample PC boards that go into the transmitters and receivers were visually inspected and the defects recorded for these parts. To record this type of data and construct p, c and u-charts, the manufacturer used an attribute data collection system. The manufacturer monitors and records the characteristics of the remote controls in each department, as well as in the insertion room, before the control goes into the next stage.

Problem

One person in each manufacturing area was designated to collect the data by hand. This person collected and recorded the data on paper and then manually plotted the charts. The procedure required sixteen man-hours per week.

The company did not want to spend this much time collecting and charting the data. The manufacturer also thought the operators should have more immediate feedback about the jobs they were doing.

Solution

The manufacturer obtained one DataMyte 769 attribute data collector for each manufacturing area and eventually plans to have one data collector for each operator. With DataMyte 769 Support Software for the IBM-PC the manufacturer saves fourteen hours per week when generating charts. Instead of recording data on paper with a pencil, the operator wands the data into the data collector. The data collector stores the data in its memory until the operator downloads the data to the PC.

See p. 20-4 for the 769 data collector.

Data collection is more convenient and accurate because the operator does not need to write down the data on paper and later hand-calculate the data for his charts. The feedback is immediate; the operator does not have to wait a day or more to look at the charts that concern his or her operations.

CASE 13-6 STAMPING PLANT FOR TELEVISION FRAMES

Plant Obtains Squareness Calculations Automatically with New Data Collection System.

A stamping plant that supplies television frames for a popular American TV brand was being pressured by its customer to implement an SPC program and show evidence of it. The plant produces several thousand frames per day. Five to ten different sizes, both square and rectangular, are made each shift.

The existing SPC program required mostly handwritten methods of collecting data. The data was then keyed into a computer for statistical analysis and chartmaking. The plant felt that by upgrading their SPC program they could not only satisfy their customers request for SPC documentation, but also obtain tangible improvements in quality.

Problem

The customer demanded statistical control of the square-ness of TV frames, as well as the overall dimensions. The stamping plant uses a fixture to check the frames. They measure the distance from a fixed reference point on one side of the frame and then the other and take the difference. By taking these measurements between the fixture and frame edge all the way around they could derive squareness.

All of the measurements had to be taken down by hand and subtracted one from the other before the squareness data could be plotted on a chart and input into a computer. Because of the volume of frames being produced, the time it took to collect the data became a burden. So did the time required to input the data into a computer.

Solution

See p. 20-8 for the 2003 data collector.

The stamping plant decided to obtain equipment that would eliminate having to write measurements down by hand. They purchased the DataMyte 2003 and an elec-tronic gap gage. Using the gap gage, they could quickly go around a fixture and take readings against the TV frame. In addition, the DataMyte 2003, with its math feature, would subtract one column of readings from another and save the result in a third column. Therefore, data on squareness was produced instantaneously. See Figure 13.6.1.

Used with their existing computer, a DataMyte software program allows them to set up the 2003 data collector for each type of frame, and program a prescribed sequence of measurements for the squareness calculation. The data col-lector is then used in the factory to record the data. At the end of the shift, the data is fed back to the computer. Charts are printed on both the overall dimensional and the squareness.

The DataMyte system provides a number of benefits, in-cluding a labor savings, reduction of error and complete documentation. The existing computer is used for long-term storage of data and for management reports. In the eyes of the customer, both the appearance of improved SPC and the actual results of better quality products went hand-in-hand.

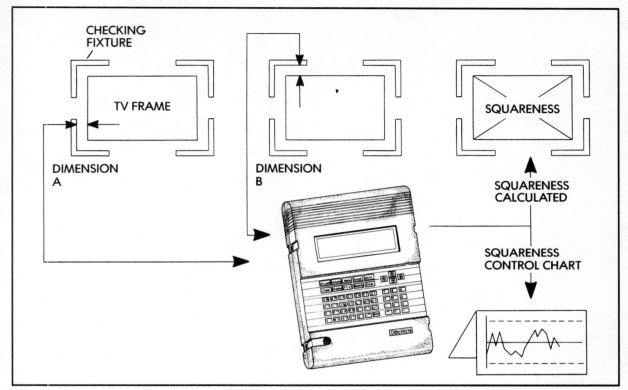

CHECKING
FIXTURE

TV FRAME

DIMENSION
A

DIMENSION
B

SQUARENESS

SQUARENESS
CALCULATED

SQUARENESS
CONTROL CHART

Fig. 13.6.1 DataMyte 2003 math option allows a squareness calculation to be derived from several linear measurements, and then plotted.

14. GLASS, PLASTICS, PAPER AND CHEMICAL INDUSTRIES

CASE 14-1   GLASS MANUFACTURER

High Speed Process Controlled with Simple Fixturing

With the increase in the cost of manufacturing, a major glass producing company wanted to cut down on the amount of scrap and rework. They needed to hold closer tolerances on a flat glass surface. By solving a measurement problem, they increased the capability of their process and cut scrap by 85 percent.

Problem Definition

Glass rolled off the assembly line too fast to manually record data. The glass was large, flat plate glass. One hundred and thirty data points were needed to perform the initial capability study. The quality control department hoped that after the process was brought to a state of statistical control, they could eliminate some of the data points required. A fast data collection and data processing device was needed. It also had to be portable and record the data at the manufacturing site. The data capture had to be accurate. Another problem was they did not want to re-design all the checking fixtures. That would be cost prohibitive.

Solution

All the existing checking fixtures were used. A common vernier caliper was selected as the measuring instrument. The checking fixtures were machined with a 3 mm reference slot. One hundred and thirty slots were machined for the initial capability study. Electronic vernier calipers were used because the information had to be collected and recorded as fast as possible. The result needed to be available at the process so decisions could be made on the spot. The caliper's data was automatically transferred to an electronic data collector. The data collector stored the data and processed it, too.

Now the measurement problem was solved. With a single test fixture modification they could use an electronic caliper connected directly to the data collector for process-

See Chapter 21 for descriptions of calipers.

14-2

Glass, Plastics, Paper and
Chemical Industries

ing and long term storage. A printer was kept at the location for the printed results. Information could also be viewed on the display of the data collector.

Some of the unseen benefits of this program were:

- Data collection time was cut by 450%.
- Only 75 of the original data points were needed.
- Scrap was cut by 85%. CPK Index approached 2.00.
- Information could be off-loaded to a host computer for long term data storage and analysis.

The equipment selected to perform the capability study and for long term process control was the DataMyte 2003 with 512K of usable data storage. No single brand of electronic verniers was selected because of the wide variety on the market today.

See p. 20-8 for the 2003 data collector.

CASE 14-2   PLASTIC CONTAINER MANUFACTURER

Operator-Oriented System Provides Better Process Control

A plastic container manufacturer had to keep track of various measurements to ensure the quality of the finished product, such as plastic milk bottles. They were looking for a system the operators of injection molding machines could use without interfering too much with their main task of making bottles.

Problem

The manufacturer measures, and compares to rigid specifications, finished thread dimensions of bottle caps and tops of plastic bottles. The manufacturer also checks the weight of the plastic containers. They need to weigh the finished containers for two reasons:

- The high-density resin used to produce the bottles is expensive. The manufacturer weighs the finished product to make sure the proper amount of resin is used when the bottle is made.
- The manufacturer also weighs the containers before and

after an appropriate liquid or powder is placed in the container. This insures that the containers hold the amounts of material that were specified.

The manufacturer measures the wall thickness and other dimensions of the containers to make sure that the bottle has been formed properly and is the proper size for its purpose.

Solution

A DataMyte 762 data collection system was the best alternative for the manufacturer, who looked at several types of computerized and handheld data collection systems. They use DataMyte 762s to record the data from electronic digital micrometers, electronic weigh scales and electronic digital calipers. The operators will take measurements and record the data at their injection molding machines. Two types of micrometers are used by the manufacturer. The operators use flat micrometers to obtain the measurements to determine the volume of a container.

The manufacturer uses a DataTruck to collect the data whenever necessary from the 762s. They use an IBM PC to store data and FAN® software to manage and report the data.

CASE 14-3   PLASTICS MANUFACTURER

Expansion Plans Require Better SQC Program

A leader in the manufacturing of several types of plastic utensils, including plates, cups, forks, spoons, etc., wanted to expand their operations. To help accomplish this, they needed to refine their SQC program. Their goals were to improve product quality, to reduce scrap costs, and to enhance data turnaround time.

Problem

The manufacturer was using manual charting methods to analyze and track the data gathered by its inspectors. Once the data was gathered, the inspectors would sit down with their calculators and manually plot x̄ & R charts. This required each inspector to spend from four to six hours per

See p. 19-6 for the 762 data collector.

day. Since the results of their analysis were generally not available for at least four hours (and often not even until the next shift), process conditions had already changed substantially and it was next to impossible for them to locate and remedy problem areas.

Solution

The SQC program enhancements were defined as follows:

- Perform data analysis automatically.
- Generate reports and charts on an "as needed" basis.
- Allow the inspectors to spend their time on the floor locating problem areas (and not involved in extensive data processing).

The equipment they chose to implement these improvements was a DataMyte 2003 data collector. As a testament to the ease of use of the system, after seeing an audit sheet and a four minute explanation of how to load all of the information into the DataMyte, an inspector proceeded to load all this information into the DataMyte and was ready to go out on the shop floor and collect data in ten minutes.

See p. 20-8 for the 2003 data collector.

Operation

The information that follows provides a brief description of how the data collector was used by the manufacturer. The data collector was loaded with the item IDs, engineering limits, and sample size required to simulate the exact audit sheet the manufacturer was currently using. See Figure 14.3.1. Each of the measurement areas from the audit sheet (12 in all) were loaded as items with 20 samples of each being used for charting. Measurements on each item were performed three times each hour on every shift.

Since the data collector was capable of generating histograms, the manufacturer decided to make use of them in their management reports. When needed, the inspector would print a histogram of the current data. These reports were then used to verify the natural dispersion of the processes. Management would then, in turn, base their decisions on how to minimize these dispersions; ultimately getting all areas to try and match the mean value.

```
                              PROCESS INSPECTION
    SUPERVISOR SIGNATURE

                               am    pm
    DATE: _2-29-84_  TIME: _7:30-->7:30_ INSPECTOR:_Chevre 1_   SETUP: _B_
```

Press 3c	Part 3c	Full Cycle	Cure Cycle	Cavity No.	Part Wt.	Mold Temp. (upper)	Mold Temp. (lower)	Cavity No.	Part Wt.	Mold Temp. (upper)	Mold Temp. (lower)
				LEFT CAVITY				RIGHT CAVITY			
64	527	196	103	21	315	350	320	22	307	340	330
76	522	191	98	32	839	330	330	31	829	340	320
75	499	222	124	12	810	330	330	11	803	340	340
63	527	208	104	12	314	330	310	11	313	340	320
73	500	208	113	92	1030	330	320	91	340	310	
82	452	208	126	211	829	340	320	151	790	350	350
83	452	209	122	102	834	330	330	112	841	350	310
84	452	206	106	342	816	320	310	312	793	340	320
85	452	238	108	132	842	350	300	141	828	330	310
86	452	173	99	481	818	350	320	482	824	330	300

Fig. 14.3.1 Sample audit sheet.

CASE 14-4   PLASTICS AND FIBERGLASS MANUFACTURER

Electronic Gap Gages Improve Tolerance Checking

A manufacturer/supplier of plastic and fiberglass piece parts needed a way to improve the overall quality of their products. The manufacturer was having too many of its parts returned. Quite often the reason stated for the return was that the part was out of tolerance.

After investigating the causes, the manufacturer determined that the methods being used to measure parts were not effective. Not only did they provide insufficient data for quality improvements, they did not address the problem of high scrap costs.

Problem

In the past, the piece part products produced by this manufacturer had exhibited a relatively high level of variability in tolerance. The problem was due in large part to the process used to measure the tolerance.

The procedure involved placing the parts in a checking fixture and then measuring their size in relation to the fixture. This was done by inserting a tapered feeler gage between the part and the fixture. When this measurement method was repeated at several locations around the part, the size of the part could be determined.

The feeler gages being used had rules etched every 0.5 mm. Thus, the operator would insert the gage, read the rule line that was closest, and then estimate down to the closest 0.1 mm. This procedure was not accurate since parallax errors were being introduced by the operators.

Solution

To facilitate their program, they purchased a DataMyte 2003 data collector and a 516 flush and gap gauge. (See Figure 14.4.1.) This provided a data collection and analysis capability that was portable, user friendly, and could be used to produce all necessary charts. The gap gage had a useable range of 12.7 mm and a resolution of 0.05 mm.

See p. 20-8 for the 2003 data collector.

Soon after implementing this system, they noticed that the variation from part to part had increased. Suspicions arose regarding the repeatability of the DataMyte and the transducer being used. This prompted further investigation of the entire system. To begin with, several parts were checked by a number of different operators using the same checking fixture and the same tool. The variation was ex-

Fig. 14.4.1 Model 516 flushness and gap gage.

cessive. By placing one part in the fixture and having several operators check it, they greatly reduced variation (approximately 0.1 mm). Thus, they had proved that operator variability in using the tool was quite acceptable and a major improvement over their previous method of estimating to the nearest 0.1 mm.

The only thing left to check was the variability of the fixture. Using one operator, one part and one fixture, the operator was instructed to place the part in the fixture and check it. The same operator then removed the part and placed it back into the same fixture to check it again. This procedure was repeated several times. The results from the checks varied greatly, proving that the variation was a result of the variability in placing the part in the checking fixture. The solution was to redesign the checking fixture so that a more positive placement of the part was assured no matter who placed the part in the fixture.

CASE 14-5 PLASTIC CONTAINER MANUFACTURER

Plastic Container Manufacturer Documents a 20% Improvement in Process Capability By Automating SPC

A plastic container manufacturer with twenty-two production lines needed an efficient way to collect all of the data for their SPC program. Manual data collection was very time consuming, as much as forty minutes per container.

Problem

If the bottle weight or wall thickness varies too much, defects occur, and excessive plastic (which is very expensive) is wasted. In addition, the thread diameter must be checked because if the thread diameter is not the correct size, the cap will not fit.

Solution

The company chose DataMyte 862 data collection systems to help them monitor their process. See Figure 14.5.1. The DataMyte 862 was chosen because of its multi-file capability. The data collector can hold data for up to ten different machines.

Fig. 14.5.1 An inspector records the diameter of a bottle opening in the DataMyte 862 data collector.

The company documented a 20% improvement in process capability, a 4.1% reduction in spoilage and a 31% reduction in downtime attributed to bottle weight variation. Their customers are receiving more containers closer to their specification targets. Container related quality problems have decreased and customers are happier.

See p. 19-4 for the 862 data collector.

CASE 14-6   PEN MANUFACTURER

Controlling Ink Filling Process Brings Substantial Savings

A manufacturer of ball point pens needed better control of how much ink went into pen cartridges. Millions of cartridges were produced each year, so reducing variation in the amount of ink could potentially save an estimated $20,000 to $100,000 a year. The amount of ink directly affected the quality of the pen, also. Too much ink in a cartridge affects the mechanical performance of the pen. Too little ink depreciates the pen in the eyes of the customer.

Problem

The ink filling process could not be measured directly. Quality control inspectors weigh a dry cartridge, have it filled, and then weigh it again. The difference would be the weight of the ink. Monitoring these weights would presumably tell them how much ink was going into the cartridges.

To do this properly at the frequency required for producing process control charts involved a lot of work. Two separate measurements plus a subtraction step were needed for each result. Measuring and recording just the filled weights of the cartridges would not establish the amount of ink actually used because of the added variation of the dry cartridge weights. A working system would involve either a lot of manual labor or a type of automated data collection system that could also derive ink weights through a subtraction calculation.

Solution

The company purchased a DataMyte 862 data collection system. The data collector could interface with their Mettler weigh scales to record weights directly—without error. In addition, the data collector had a math function, which can subtract one column of numbers from another and put the results in a third column. See Figure 14.6.1. The data in the third column, in this case the ink weights, could be printed out in the form of an $\bar{x}$ & R control chart or histogram.

See Chapter 19 for data collectors and Chapter 21 for weigh scales.

Although two measurements were still necessary for each result, since the readings were input directly from the scale via an RS-232 cable, there was much less error. Less time was needed to record the measurements, and since the subtraction step was handled internally in the data collector, there was a substantial reduction in time and error overall in the data recording process.

Data processing was instantaneous. The data collector was simply connected to a printer to produce the control charts. Charts on cartridge dry weights were also available as an added benefit.

Overall, the data collector system provided a savings that would pay for its cost in an estimated two to three months.

CASE 14-7 BATTERY MANUFACTURER

Battery Manufacturer Automates SPC

A manufacturer of six and nine-volt batteries had been practicing SPC for many years when they decided to auto-

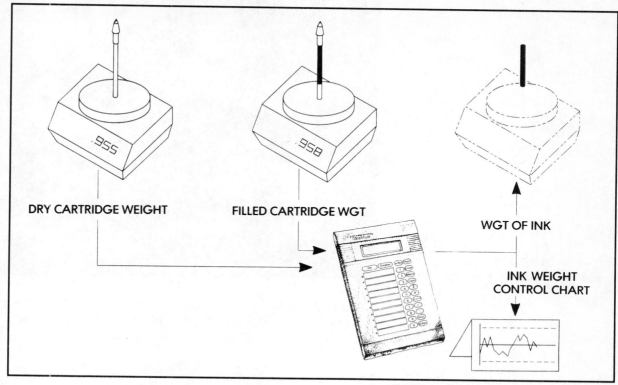

DRY CARTRIDGE WEIGHT

FILLED CARTRIDGE WGT

WGT OF INK

INK WEIGHT
CONTROL CHART

Fig. 14.6.1 DataMyte 862 math option allows
the ink weight to be derived and plotted.

mate their data collection. One reason they choose
DataMyte data collectors is because they are able to withstand the harsh environment.

Problem

The battery manufacturer tested the data collectors in the
harshest environment first, figuring that if they could withstand the fine powder so detrimental to electronic components, the data collectors could be used anywhere.

Several hundred tests or measurements are performed on
the battery or its components between the beginning of its
construction to the moment you buy it. It was nearly impossible to manage that much data.

Solution

The company chose DataMyte data collectors to help
them practice SPC. See Figure 14.7.1. They chose Data-
Myte data collectors for three reasons: they passed the test
of working in a harsh environment, they could continually

Fig. 14.7.1 A DataMyte 762 being used with a weigh scale, caliper and digital indicator.

upgrade the data collectors, and they would help them collect and analyze the enormous amount of data generated daily.

CASE 14-8   PRINTING COMPANY

Printer Collects Ten Times More Data

A supplier of corrugated packaging containers principally serving the food processing industry was committed to supplying the highest quality product to their customers.

The company's principal manufacturing processes included cutting, printing and wax coating operations. This company is one of the largest suppliers of this type of product with numerous manufacturing sites throughout the United States. Although the company would like to reduce manufacturing costs, their principal motivation to begin an SPC program was to demonstrate to their customers their commitment to quality. Their customers were not demanding documentation of conformance to specifications, nor were they mandating SPC techniques.

Problem

The biggest obstacle to this company was converting their existing inspector based quality control program to operator based data collection and analysis. Their primary target was the printing presses where they used MacBeth Densitometers to record color density and percent dot concentration. See Figure 14.8.1.

Time was a critical concern from two aspects. First, the operator was engaged in numerous activities on the press and could not be burdened with excessive data collection tasks. Second, responding to out of control conditions needed to be rapid to ensure consistency of finalized products.

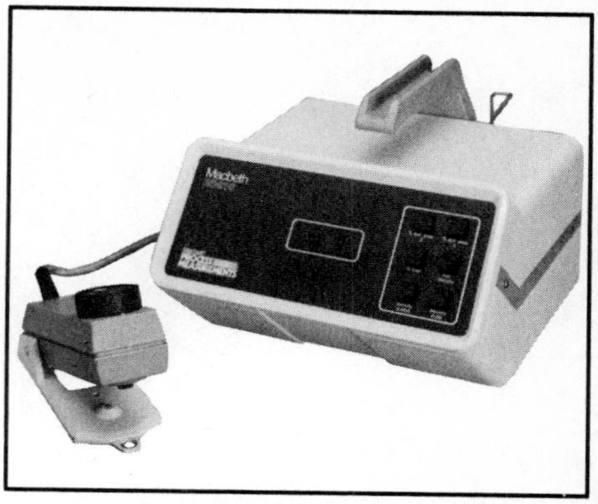

Fig. 14.8.1 MacBeth densitometer.

Solution

The company elected to use the DataMyte 762 in conjunction with their MacBeth densitometers. By using the DataMyte, operators were capable of collecting up to ten times more data than in manual methods. Since graphs were generated immediately, corrective action responses were implemented in process. Before using DataMyte data collectors, in-process corrections were rarely made due to the short cycle runs and length calculation requirements for process control parameters.

See Chapter 19 for the 762 data collector and Chapter 21 for the densitometer.

Before using DataMytes, operators would depress the densitometer head and handle assembly to record a value. Before proceeding to the next value they would stop to record the data on a tally sheet and then proceed again. With DataMytes brought into the operation, the operator depresses the densitometer and upon releasing the head and handle assembly the reading is automatically recorded. At the end of each subgroup, an $\bar{x}$ & R chart appears automatically for each color on the product. At the end of a run, the operator unloads his data to the DataMyte FAN II software program and downloads his product file.

The operation on the printing presses was so successful that this manufacturer expanded the use of DataMytes to the wax coating application where electronic balances measure the amount of wax applied and DataMytes monitor processes for control. Additional DataMytes are used in conjunction with micrometers to measure the thickness of raw materials. DataMyte 769s are used extensively for visual defect audits. Today, over 50% of all processes in this company are controlled using DataMyte products.

CASE 14-9 MAIL SERVICE

Automatic Data Collection Zips Out Case Density Studies

The Eastern Region of a large mail-handling concern performed case density studies to determine how manual letter and flats diagrams should be designed. The diagrams minimize the distribution required in subsequent operations and arrange the case separation for maximum distribution efficiency. Performing this task manually was labor intensive and therefore very costly. Also, the chance for introducing errors into the data during transcription and key punching operations was high. In an effort to upgrade their capabilities and because of the high volume of mail passing through the eastern region on a daily basis, the mail service determined that they needed to implement an improved system as soon as possible.

Problem

The Case Analysis System (CAS) required a daily 500 piece sample of ZIP codes. To ensure an accumulation of representative data for designing the case diagram, special sampling techniques were used. The procedure for sampling was performed as follows:

1) A 500 piece sample consisting of letters and flats was taken daily at each test point. This sample was actually divided up into five 100 piece sub-samples, which were taken at randomly selected hourly increments. The time intervals chosen for sampling were based upon the probability of mail availability during the hourly increment.
2) Next, the sample ZIP code data was recorded manually on a density sampling work sheet. Each work sheet could accommodate 250 ZIP codes for each operation, so two coding sheets were turned in. Note that this method required the manual counting and recording of the mail in these separations.
3) Finally, the collected data was converted into machine language for processing by the host computing system, which was located on the west coast. This process involved transcription, keypunching, editing, and re-key-

punching operations before the actual processing could begin.

Solution

The eastern region determined that the solution for reducing costs and for increasing the data turnaround time was to close the loop in their data collection/processing system by fully automating the procedure. Handheld microprocessor-based data entry terminals were used by the auditors for recording the ZIP code data. There were basically three advantages to automating the data collection/processing procedure:

1) The rapidity at which data can be recorded.
2) The immediate availability of the data in computer processible form.
3) The increased reliability of the data due to the elimination of transcription and keypunching operations.

Once the mail service had defined the known problems and areas of concern, they began to investigate the best way to implement their program. They determined that the program implemented must be able to:

- Perform data collection and analysis automatically.
- Interface with the various types of existing computing terminals that are used to access the central mainframe on the west coast.
- Ensure the integrity of data. (This was necessary since the data was being sent via telephone transmission lines.)

Based on their analysis, they selected a DataMyte 1000 handheld data collector. The DataMyte 1000 provided data collection and analysis capability that was portable, user friendly, and could be used to interface with all their existing computing terminals and mainframes. The data collector could be programmed to transmit using sophisticated communications protocol, providing error checking to ensure data integrity. The data collector was rugged and had proven reliability as a field terminal.

By using the DataMyte 1000 for data collection, analysis, and reporting purposes, the mail service was able to improve their case density studies program. This unique system was able to speed up the whole procedure for performing case density studies. Further, because of the high degree of automation, the possibility for human error was greatly minimized. The savings in keypunching costs alone were substantial. The DataMyte, after just one use, saved the eastern region 144 manhours at $15.00 an hour, more than justifying its purchase in a single use.

Operation

The information that follows provides a brief description of how the DataMyte 1000 was used by the mail service.

The DataMyte 1000 was installed at the user's facilities, it was set to IN-Mode, Section of Record 5. The unit was configured with an entry counter feature that audibly alerted the operator when the 500 piece sample was completed. As each piece of mail was sampled, a five digit code was entered into the DataMyte's memory via the keyboard. Immediately following data collection, the DataMyte was linked to the central mainframe and the data was transmitted. At the mainframe, the data was written to a sample ZIP code file, where it accumulated until Case Analysis System (CAS) programs could analyze and collate the data. Once the data was processed by the CAS programs, it could be used to design a new case diagram, analyze an existing case diagram, or to generate ZIP code reports for further analysis.

CASE 14-10   PHOTOGRAPHIC EQUIPMENT MANUFACTURER

Latest Short-Term Payback Installation Eliminates Manual Setup Changes

A major photographic equipment manufacturer uses DataMyte data collectors throughout its corporation. When this manufacturer makes a capital equipment purchase, it assigns an industrial engineer to analyze the return on investment. The company's first DataMyte purchase was for the film capsule line. The accompanying engineering study compared the automated data collection, with real-time feedback, to the manual data collection the company had conducted previously. The study justified buying 40 more systems for the plant.

The manufacturer has since installed many more DataMyte systems throughout their plants. In all major system installations, in their tape, film, film cartridge and camera divisions, the manufacturer has documented a payback in less than six months.

The manufacturer's latest installation incorporated the Allen-Bradley VISTA LAN/1 network.

Problem

In many of the company's manufacturing processes, part changes on machines occur several times in one day. Each machine may be required to run hundreds of different parts each year. When this condition is multiplied by hundreds of machines in the department, data management becomes a severe problem. Consistency in part quality between runs is difficult to correlate.

Solution

As a result, this firm looked to hardwired networks to speed up the setup changes in the data collectors between parts. Using a VISTA LAN/1 network eliminates the problems described above. The network incorporates data collectors, line drivers, a host computer (in this application, an HP 1000) and customer developed software.

The hardwired VISTA LAN/1 network is a true two-way communication system. The host computer serves as an enormous data base of part setups. Literally thousands of electronic files are stored in the computer. Each file fully describes the various parts and process parameters. These electronic files are accessed by operators on-line through a menu selection in their DataMyte data collectors. When new part setups are required, operators keypunch in the desired part number, which signals the host computer to pick up the file resident in the data collector. Upon uploading the resident file from the data collector, all new data is filed with historical data archived in the host computer from previous runs. The archived data at the host computer can be accessed via a PC network by various departments within the company. The computer then sends the requested file to the data collector.

The individual requesting information from the host computer can call up a file by part number and review some or all of the historical data and perform post process analysis of the data in a variety of third party statistical software programs.

Fig. 14.11.1 An inspector looks for visual defects and records the results in the DataMyte 769 data collector.

CASE 14-11 GLASS CONTAINER MANUFACTURER

Glass Company Automates Data Collection With A DataMyte 769 and 862

A glass container plant with six production lines that have the capability of running between 370 and 420 bottles a minute needed a way to practice SPC real-time. With that fast of production, a lot of bad product could be produced very quickly.

See p. 20-4 for the 769 data collector.

Problem

The company needed a way to practice SPC real-time. They were producing far too many rejects that had to be reprocessed because they were too heavy or light or had a visual defect such as fused glass, thin glass and out of round. The bottles have stringent weight requirements. If the bottle is too heavy or light, the operator can adjust the amount of molten glass that is dropped into the molds.

Solution

The company chose to automate their SPC system using DataMyte 862s and 769s. Every fifteen minutes an operator uses a DataMyte 862 and weigh scale to record the weight of three bottles. The operator can adjust the amount of molten glass that is dropped into the molds if the bottle is too heavy or light. If the bottle does not meet the weight requirements, it is reprocessed as cullet (recycled glass). The inspector uses the 769 data collector to track visual defects on the bottles and cartons. See Figure 14.11.1. If any defects are present, the inspector records both the defect and the mold number. The mold number allows the quality department to track which mold or line is producing defective bottles.

CASE 14-12   TEFLON SEAL MANUFACTURER

Company Automates Gaging Stations to Maintain Production Flexibility

A company manufacturing pressure molded teflon seals needed to expand its implementation of SPC.

Problem

The company had 30 different molding machines producing a wide variety of teflon parts. Operators were taking time to gather dimensional measurements with a vernier caliper or micrometer. After each subgroup was collected it would be manually averaged and plotted on an $\bar{x}$ & R control chart. Some operators are responsible for several machines and not all molding machines may be in operation at any given time.

The company wanted to reduce the time involved in operator based collection as well as improve the accuracy and reliability of the charts.

Solution

The company set up their gaging stations to allow machine operators to take measurements faster, keep a number of control charts and have the most amount of time possible to maintain production on the molding machines. Each station used a DataMyte 862 data collector, video monitor and electronic micrometers and calipers.

The DataMyte 862 could handle a number of independent jobs in separate files. This allowed data collection and

See Chapter 19 for a description of the 862 data collector.

SPC on several machines by one operator, and the ability to start or stop data collection to match the production.

CASE 14-13   PLASTIC BOTTLE MANUFACTURER

The DataMyte 862 Reduces Time Spent Collecting, Storing, and Reporting Measurements

A major manufacturer of plastic bottles for the beverage industry needed a way to reduce the labor intensive task of collecting, storing, and reporting measurements. The measurements help them to keep product quality up and are also required by the companies that they supply.

Problem

The company needed to keep track of several measurements:

- The weight of an empty container
- The volume from the weight of the container
- Concentricity studies on their containers (see Figure 14.13.1)

Fig. 14.13.1 Critical dimensions on a beverage bottle.

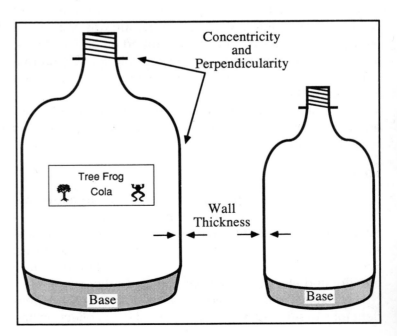

- The wall thickness of the containers

Solution

The multi-file DataMyte 862 is able to record all of the measurements needed. The company received payback in less than three months.

See Chapter 19 for a description of the 862 data collector.

CASE 14-14   INDUSTRIAL CERAMICS

Improved Monitoring of Die Wear Paid for Data Collection Systems in One Month

A manufacturer of industrial ceramics, including ceramic seals, ceramic ball bearings, and ceramic grinding wheels, faced tremendous pressures from off-shore competitors, both on quality and price. By improving the measurement and monitoring of dies they are reducing die change intervals and saving costs.

Problem

The most glaring problem was on the six to eight hundred ton press that takes powdered ceramic material and presses the material into a form. Each die costs up to $10,000. The previous method of predicating die change was with an arbitrary number of 1500 strikes of the press. After 1500 strikes, the $10,000 die was discarded.

Solution

Using the DataMyte 762 data collector along with calipers measuring wall thickness, ID, OD of the final product, they not only retained customer certification, but more importantly gained the ability to predict the change precisely. If they find they can go more than 1500 strikes of the press, they save money by not having to buy a new die so soon. If they find they have to replace a particular die before 1500 strikes of the press, they save money by not making a bad product.

The payback on this system was less than one month due to the frequency of die changes and the ability to precisely predict die changes. In addition, for customers that required certification of specs on the products they ordered, the data was generated as the parts were being manufactured This eliminated the task prior to shipment of generating a report of certification of process which was shipped with the product to the customer.

CASE 14-15 PLASTIC BOTTLE MANUFACTURER

Company Uses Data Collection to Check Bottle Forming Process

A large maker of plastic bottles used for motor oils found that they needed better process controls to meet the demands of the oil companies. By using DataMyte data collectors the company gained better control and quickly reduced the scrap and rework costs.

Problem

The company needed to monitor critical characteristics in the blow molding of plastic bottles. They needed to make bottle neck diameter, thread diameter, height and thickness measurements. They were also checking weights to control the amount of plastic used. Other applications included attribute data such as bottle defects and label positioning.

Solution

After looking at several methods of automatically collecting the data, they determined that a combination of DataMyte 2003, 769, and 862 data collectors could work successfully in their application. Using the DataMyte 2003 they would check bottle defects and thread diameter. The 2003 data collector was also used to audit the accuracy of their vision system for labels positioning. Rejected bottles were studied and causes were entered into the 769. The DataMyte 769 would then provide a Pareto chart listing the most frequent problems, which the company used to make adjustments to their label application machines.

See p. 20-4 for the 769 data collector.

The largest cost savings came from using a DataMyte 862 data collector to monitor weights and volumes. The data collector produced control charts helping them better control the amount of high density resin used in the blow molding machines. The data collector also tracked trends and alerted them in advance of a process problem.

CASE 14-16 ALUMINUM CAN MANUFACTURER

Major Beverage Producer Uses DataMyte Data Collectors for Seam Checks

A major soft drink manufacturer determined that they needed to find a better way to collect data on their can sealing process. When a top is connected to an aluminum can just after filling, it is sealed by wrapping the edges of the can cover around the flange of the can body. This process is known as seaming. The quality checks made on this process are known as the "seam check." The quality assurance manager found the automated computerized seam check systems were too expensive and decided to look at DataMyte data collection systems as a possible solution.

Problem

Their operators measured each characteristic and wrote each data point on a check sheet. With twelve heads on the seamer and nine characteristics to monitor for each head, a minimum of 108 charts were required. Producing that many charts was impractical without the use of automated data collection.

Solution

The company decided to use the DataMyte 2004 data collector. The 2004 was chosen because of its large memory capacity (512k) and special features that would allow all seam check characteristics to be monitored for each head of the seamer. By using the 2004, they not only solved the company's immediate needs but also obtained a system that would be flexible enough for future applications such as attribute data collection and analysis.

The amount of time it takes to do each seam check has been reduced by ten minutes for a total time savings of two hours each day. The process is analyzed by seam head for each characteristic during any specified range of time. For example, a capability report is generated each week for each characteristic to determine if any special attention is needed on a particular head or heads. This same type of analysis is done on a monthly basis. A control chart may be displayed at any time for a particular head and characteristic to determine if a significant change in the process has taken place.

See p. 20-10 for the 2004 data collector.

15. METALWORKING AND MACHINERY INDUSTRIES

CASE 15-1 ALUMINUM EXTRUSION AND FABRICATION MANUFACTURER

Use of Handheld Data Collectors Cuts Scrap, Speeds Up Reaction Time, and Verifies Process Capability

An aluminum extrusion and fabrication manufacturer was practicing SPC by collecting data and plotting charts manually. Manual data collection and charting caused their reaction time to be too slow. Therefore, they had a lot of scrap. The company decided to try automating data collection to increase their speed and cut down on the amount of scrap.

Problem

The manufacturer was measuring several different parts in two areas of the plant. The time that it took to collect all data and then plot the charts severely delayed their reaction time. They needed a fast and accurate method to collect data and plot charts.

Solution

Since the manufacturer already had an SPC program, it was readily apparent what kind of data collection system would work the best. Six DataMyte 762s were selected to be used at operator stations.

See p. 19-6 for the 762 data collector.

In the fabrication area, data is collected on piercing, sawing, CNC milling, and downstream assembly operations. The DataMyte 762 is used with a multiplexer interfaced to specially constructed digital indicators, a micrometer and caliper at each cell.

In the extrusion areas, the DataMyte 762s are used for monitoring critical features during "stretching" operations to provide dimensional certification. Characteristics are also monitored on mill and drill operations in this department. The DataMyte 762s are used with micrometers and calipers. See Figure 15.1.1.

Previously, parts were dimensionally certified using a CNC machine. DataMyte data collectors were able to significantly reduce "verification" time on parts. More accu-

Fig. 15.1.1 Calipers and data collector for checking stretching operations.

racy has been obtained with DataMytes, and faster reaction time has led to a reduction in scrap.

CASE 15-2   MACHINING COMPANY

Use of Handheld Data Collectors Cuts Scrap by 31%

A machining company was having many problems collecting data manually in their milling and shaping areas. The process was too slow. DataMyte data collectors helped them to cut down their reaction time and cut their scrap by 31%.

Problem

In the milling area, there are two operators, one responsible for each line of 18 machines. In the shaper area, there is one operator running one line of 18 machines. Each operator monitors 18 machines for SPC.

In either department, the operator is responsible for bringing parts from each machine hourly to the central gaging station.

The problem with doing this manually was that data collection took too much time and the operator lost productive time charting data rather than running the machine. All data was calculated by hand for control charting and then

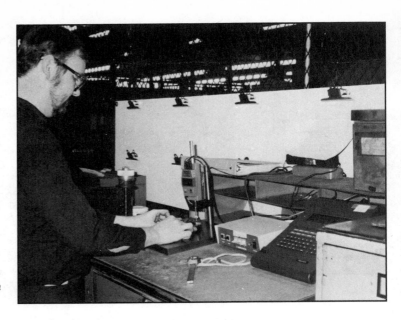

Fig. 15.2.1 Multi-part gaging station using a DataMyte 2003 data collector.

See p. 20-8 for the 2003 data collector.

keypunched again into the computer for capability studies and monthly reports.

Solution

The DataMyte 2003 was selected in each area because of its multi-part capability.

In the milling area, 36 part setup files are stored in the data collector. The DataMyte 2003 is used with a Mitutoyo MUX-10 and interfaced to a Mitutoyo caliper and Mitutoyo 543-423 indicator. See Figure 15.2.1. Every hour the operator pulls four parts from each machine, takes them to the data collector and measures the parts for overall height and tooth width. On-screen graphs provide real-time feedback on process control.

In the shaper area, 18 part setup files are stored in the DataMyte 2003. The data collector is used with a MUX-10 and three Mitutoyo micrometers. Each hour the operator pulls two parts per machine, and uses the micrometers to check pitch diameter. On-screen graphs provide instant feedback using $\bar{x}$ & R charts.

In both departments, hard copy reports are generated weekly off the DataMyte 2003.

A scrap reduction of 31% was realized because the use of DataMyte data collectors provided for faster reaction time to problems. More accuracy was achieved also using automated data collection. Staffing in the SPC department was reduced by 75% by going to automated data collection.

CASE 15-3   SPECIALTY STEEL MANUFACTURER

Combining Data From DataMyte 769 and Lotus 1-2-3

A manufacturer of specialty steel was having problems correlating ultrasonic testing results with production records in order to spot trends and take corrective action. The company is a manufacturer of specialty steel and steel alloys. The highly competitive nature of the world steel industry as well as the high cost of materials involved required that problems with billet quality be tracked back to the source as efficiently as possible.

Problem

Due to the amount of data needed to track billets through melting, casting, cooling and testing, collecting it manually proved cumbersome. The company needed to record furnace numbers, ingot location, billet size, surface condition, operator names, sonic testing equipment used, and sonic test data including type of defect, its location in the rod and severity. Collecting this information on paper and keypunching into Lotus 1-2-3 for correlation with data on furnace temperature was not acceptable.

Solution

The solution was to provide a system that eliminated the need for keypunching by allowing operators to collect the required data efficiently and down load to Lotus 1-2-3. Once collected, it could be readily available to quality personnel for production analysis. The system that solved this problem was the DataMyte 769 and support software.

At the end of a shift data recorded in the 769 is downloaded to Lotus 1-2-3 and combined with furnace temperature data. The capacities of the 769 to accept data on 250 different attributes, interface with software such as Lotus 1-2-3 and dBase III and prompt operators for entries and

See p. 20-4 for the 769 data collector.

dBase made in the 769 an ideal solution.

The DataMyte 769 has allowed this company to collect the type and quantity of data needed to better understand their production process. It is now possible to pinpoint the cause of quality problems as well as provide data to customers on specific products. In addition, data is now available to help determine the costs associated with specific failures.

Choice fields are set up in the 769 for furnace numbers, ingot location, billet size, surface condition, operator names and sonic testing equipment. Attribute fields are used for type of defect, its location in the rod and its severity. Quantity fields are also used for entry of data recorded on shop sheets at the end of a shift. When the shift is over, data from the 769 is transferred to Lotus 1-2-3 and combined with furnace temperatures and data from other sources.

The DataMyte 769 data collection system has permitted this company to combine information from different sources permitting meaningful analysis of relationships among different processes in terms of final product quality. The company has built a data base of quality and production data that can be used to track defect causes and improve quality and productivity.

CASE 15-4   HYDRAULIC VALVE MANUFACTURER

Valve Manufacturer Produces Charts to Send With Parts Shipments Using New System

Hydraulic valve manufacturer had a problem: Although they had a totally automated line using programmable controllers making automatic adjustments, they were producing marginal parts barely in specifications. Due to the fact that each piece was measured in-line, the grinder would make constant adjustments creating tremendous variability. The plant needed a way to monitor process performance and record data for reporting. A DataMyte 762 was selected.

Problem

Every valve which came down the track would be gaged

with in-line transducers. Data would then be sent to a programmable controller which evaluated the information and sent a signal to the machine to make an adjustment. Much of this passed an individual piece inspection rather than statistical control limits of the process.

Solution

A programmable controller was equipped with an RS-232C interface which would allow it to send data to the DataMyte 762. Although each piece is inspected for multiple characteristics, the DataMyte "skip counting" function allows a sample or recording of every hundredth piece, which is five times per hour. Automatically after each subgroup the DataMyte would updates its $\bar{x}$ chart through its graph logging feature. This allowed supervisors to see how the process was running. It became clear that many erratic adjustments were made when the machine should have continued to run. This created a large variation in the process and a low or poor CPK value. The CPK was the main index in which the manufacturer's quality was decided by their customers.

Once a day, data in the DataMyte 762 was transferred to a DataTruck and sent to a PC using the DataMyte FAN II software. Long-term analysis could then be made. This eliminated the need for a final inspection step of taking 50 pieces out of a box and running a capability or CPK analysis (histogram) to be sent with each shipment saving 4-5 man-hours per week. The payback on the system was considered to be 7 months.

See p. 19-6 for the 762 data collector.

CASE 15-5   HYDRAULICS MANUFACTURER

Fifty Percent Labor Savings Gained With New Data Collection System

A leading manufacturer of tube fittings for hydraulic and pneumatic applications required that all SPC data be instantly available to their customers for liability reasons. This division of a billion-dollar corporation is extremely concerned with quality control, since the forces involved in hydraulics and pneumatics can lead to many liability problems. If the product should fail. In order to ensure the best product, the company insisted that all SPC data be instantly available to their QC manager as well as to any current and potential customer.

Problem

Instant data collection was hindered by the fact that the QC manager required this data from two separate plants, located thirteen miles apart. The two plants consisted of 160 separate machining operations, with SPC being conducted at each machining operation. The SPC program collected both variable and attribute data in large amounts, and this data is being keypunched into an IBM AT. In order to make an automated system work, they needed a data collector with networking capabilities for instant data reports, and the system had to be economically feasible.

Solution

See p. 20-8 for the 2003 data collector.

In order to meet the economic needs and greatly reduce the time spent on an SPC program, DataMyte recommended a 2003 data collector capable of recording both attribute and variable data electronically. Since the 2003 also has 512K memory, the unit could be shared among several machining operations. This would reduce both the number of data collectors and gages required to handle both plants.

Using a networking system designed in the plant, the customer was able to hardwire all of the units to the IBM AT through the use of short call modems. Used in conjunction with the DataMyte FAN II software, this system allows the QC manager to collect data from various operations without leaving his office. Data can then be sent by modem to the requesting source, whether within their corporation or from one of their customers.

By using a single DataMyte 2003 to cover four separate operations, the customer saved over $135,000 in gages alone that would normally have been required to accomplish this task. In addition, the customer estimates that 50% labor savings has been achieved over a manual SPC program. By networking the 2003s together, the customer is able to provide instant data whenever it was requested. Because of the abilities of the units, only 40 of the 2003s were required to handle 160 machining operations.

CASE 15-6   SCREW MANUFACTURER

DataMyte System Helps Solve Plant-Wide Scrap Problem

A medium size screw products manufacturer wished to reduce scrap by improving quality on their multiple spindle screw machines. They elected to use SPC techniques.

Problem

Six to eight dimensions were critical to the part. All had to be held to ± .0002 inches. Operators were having to maintain a minimum of six $\bar{x}$ & R charts by hand. Productivity was cut drastically as a result. Yet, at 1000-2000 pieces per hour, scrap could add up very quickly.

The company determined they needed a faster way to get information to the operator. They considered computerized data collection as a possible solution. The company then set certain criteria for evaluating new equipment for SPC.

Solution

Much of the success the company had in solving their problem was due to their careful evaluation of needs. With fifty machines turning out a high volume, they realized that scrap could not be reduced without close control. The machine operators were the key players. An improved SPC program must focus on the operators, providing them with a way to input data and get results quickly. This would help them make good decisions, react faster, and ultimately reduce scrap.

Equipment for their SPC program had to have these features:

- It must be rugged, built for continuous usage in oily, dirty environments.
- In order to gain operator acceptance, it must be simple to use.
- It must have immediate, clear graphic display of process informat right at the machine.
- It must be low-cost, so it is economical for fifty machines.

The company decided to install a DataMyte system, consisting of these components:

- DataMyte 762 data collectors, one for each operator station. The model 762 was chosen, having inputs for up to three gages and capacity for up to 4000 readings. Different gages were needed to measure the various dimensions. 4000-reading memory capacity was needed to store multiple samples of up to eight dimensions.
- Electronic micrometer, bore gages and dial indicators.
- Video monitors for each station.
- A DataTruck to harvest accumulated data from each 762 on a route basis.
- FAN software for their IBM PC, to archive data and do management reporting.

This system satisfied all their criteria, increasing the speed of data collection and analysis, increasing worker productivity with regards to SPC, improving data management and thereby reducing scrap on a plant-wide basis.

In addition, they found that operator training was made easier. All data was recorded in the same manner. Charts on the video monitors provided analysis in uniform fashion. Gage reading and recording error was virtually eliminated, which was especially important when working to such tight tolerances.

All data was recorded and archived in the same format. This allowed management to do cross comparisons on their machinery. Using Pareto analysis, the least capable machine was identified for troubleshooting. Jobs that de-

Operation

What follows is a brief description of SPC activities using the DataMyte system. Each half hour, an operator would sample one piece per spindle and examine six to eight dimensions. The DataMyte 762 at the machine was connected to a micrometer, bore gage, and dial indicator simultaneously, with the micrometer being used to check most of the dimensions. The operator would then touch the $\bar{x}$ key to display the $\bar{x}$ chart, and the R key to display the R chart. When using manual methods, it took 30 to 35 minutes to make a decision. Now decisions took less than ten minutes. Out of control points were flagged with a big "0" on the video monitor. Two times each day, a supervisor came by and dumped accumulated data from the 762 into the DataTruck. The DataTruck was then connected to the IBM PC, and all the data was put on disk for permanent storage.

manded higher precision were shifted to machines that proved to be the most capable at holding tolerances. In this way, management and workers together helped reduce scrap significantly and boost productivity at the same time.

CASE 15-7   RIVET AND FASTENER COMPANY

DataMyte System Boosts Operator Effectiveness in SPC Program.

A manufacturer of rivets, bolts, pins and threaded fasteners has an SPC program in place to monitor their manufacturing processes. The operators take measurements, collect the data, calculate averages and ranges for the subgroups and plot $\bar{x}$ & R charts. One of the critical measurements is the shoulder diameter on an expensive rivet. The diameter must be between 0.5851 and 0.5858 inches.

Problem

The manufacturer wanted to institute an automated SPC system because the operators did not like to collect the data by hand and then spend time plotting the data points on the control charts. The data collection and calculations were viewed as an additional chore instead of an integral part of the manufacturing task; the operator's part in the SPC program was directed too much towards the data collection and control charting. The operators work with more than one machine, so they cannot devote all of their time to monitoring a single part.

The operator had to plot the calculated data points on a paper chart that quickly became soiled and oily because the operator had to handle it every time a data point needed to be added. With charts that were hard to read and plot points on, long-term data storage was out of the question.

Solution

The operators now use electronic digital micrometers and calipers to take measurements. The data is automatically collected by DataMyte 762s. A DataMyte CRT is connected to each 762; all the operator has do is press a button on the

See p. 19-6 for the 762 data collector.

762 to see a control chart. An operator does not have to worry about getting the 762 dirty because the 762 is made to withstand the factory environment, having a sealed membrane keypad and hardened plastic shell.

Operators are glad to use the DataMyte system because it has made their jobs easier. Data collection is a less cumbersome task and they have more time to analyze the control charts — and work on the other machines. The manufacturer is also using the FAN® software instead of oily paper charts to archive the data and track the operations over time.

CASE 15-8   INDUSTRIAL SPRING MANUFACTURER

DataMyte System Increases Capability of SPC Program

"How springy are your springs? A little squishy? Too stiff?" These are the questions that a manufacturer of industrial springs faced. The need for ultra-high precision springs in consumer products, missiles, computers, automobiles, electronic instruments and a host of other applications were growing rapidly. To get the answers to the questions his customers were asking, this manufacturer turned to a Carlson Electronic Digital Spring Tester. See Figure 15.8.1

The tester can be used on either extension or compression springs. Extension springs are loaded by pulling them apart, while compression springs are loaded by pushing them together. See Figure 15.8.2.

Problem

The manufacturer was doing manual charting at the time of introduction to the DataMyte system. The major problem areas with the manual method were:

- Time consuming
- Error prone
- Charts existed on only one piece of paper

The manager was not able to observe the process without picking up the operator's charts, which also meant that the

Fig. 15.8.1 Carlson Spring Tester

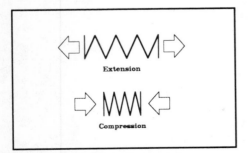

Fig.15.8.2 Spring test characteristics.

chart had to be photocopied and returned at once.

Solution

Since the operators were paid by the hour, it was felt that paying for testing and analysis was a better buy than paying for pencil and paper math errors. The DataMyte 762 with a DataTruck offered some very real advantages.

See p. 19-22 for the DataTruck.

- Instantaneous data recording. The RS-232 output of the spring tester is transmitted directly to the DataMyte 762. No stopping to read the gage and write the value.
- Error free. The data is transmitted directly to the DataMyte 762, which calculates average, sigma, and range automatically.
- Management reports are available at any time. The DataTruck can pick up a copy of all the data in up to twenty 762s and generate statistical reports to a monitor or printer.
- Setup information can be saved in the DataTruck or by FAN® software. No more rummaging around file drawers looking for the chart for the current part. The setup can be down loaded with or without data. The operator can either start a new chart or pick up where he left off.

CASE 15-9   MACHINE SHOP

Shop Finds Faster Data Input Improves SPC Program

A small machine shop has had SPC implemented for three years. All the gaging, data collection and report generation was performed manually by the machine operators. The QC manager would periodically audit the operators by taking samples, manually inputting the data into a portable lap computer and generate statistical graphs for filing.

Many SPC benefits have materialized over the three years, such as a 35 percent reduction in scrap, 80 percent reduction in rework, and every process having a CP index of two or greater.

Problem

The problem that needed to be addressed was how to increase productivity. If they could eliminate manual data collection they could increase the productivity of their operators.

The solution was to find a data collection device that each operator could have at his machine. The device should also produce reports in real-time and have the ability to interface to a computer for storage and analysis.

Some electronic gages could be purchased, but the budget did not allow for a total revamping of their gaging. Therefore, whatever device was implemented, keyboard input of data was a requirement of the system.

Solution

A DataMyte system was purchased for the following reasons:

- The fixed station data collectors could tie into gages, or accept keyboard input.
- Real-time statistical graphs were available for the operators, which would allow for even better control of the process.
- When the Q.C. manager wanted to audit the operators, he could now use a DataTruck to harvest data from the operators' data collectors.
- An IBM PC was purchased for more efficient data report generation and its ability to run FAN® software.
- When the budget allowed for more sophisticated electronic gaging, DataMyte data collectors had the flexibility to connect to several different types and brands of gages. This way, a logical choice of gaging solutions for each process could be addressed.

See p. 19-4 for the 862 data collector.

Each 862 was set up differently, depending on the process at each location. SPC benefits now included increased productivity from the workforce, without all the errors associated with manual data collection/report generation. Instead of hard copy filing, which ate up file cabinet space, the data was stored on floppy disk and could be easily recalled at any time via part number. If changes were required in data collection routines, these changes could be done electronically from the DataTruck or the IBM PC. If customers wanted to see the SPC program at this organization, one trip out onto the shop floor would convince them of an efficient, results orientated system.

CASE 15-10   TURBINE MANUFACTURER

Micrometer and Calipers Used for Critical Measurements on Stator Bars and Field Slot Widths

A large steam turbine-generator manufacturer determined that their present methods for taking critical measurements were too slow and costly. This manufacturer, a very large and well-respected leader in its field, recognized that those areas must be brought under control.

Problem Definition

The problem involved the measurement of the stator bars down the length of a generator (from 100 to 400 feet). These measurements were critical since they were needed to determine the amount of insulation tape required for wrapping around the bars. (If not enough tape was used, a voltage breakdown condition could result; if too much tape was used, clearance problems could be encountered.) To compound the problems inherent with such measurements, consider that seven measurements were required per bar and that anywhere from 36 to 72 bars could be used in a generator.

A similar problem was encountered when measuring field slot widths. Once again, the amount of insulation to be used was the critical factor to be determined by the measurements. The number of measurements required for the field slot widths was also staggering; there could be an average of seven measurements in each layer, with six to eight layers per slot, and up to 30 total slots.

Once all the measurements were taken (for both the stator bars and the field slot widths), the data was keypunched and then fed into a minicomputer to determine the correct amount of insulation. As can be expected, keypunching alone required a lot of time.

Solution

The company recognized that efficient data collection was necessary and began to plan some program improvements. They determined that the program implemented must be able to:

• Significantly reduce data acquisition time.

- Eliminate the time and costs involved in keypunching the measurements.
- Determine the inherent capability of the machining processes by identifying, isolating, and controlling key parameters that affect critical dimensions.

See p. 20-8 for the 2003 data collector.

The equipment selected was the DataMyte 2003 hand-held data collector, interfaced to a digital micrometer and caliper for input. The DataMyte 2003 could record the measurement in an organized, machine-readable form. Further, the data could be sent directly from the data collector to the minicomputer without the need for keypunching.

CASE 15-11 ELECTRONIC CONNECTOR MANUFACTURER

An electronic connector manufacturer found that their data collection system was too slow and didn't allow them to get real-time feed back on their process. This company makes electronic connectors consisting of a mixture of brass and copper.

Fig. 15.11.1 An operator displays an x̄ chart on the DataMyte monitor to review the performance of his machine.

Problem

The companies method of collecting data was too slow and error prone. The contacts come off the screw machine fast enough so that there wasn't time for the operator to inspect the parts and get real-time feed back on the process.

Solution

The company knew that they needed to automate their data collection system. They choose to automate their data collection with the DataMyte 862 data collection system including DataMyte FAN II software for the IBM PC. See Figure 15.11.1. The operators use Mitutoyo micrometers to measure the contacts. The readings for length and width are stored in the data collector, and the operators are able to display charts within minutes.

CASE 15-12   BRASS MILLING OPERATION

Automatic Data Entry Speeds Statistical Reporting.

A large Midwestern brass company had been looking at ways to implement statistical process control on its operations. Many of the company's customers were demanding that the company supply evidence that its materials meet specifications. A commitment was made by management to implement SPC throughout their plant in an effort to control costs and improve overall quality.

Problem

One particular area that had been a continual concern for the company was its rolling mill operation. In this process, rolls of brass plate are reduced in thickness. Large, 1000-5000 foot rolls are compressed from 60 gage thickness to as low as 8 gage utilizing a Z-Mill. The gage thickness is measured by a contact ball micrometer mounted on the Z-Mill. Output from this measuring device is directed to a large, free-standing control panel located 20 feet away. Operators watch the output from the micrometer on a needle gage in the control panel and adjust the Z-Mill as needed. The problem that the company faced was that they had no way of effectively monitoring how consistent the final gage of

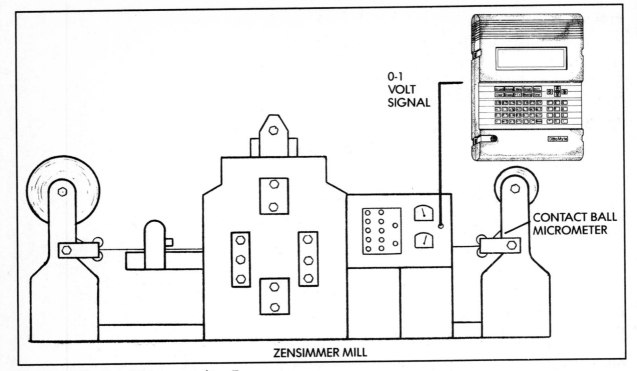

0-1 VOLT SIGNAL

CONTACT BALL MICROMETER

ZENSIMMER MILL

Fig. 15.12.1 DataMyte 2003 connected to a Z-mill.

the brass roll was. Their only permanent record was a strip chart recording of the controller's meter, and this was almost useless since it was usually over 15 feet long and often illegible. It was not possible to identify areas in the roll that fell outside specifications, and so no corrective actions could be taken.

Solution

See p. 20-8 for the 2003 data collector.

The solution to the problem involved the use of a DataMyte 2003. See Figure 15.12.1. The company's engineers provided a 0-1 volt output signal that was tied into the data collector's analog input port. They used the controller's linear metering capability to trigger thickness readings at specified length increments along the full length of the brass roll. In this manner, equally spaced thickness readings could be stored in the data collector.

Each brass roll in the plant is uniquely identified by its heat treat number. Therefore, each roll (heat treat number) is assigned a file in the DataMyte 2003. As the operators load the brass rolls into the Z-Mill, they pull up the correct file in the data collector. As the roll is processed through the

Z-Mill, thickness readings are recorded directly into the data collector. Once roll processing is completed, statistical information is immediately available. Capability reports are generated and made a part of the permanent record. In addition, they produce run charts of the data which are used to indicate areas of the roll that fall outside specification limits. These areas are cut out of the roll and reprocessed.

With the data collector, the company has solved a major collection and analysis problem in a critical area of their operations. They are now able to provide real-time analysis of their data as well as supply useful records to their customers. They have been able to improve their overall quality and to assure their customers that their materials are meeting specifications.

CASE 15-13   STEEL MILL

Pyrometer and Handheld Data Collector Measures Bar Stock Temperature

To eliminate scrap and rework, a large steel manufacturer needed a fast, accurate, and easy-to-use method of measuring the temperature of hot bar stock as it emerged from the furnaces. Since molten steel is too hot to be safely measured by traditional methods, the manufacturer needed a solution to their problem that was based on state-of-the-art technology. By implementing a statistical process control (SPC) program that included the use of a DataMyte 2003 data collector configured with an optical pyrometer, the manufacturer was able to solve the problem.

See p. 20-8 for the 2003 data collector.

Problem Definition

The problem involved measuring the temperature of hot bar stock as it emerged from the furnaces. Any manual method posed an extreme safety hazard to the inspector. Furthermore, a fixed station pyrometer did not offer the mobility necessary to follow the hot steel through its various forming and shaping stages. Without knowing the temperature of the bar stock, statistical process control was limited. This, in turn, resulted in unnecessary scrap and rework.

Solution

They determined that what was needed must be able to:

- Allow the temperature of the bar stock to be measured without posing any safety hazards to the inspectors.
- Accurately measure the temperature of the molten steel.
- Measure temperatures at various locations during the forming and shaping stages.
- Analyze and collate raw data as it was input by the inspectors.
- Interface with the manufacturer's main computing system (in this case, an HP3000 mainframe).
- Generate x̄ & R charts, capability reports, and histograms on an "as needed" basis.

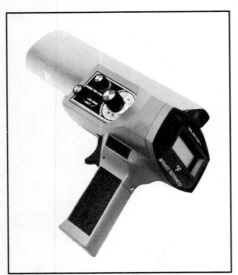

Fig. 15.13.1 Optical pyrometer.

In addition to the above requirements, the manufacturer did not want the system installed to require additional manpower or the use of expensive lab-type equipment.

Based on their analysis, they chose a DataMyte 2003 and a handheld optical pyrometer. See Figure 15.13.1. The optical pyrometer, which is a lightweight, handheld device that measures heat via infra-red reflections from the hot metal, allowed the manufacturer to safely and accurately measure the hot steel at a distance (in this case from 10 to 30 feet). The DataMyte 2003, which was attached directly to the pyrometer, received analog signals from the pyrometer, digitized the signals, and then stored the data into cell memory.

The operator could then go to the shop office, connect the data collector to its printer, and print out the basic x̄ & R charts. The data was also transmitted through an interface cable to a terminal, and on to the mill's central HP3000 for further analysis and permanent storage.

CASE 15-14 CYLINDRICAL CONNECTORS

Manufacturer Reduces Scrap Costs With Improved Monitoring and Data Collection

This cylindrical connectors manufacturer needed to reduce its scrap costs so it could increase its competitive

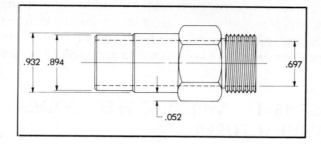

Fig. 15.14.1 Example connector.

edge. The firm was an industry leader but several smaller firms were producing similar products and providing stiff competition. A reduction in waste and scrap costs would let this world-wide supplier of connectors trim its pricing and maintain its level of competition in the industry.

Problem

With several machines producing the same connectors, it was very difficult to pinpoint specific areas where excess waste was generated. Since dimensions were extremely critical on each connector, it was necessary to scrap a connector if its dimensions exceeded specifications. Each machine and machine operator were responsible for maintaining dimensions within specifications.

Solution

Each machine/machine operator combination needed to be monitored independently. A costly option would be to install a fixed station system to monitor each system and the connectors it produced. A better option for this manufacturer was one system which could monitor all the machines and correlate the accumulated data to each machine. Since the company used micrometers and calipers for measurement of critical dimensions, it needed a portable collection process which could collect data from such mechanical devices.

The company found the DataMyte 2003 an excellent tool for its needs. It could interface directly with micrometers and calipers so it would eliminate one step in the data collection process and use the gage itself for automatic data capture. The data collector could be set up to identify the data collected from each machine.

See Chapter 20 for a description of the 2003 data collector.

One operator and one data collector/caliper combination can now monitor each machine independently, determine which machine and operator are having difficulty in producing quality products, identify possible solutions to the problem and help keep scrap to an absolute minimum.

CASE 15-15   WELDING ELECTRODES MANUFACTURER

Better SPC Techniques Help Reduce Scrap

A large Eastern manufacturer of welding electrodes was experiencing difficulty in controlling some critical parameters on its product. The coating on a welding electrode is critical to the integrity of the weld since it is this coating which controls the gaseous envelope that forms around the welding arc. If the coating on the welding electrode is not properly deposited, gas bubbles and slag can become entrapped inside the weld, adversely affecting the integrity of the welded joint.

Problem

Two critical parameters needed to be controlled in order to reduce the amount of scrap being produced by this process. The first parameter was the concentricity of the coating. If the coating was not deposited concentrically to the rod, the arc could burn through, resulting in a faulty weld. This parameter had to be checked when the coating was in a newly deposited, semi-soft condition since there was no practical way to determine concentricity after the coating was baked hard. Also, by identifying defective rods prior to baking, the wire could be cleaned and recycled and the cost of baking a defective rod eliminated. The second parameter was the overall thickness of the coating, which was measured after the rod was baked.

Solution

The company contacted DataMyte for two reasons. First, they were interested in decreasing their scrap rate by more effectively controlling the process. Secondly, they wanted to institute a less time consuming method of producing control charts and other statistical reports which would be provided to their customers.

DataMyte provided a two-fold solution to the problem. Measuring the overall coating thickness on a hard-baked rod was accomplished with a caliper and the DataMyte 862. Monitoring the concentricity of the coating was a more difficult application. Prior to installing the DataMyte system the concentricity was measured by placing the rod in a knife-edged jig and turning the rod until the jig contacted the bare metal. A dial indicator was then placed against the coating as the rod was rotated. The operator observed the high and low readings, subtracted the two and entered the difference. The DataMyte solution was to install a digital indicator which transmitted readings directly to the 862 for a total-indicator runout (TIR) measurement. The 862 records the high and low readings, subtracts the two automatically, and plots the difference on a control chart or histogram.

See p. 19-4 for the 862 data collector.

By installing a DataMyte system, the company was able to have the operator closely control his process and reduce the amount of scrap produced. The time spent in recording readings taken on his samples was cut in half, giving him more time to monitor his process.

CASE 15-16   WIRE ROPE MANUFACTURER

Handheld Data Collector Aids Machine Tool Selection

Buyers of machine tools face a difficult task when evaluating machinery to be used in their plant. Machines can be purchased from a number of international sources; machines that do basically the same thing. A single machine can cost $100,000 or more, and it must operate profitably for a number of years. As age works against the machine tool, more often than not, higher quality and tighter tolerances are demanded of it.

Faced with many choices, but only a few right ones, industrial engineers at a large cable company are using sophisticated methods for machine tool evaluation. To increase the confidence level of a purchase, the company conducted a process capability study at the machine tool maker's shop, using electronic gaging and a portable data collection system.

Problem

The company continually needs to evaluate new machinery for its plants. The machines include extruders, cablers and braiders, all available from a number of sources. To select a tool they use a test production run, and look for the least natural variation in a machine. They employ the methodology of statistical process control (SPC) to find proof that a machine is stable during a production run. This requires measuring critical characteristics continuously during the run and analyzing the data.

Solution

Compared to conducting a capability study in one's own plant, a machine tool evaluation requires using equipment that can be carried around. Still, the equipment must provide accurate repeatable analysis, so that test runs on similar machines in different locations can be compared. The company had been using hand micrometers and handwritten methods of recording data. They have since adopted the use of a handheld data collection system.

See Chapter 20 for a description of the 2003 data collector.

The data collection system, made by DataMyte Corporation, consists of a DataMyte 2003, an Epson printer, and gages having data output capabilities. Gage readings are automatically recorded in the data collector, allowing an engineer to concentrate on taking accurate readings without having to stop to record them. Data from gages that cannot be interfaced are entered on the built-in keyboard.

When connected to the Epson printer, the data collector provides instant hard copy results. The test run documentation, can include a real-time analysis of operating level, dispersion, trends and required adjustments ($\bar{x}$ & R chart), and a graph showing the distribution of all data for each characteristic such as Cp, CpK and estimated percent out of specification and histogram (capability report).

When evaluating a machine such as a cabler, which makes braided cable wire, several characteristics are monitored. The diameter of the cable coming off the machine is sampled continuously, with least variation the criterion. To do this, a LaserMike optical micrometer is positioned in-line with the machine. An RS-232C output cable connects the LaserMike to the data collector, and readings are recorded at various intervals.

Another characteristic is the lay of the wire, which is the distance between braids. This is measured by hand, using a Fowler Max-Cal caliper, which transmits the readings to the data collector. A tension meter is used to measure let off tension, which is being maintained at each wire spool feeding the cabler. Other characteristics are uniform flyer rpm, taken directly from the machine readout, and bearing temperature. Bearing temperature is expected to rise from a cold state and then level off. The readings can be taken with a handheld pyrometer, and plotted in the form of an $\bar{x}$ & R chart.

The company finds an evaluation against a criterion, such as least normal variation, easier with the use of direct connect gaging and a handheld statistical computer. The large amount of data needed for a capability study is obtained more quickly with no loss of accuracy. Getting hard copy results instantly gives the evaluation engineer the opportunity to share results with the machine tool maker. Evaluations are more fair.

CASE 15-17   FOUNDRY OPERATION

Company Uses Attribute Data Collector for Scrap Reporting

A foundry dealing in small-to medium-size castings wanted the ability to categorize the scrap that was being produced in their plant. They also wanted the ability to detect which part, what casting and which cavity of a particular mold the scrap came from. The system had to be used on the floor in an extremely hostile environment and be easy enough to use by non-technical people.

Problem

The major problem they were experiencing was an extremely high scrap rate and they had no way of determining where the scrap was coming from or when the scrap was actually cast. By automating the system they hoped to document all the casting parameters, including the part number, cavity, cast date, and the number of defects on any one casting.

Solution

The approach they decided to take was to use a DataMyte 769 data collection system. This system would provide the ability to collect all the required information, give real-time feedback to floor supervisors and also long-term data storage on the computer.

With the large amount of information that was processed every day they determined their best solution for software would be to use one of the popular data base programs available on the market and use that for archiving purposes.

Over the course of a three month period they were able to reduce their scrap by 30% with the use of this system.

See Chapter 20 for a description of the 769 data collector.

CASE 15-18   PRESSURE VALVE MANUFACTURER

Company Catches Assembly Defects in Valves

A manufacturer of complex pressure valve assemblies wanted to perform an outgoing audit to ensure proper assembly of various valve assemblies. The way they were currently doing it was slow and not very efficient. A lot of errors showed up in the data that were errors due to data collection and not real assembly errors. So they couldn't distinguish between assembly errors and operator recording errors.

Problem

This particular plant manufactured approximately 200 different valve assemblies. The need for this system was twofold. One for internal defect analysis and also to provide a report that could be supplied to the customer on an outgoing audit basis.

Solution

To help collect and analyze the large amount of data, they started using a DataMyte 2000 data collector. Each inspection procedure was recorded in the data collector and stored under a file which referred to the actual part number. A directory of all the files was printed out so that the final inspector could call up the appropriate file, perform the required set of inspection points and then print the information out for hard copy. The reports were attached to the shipment and sent out with the proper valve assemblies.

By automating the final inspection sequence they were able to catch defective assemblies before they left the manufacturing facility.

CASE 15-19   STEEL COMPANY

Company Controls Costs By Holding Tolerances to Two Percent of Nominal

A steel company wanted to improve the methods used to collect and chart the measurements being used to control the rolling process in production of steel beams. By automating the analysis of steel beam dimensions, the company could avoid "giving away" too much steel.

Problem

As steel beams are produced, a one-foot section is cut from an end and used to calculate the footweight. This involves taking dimensional measurements from the flanges and web of the beam section, averaging them out and multiplying this result by a known density factor. The calculations are made separately for the flange length, flange thickness and web thickness. The individual results are then added up to produce a footweight, the weight of a one-foot section of a steel beam. Due to the number of measurements and the calculations involved, the company wanted to automate this function, which would allow them to react more quickly to size variations. Specifications call for the finished beam to fall within ± 2% of expected footweight. The company attempted to stay within ± 2% of tolerance.

Solution

A DataMyte 2000 was chosen to solve this problem. The measurements and attendant calculations needed to generate footweights needed a file consisting of 12 items which the 2000 could handle. The footnote capability also allowed them to record the heat number (lot no.) and tag it to the measurement data for identification. The DataMyte 2000 could also store the file setups for many different foot-

See p. 20-6 for the 2000 series data collectors.

weights, since the rolling mill operators may produce many different beam sizes in one shift or in one week. Because of the concern for staying within ± 0-2% of nominal size, an additional item was added to each file to look at the final footweight calculation as a percentage of nominal (100%).

By being able to easily calculate footweights, the company was able to get this value produced on a constant basis. Knowing quickly what the mill is producing allows the operators to control the process to stay close to 2% over nominal and avoiding undersize beams as well as beams significantly larger than ± 2%.

The dollar savings are projected to be significant if beams can be held to PM2% over nominal. They can thus avoid "giving away" one or two percent or more of their production.

CASE 15-20 MACHINING SERVICES SUPPLIERS

General Motors Supplier Achieves Certification Status Using DataMyte Data Collectors

A supplier to General Motors and Hydra-Matic needed to supply both of them with prove of their quality to be certified. The company was already practicing SPC but the methods they used weren't effective enough to get all the data processed to meet the certification requirements.

Problem

The company needed to automate their data collection system so that they could process the data to meet GM's certification requirements. Hundreds of tests are done on the products from the time they are made until they are shipped. Each of those tests must be documented. It was nearly impossible to keep the paperwork on those tests.

Also, the company was finding dimensional problems on the parts but after the products were manufactured since it took an operator 20 minutes each hour to produce a chart.

Solution

The company decided to automate their data collection system using DataMyte data collectors. See Figure 15.20.1.

Their reject rate for GM products has dropped from above the 50 parts per million standard down to 13 parts per million. And they are able to prove their quality, an act essential to the certification status that they have with Hydra-Matic and General Motors.

CASE 15-21   PRECISION SPRING MANUFACTURER

A Spring Manufacturer Found That Using DataMyte Data Collectors Made Inspection Three Times Faster

A precision spring manufacturer needed to prove themselves and their customers that they were making quality parts. They were using a lot of pencils and doing a lot of arithmetic to prove their quality.

Problem

This spring manufacturer spent far too much time calculating their quality. And their was a time delay between the inspection and the process. The company was also aware that every time someone handled data, they could make mistakes.

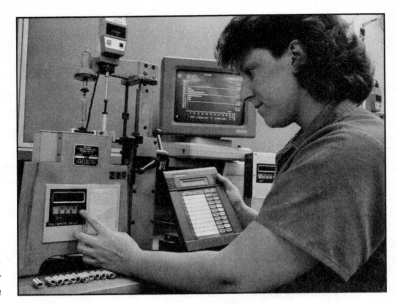

Fig. 15.21.1 A technician uses a Carlson Spring Tester and a DataMyte data collector to measure the tension and compression of a spring.

Solution

The company decided to automate their data collection system. They purchased several DataMyte data collection systems. See Figure 15.21.1. The company ran an experiment to see how much time the data collectors could save them. They found that the inspector using the DataMyte data collectors could check three times as many parts as he could using their manual method. DataMyte data collectors have allowed this spring manufacturer to be faster, get real-time feedback and not worry about errors slipping in when they handle data.

16. TRANSPORTATION INDUSTRY

CASE 16-1   STAMPING PLANT

Critical Dimensions Checked on Vehicle Vent Windows

An automotive stamping plant that provides metal stampings and fabricated parts to their assembly division found that manual methods for data acquisition were too slow and, more importantly, unreliable. It was critical that the vent windows be uniform in nature. This manufacturer soon determined that what they really needed was to enhance their data acquisition capabilities.

Problem Definition

In the stamping process, vent windows go through seven different operations, starting with the initial stamping operation to final inspection and shipment. At each stage of the process, data must be gathered on critical characteristics of the metal frame, such as length, width, contour, and sweep. A checking fixture with marked measurement points was used. See Figure 16.1.1. The measurements would then be recorded on a sheet of paper. After recording a number of measurements, the data would then be sent to another department where it would be analyzed (provided, of course, that there was enough time to perform the analysis). The problem faced by this manufacturer was directly related to the sheer number of measurements that had to be made and the time required to process the raw data into usable chart form.

Solution

Rather than cut back on the characteristics which were measured, which they felt they could not do, they looked at ways to speed up data collection. They determined that the program implemented must be able to:

- Allow data to be collected automatically.
- Create information files.
- Process the data collected.
- Generate charts and reports.

- Reduce the number of man-hours required to collect, analyze, and process the data.

Based on their analysis, they chose a DataMyte 2003 data collector and gap gages. The DataMyte provided a data collection capability that was portable, user friendly, and could be used to display the necessary charts and reports whenever they were needed.

Once the DataMyte was implemented as part of the SPC program, the manufacturer's roving auditors were able to automatically capture data with an electronic gap gage at each step of the vent window manufacturing process. Immediately following data collection, statistical summaries were displayed on the LCD display. When desired, the auditors could simply press the graph key on the DataMyte to display the required charts and capability reports.

Use of the DataMyte in this application had several positive consequences. First and foremost, it increased the validity, and therefore the believability of the data. Since data was captured automatically with a transducerized measurement probe and stored in DataMyte's solid state memory, transcription errors were completely eliminated. There was always a rush when data was collected manually and sometimes an auditor would massage the data to avoid the paperwork generated when a lot of units were out of spec. This problem was also eliminated, thereby increasing data validity. Secondly, the data being captured by the DataMyte is now being utilized fully. Prior to using a DataMyte, the data was seldom analyzed. This was due to the sheer volume generated and the lack of staff to manually summarize it.

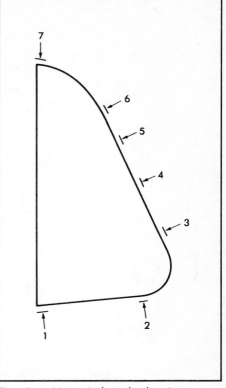

Fig. 16.1.1 Vent window check points.

See p. 20-8 for the 2003 data collector.

CASE 16-2   FINAL ASSEMBLY

Margin and Flush Checks Made During Final Assembly

A leading automobile manufacturer needed a data collection and analysis device that was capable of performing critical margin and flushness measurements at several locations on the perimeter of the hood, fenders, doors, quarter panels, deck lid, cowl and package tray.

Problem Definition

To form the body of the finished vehicle, a number of sheet metal parts and assemblies must be joined together during assembly. Each part is subject to dimensional variations which affect the outcome of the final assembly. This is also true of the process used to join the parts. To eliminate the necessity of selectively mating parts for fit during assembly, the dimensional variation of the sheet metal parts must be controlled during the stamping process.

Assuming that there is a minimum of piece-to-piece variation in the sheet metal parts, the "build-up" process, in which the assembly of those parts takes place, is controlled by measuring their fit after assembly. The measurements performed are critical to the overall process of producing an automobile.

Solution

The company chose a model 516 gap and flushness probe, used in conjunction with the DataMyte 2003 data collector. One end of the probe had a set of fingers, that when inserted in a gap and separated, would measure the distance. Both curved and straight margins could be measured. The other end of the probe had a shoe and movable rod to measure flushness (see Figure 16.2.1)

Using the data collector and probe, the inspectors could take a series of measurements on each vehicle in a sample group. The measurements were taken at predetermined check points on the finished vehicle. Margin and flushness were recorded at several locations on the perimeter of the hood, fenders, doors, quarter panel, deck lid, cowl and package tray.

Several unit samples (typically 25 to 30 per inspection) were taken twice each shift. As the measurements were made with the gap and flushness probe, the data was fed directly into the data collector. When the inspection was complete, the data collector was connected to a desktop computer and the data transferred.

Once the data was transferred to the computer, it was analyzed using DataMyte FAN II software. Once the data was analyzed, a capability analysis report was generated. See Figure 16.2.2. This report was reviewed immediately by the plant's Quality Control and Process Engineering personnel to determine the overall operating conditions of the

See Chapters 19 and 20 for data collectors and Chapter 21 for gap gages.

assembly process. If a problem condition in the fit of any body panel was identified, the quality control function responsible for that part was immediately notified and a corrective action initiated.

Case 16-3   PAINT AUDITING

Computerized Paint Audit System Creates Graphic Reports

The U.S. manufacturing facility of a Japanese automobile manufacturer wanted to put their entire paint audit program onto a computer system. Computerizing the program would provide both graphic reports and faster turnaround of the data. Summarized data could be used to alert the front end of the paint line of the current status of defects occurring at the end of the paint line. With the information, adjustments could be made immediately to correct the problem.

```
            DataMyte
     FAN II Support System
   Model 1049-05  Version 4.2
       Capability Report
Electronic Dimensional Audit System


07/10/89 14:26
```

Capability Report

```
USL ----------------------------------------------------

CL  ----------------------------------------------------

LSL ----------------------------------------------------
     POINT 01          POINT 03          POINT 05          POINT 07
          POINT 02          POINT 04          POINT 06
```

Unit ID: L!H-DOOR

Item ID	POINT 01	POINT 02	POINT 03	POINT 04	POINT 05	POINT 06	POINT 07
Nominal..........	N/A	N/A	N/A	N/A	N/A	N/A	N/A
Eng Hi/+Tol.......	5.5000	5.5000	5.5000	8.0000	8.0000	7.0000	7.0000
Eng Lo/-Tol.......	3.0000	3.0000	3.0000	3.0000	3.0000	4.5000	4.5000
Mean.............	3.9813	4.9950	4.0431	3.7588	4.1606	5.6450	5.6213
Range............	2.1800	2.8000	1.4000	3.5000	2.3000	2.9000	2.5100
Sigma............	0.4103	0.5222	0.3053	0.8020	0.4313	0.5317	0.4612
6 sigma..........	2.4617	3.1330	1.8316	4.8121	2.5878	3.1904	2.7669
Mean + 3 sigma....	5.2121	6.5615	4.9589	6.1648	5.4545	7.2402	7.0047
Mean - 3 sigma....	2.7504	3.4285	3.1273	1.3527	2.8667	4.0498	4.2378
Start Date........	07/10/89	07/10/89	07/10/89	07/10/89	07/10/89	07/10/89	07/10/89
Start Time........	14:51	14:52	14:53	15:09	15:09	15:10	15:11
End Date..........	07/10/89	07/10/89	07/10/89	07/10/89	07/10/89	07/10/89	07/10/89
End Time..........	15:15	15:15	15:15	15:16	15:16	15:16	15:16
Cp(3).............	1.0156	0.7979	1.3650	1.0391	1.9321	0.7836	0.9035
Cr(3).............	0.9847	1.2532	0.7326	0.9624	0.5176	1.2761	1.1068
Cpk(3)............	0.7972	0.3224	1.1391	0.3154	0.8970	0.7178	0.8105
Zmax/3............	1.2339	1.2735	1.5909	1.7628	2.9673	0.8494	0.9966
Estimated % Out...	0.8530	16.6087	0.0326	17.1066	0.3581	2.1160	0.8951
Samples..........	16	16	16	16	16	16	16

Fig. 16.2.2 Example EDAS report.

Problem

One of the first problems they had to overcome was the large amount of data that had to be collected over a relatively short period of time. The data was composed of:

COLOR
BODY STYLE
PANEL
SPRAYBOOTH
DEFECT TYPE

The intent was to have one inspector on each side of the vehicle recording defects. Unfortunately, the line speed was such that the operator would soon lag behind and therefore was not able to perform a thorough inspection. They began a search for a data collector that would allow fast data entry and still offer the portability required in a large paint department.

Solution

They settled on the DataMyte 769 because bar code entry would provide the fastest means of inputting the data. The 769 data collector attached to a printer automatically generates a bar code data collection sheet. The bar codes were simply three digit numbers, with a human readable description printed below each code.

The 769 data collector could then display charts and statistics on a monitor that included the following information:

Total number of defects
Average defects/vehicle
Total dirt defects
Average dirt/unit
Average mottle defects
Top three defects
and Pareto charts based on:
Color
Side of vehicle
Spraybooth
etc.

The implementation of this system resulted in a reduction of manpower in repair of defects, reduced warranty costs,

See Chapter 20 for a description of attributes data collectors.

reduced man-hours in repair of defects, reduced man-hours in data analysis, and better customer perception of their products.

CASE 16-4   CAR BODY PLANT

Adding SQC to Incoming Inspection Eliminates Faulty Moldings

A car body plant in Detroit discovered a great deal of variability in the linear dimensions of window channel moldings produced by outside suppliers. The plant manufactures decorative trim moldings, sunroofs and other trim parts for use in many of the car company's auto and truck assembly plants. These moldings have a significant effect on the appearance of a vehicle. For many years the plant and suppliers checked the linear dimension of incoming moldings in the traditional manner, but this was not effective in culling out all faulty moldings prior to installation.

Problem

During the assembly of sunroofs, there were problems with the channel moldings that formed the protective edge of the sunroof. These moldings, produced by outside suppliers, exhibited a high degree of variability in length. When the molding was too long, there was an unsightly gap on the edge of the sunroof. When moldings were too short, there was a gap where adjacent moldings came together.

Both the plant and the supplier inspected the molding dimensions in the traditional manner, using a "GO" or "NO GO" checking fixture, but it was not possible to check each molding prior to installation on the glass. As a result, it was a normal daily occurrence to have sunroofs rejected for rework because of problems with channel moldings. An additional hourly person was often employed on a full time basis just to inspect and rework the sunroofs after they were assembled because of moldings being too short or too long.

This situation existed in the plant for several years and there seemed to be no clear cut method to monitor and accurately control the length of these channel moldings from vendors. The costs of this problem were considered excessive.

Solution

Several solutions, ranging from changing vendors to 100% incoming inspection, were considered. The practical solution appeared to be the proper utilization of statistical quality control (SQC) procedures for incoming inspection. It was hoped that this approach could demonstrate to suppliers that some of their production procedures were not in statistical control. A planned program was developed and implemented utilizing available plant personnel.

Quality control personnel used DataMyte 2003 data collectors and 514 gages in a checking fixture to measure and record deviations in lengths on incoming channel moldings.

See Chapters 19 and 20 for data collectors and Chapter 21 for the 514 gap gage.

The first control charts indicated that the length dimension from vendors was out of statistical control in several cases. The vendor's quality control personnel were called in and most were surprised by the findings. Most felt that their production could control the linear dimension. Soon, some of the vendors began to monitor their length dimension with their own DataMytes. As a result, their moldings became less variable and began to center on a nominal value. This reduced scrap and rework for the body plant, as well as reduced the need for continued incoming inspection.

The implementation of SQC procedures during incoming inspection at the body plant has contributed to an improvement in the overall quality of channel moldings used in the assembly of sunroofs. The sunroof assembly department has not been forced to rework a sunroof due to faulty channel moldings since shortly after the SQC program was initiated.

Operation

Because of the relatively simple operation of the data collector, it was decided that the best way to implement an SQC system in incoming inspection was to involve the hourly inspection workers. They had the ability to plot and monitor control charts as well as act on the information produced by the charts. An hourly employee was trained in SQC calculations and control chart plotting and then he helped train other hourly workers.

Incoming inspection personnel take samples out of every shipping container. These samples are placed on the checking fixture and measured with a 514 gage for linear dimension. The data are collected using a DataMyte 2003. Each inspector follows instructions kept by the DataMyte to record these measurements. The data collector is set up for 1 item by 10 samples per subgroup.

The inspectors generally take 10 moldings out of each shipping container. Then, he prints a $\bar{x}$ chart and histogram.

CASE 16-5   STAMPING PLANT

Manufacturer Cuts Waste in Bar Stock Blanking Stage

An automotive stamping plant that specializes in the manufacture of decorative trim moldings found that traditional methods of checking linear dimensions was causing excessive scrap and rework. Since the moldings have an impact on the appearance of a car or truck, the company felt they contributed to the overall image, good or bad, of the entire corporation. The ability to manufacture moldings of high quality was therefore very important.

In their efforts to improve the overall quality of their trim moldings, this manufacturer decided to improve their statistical quality control (SQC) program. As by-products of improving the SQC program, the manufacturer also wanted to be able to increase productivity and to increase customer awareness of their manufacturing capabilities.

Problem Definition

In the past, moldings produced by this manufacturer had exhibited a higher level of variability than what was considered acceptable. One process that was very critical to the ultimate quality of the moldings involved the cutting of metal bar stock to a specific length. This process, which needed to be performed often, required a high degree of accuracy. This cut was accomplished on a rolling mill during an initial forming process. For many years, this dimension was monitored in the traditional manner by using a "GO" or "NO GO" checking fixture. If the linear dimension couldn't be maintained within a tolerance of plus or minus 1.5 mm, problems would arise during later production processes.

As a result of not having a clear-cut method to monitor and accurately control the length of the trim blanks, and since traditional measuring methods were being used, it was considered a normal occurrence to have a basket of parts (1,500 to 6,000 pieces) returned each day for rework or disposal as scrap. In fact, an additional hourly person was often employed on a full time basis just to rework the materials that had been cut too long. The short pieces normally had to be scrapped. It was sometimes even necessary to have a mill in operation just to rework the defective oversize parts. Actual cost figures for rework and other expenses resulting from their problems were not available; however, the company did consider them to be excessive. It was thought that this situation had existed for fifteen to twenty years.

Solution

Several solutions were considered, ranging from acquiring new rolling mills, to increasing the production by using outside vendors. It was felt that tightening the engineering specifications would not solve the problem because of the inability to maintain the existing tolerances. A more practical solution seemed to be the proper utilization of statistical quality control procedures.

A planned program was developed and implemented using available plant personnel. They purchased a DataMyte 2003 data collector and a 514 gap gage to measure and record deviation from nominal cut lengths. The data was plotted on $\bar{x}$ & R and sigma charts to identify the source and magnitude of variation.

See Chapters 19 and 20 for data collectors and Chapter 21 for the 514 gap gages.

The first control charts generated indicated that the cut-off process on the rolling mills was out of statistical control. Further analysis indicated that there were at least two assignable causes for this variability.

First, the operators were "tweaking" the length setting on the mills. That is, they would make a length adjustment whenever they had a sample that did not check as a "GO" on the checking fixture. To counteract this problem, the operators were instructed to change the cut-off setting only when it was indicated by the control chart.

Second, the mechanically actuated cut-off blades on the mills were not operating in a consistent manner. Due to wear, the setting would drift over a period of time. The company made plans to replace these blades with pneumatic cut-off blades that would not drift.

Shortly after implementing the SQC program using the data collectors, the length dimension became less variable and began to center on a nominal value. This helped reduce scrap. Instead of receiving a daily basket of parts for rework or scrapping, the rolling mill department had returned only one basket of materials during the first six months (this basket was later discovered to have been from a mill whose operator had neglected to keep a control chart for his machine that day). Other benefits realized by using the data collector included:

- An increase in the production rate, due in part to the decreased operator time spent on inspection and machine adjustments.
- Increases in linear dimension performance that included: a decrease in variability ($\pm 3\sigma$ and $\bar{x}$ in control); achieving a capability ratio of less than .75; and a return to processes centered on nominal values.
- Increases in productivity gains that included: reduced machine downtime, reduced labor for sorting out defects, reduced need for inspections, elimination of extra labor for rework, almost complete elimination of scrap, and improved relations and image with customers.

The use of data collectors in the rolling mill area of the plant will be expanded in the near future so that each operator will have his own unit to enable more frequent sampling. The success at this company illustrates the role hourly workers can play in helping to build quality products in any manufacturing environment.

CASE 16-6 AUTO MANUFACTURER

Dynamic Torque Measurement Reduces Rework

A large automobile manufacturer needed a method for measuring dynamic torque. As an auditing procedure, this manufacturer had monitored static torque for some time; however, they felt by auditing dynamic torque they would be able to better analyze the fastening process, and reduce rework.

Problem Definition

The problem faced by this manufacturer involved the measurement of dynamic torque. Their engineers felt that by auditing dynamic torque, they would be able to more fully understand what happened to a fastener as it was being torqued down. Dynamic torque measurement could capture data related to the fastener itself, the tool being used, and the force being applied by the operator.

The problem involved not having a versatile method of

hooking up a tool to the data collector that would allow dynamic torque to be measured.

Solution

It was determined that the best solution would be to place in-line torque transducers on the assembly tools and set a DataMyte 2003 data collector to read the peak torque signal generated. See Figure 16.6.1. In addition to being able to measure dynamic torque, other benefits realized by using the DataMyte 2003 included:

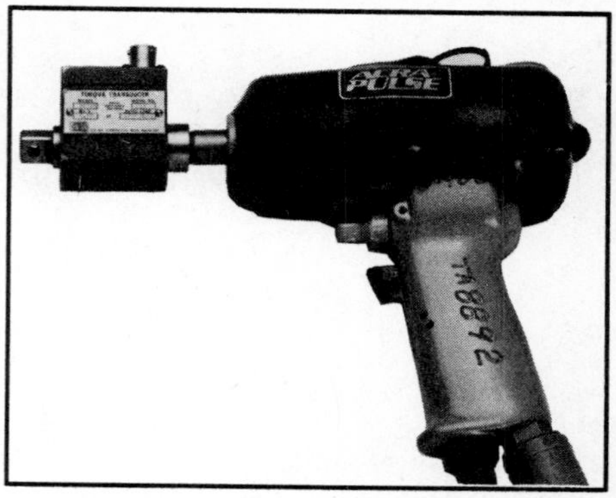

- A significant reduction in data acquisition time.
- Elimination of the time and costs involved in keypunching the measurements.
- The ability to have charts and capability reports generated on demand.
- Added flexibility. The data collectors could be used to measure both static and dynamic torque on a variety of fasteners.

By using the DataMyte 2003 for data collection purposes, the manufacturer was able to implement a program that provided them with the ability to perform dynamic torque measurements.

The data collector was flexible and handheld, which provided advantages over most torque data collection devices. The air stall tools they used required no special settings, other than the installation of in-line transducers.

Fig. 16.6.1 In-line torque transducer mounted on a torque tool.

See Chapters 19 and 20 for data collectors and Chapter 21 for torque tools.

CASE 16-7 OUTBOARD MARINE ENGINE MANUFACTURER

Capability Studies Improve Head Cover Torque Reliability

A midwestern marine engine manufacturer was using handheld dial indicator type torque wrenches to set final torque on the six fasteners holding cylinder head covers. This was a time consuming process, but these fasteners were critical to the operation of the engine. Although air driven nut runners were used to run down the nuts prior to their final torque, the requirements of virtual 100 percent compliance to specifications seemed to preclude their use for final torque setting.

Problem

Air driven, torque-controlled nut runners were available, and the quality control manager felt that significant savings could be realized by using them for final torque setting. The problem seemed to be with how to make sure torque specifications were being met. The solution seemed to lie in implementing a statistical monitoring program to ensure the highest quality was maintained at the highest confidence level.

Solution

A two part program was proposed. First, capability studies would be performed to establish the capability of both hand torquing and air driving the six critical fasteners. These studies were to be performed with static or breakaway type torque checks. Secondly, if as expected the air driver was more capable than hand torquing, a monitoring system based on dynamic torque would ensure continuing adherence to specifications.

The capability study, performed with the DataMyte 2003, showed that the air gun was far more capable than torquing by hand. See Figure 16.7.1. The study also showed a strong tendency of the manual method to over-torque the fasteners.

The second part of the program was the daily monitoring of the air driver. Five samples were taken daily from an in-line torque transducer. These readings were recorded as the fastener was torqued. The gun appeared to have more variation than expected. An alert operator noticed that although the cadmium plated fasteners were purchased to the same specification, some batches were bright and shiny and some were dull and blackened. The operator also observed that there was a consistent 10 percent variation between torque readings that was dependent on fastener finish.

Having determined that a large part of the variability in the process was caused by raw materials, the Q.C. manager alerted engineering, purchasing, and incoming inspection to the problem. A close examination of the fasteners showed that both types met the specification as written! At

See Chapter 20 for a description of the 2003 data collector.

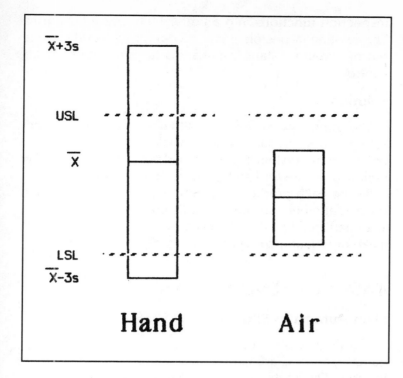

Fig. 16.7.1 Distribution of torque measurements made by hand as compared to those made by a torque controlled nutrunner.

this point engineering issued tighter specifications to ensure uniform fastener procurement.

CASE 16-8   AUTOMOTIVE GLASS DIVISION

DataMyte 769 Data Collector Helps Automotive Glass Division Maintain Quality and Delivery Schedules With Automated Attributes Inspection

Divisions of major automotive manufacturers are moving toward just-in-time production, with their assembly plants as customers. By using in-process attributes inspection systems, an auto glass division maintained close control over quality, which kept their production on schedule.

Problem

Glass inspection requires detailed visual inspection and rapid attributes analysis to provide advance warning of process problems. The company needed a way to turn the inspection functions into a better feedback mechanism. The existing inspectors made accept and reject decisions, but provided no statistical analysis for preventive quality control.

Solution

The company decided to install a data base-like attributes system out in the factory floor. They chose the DataMyte 769 attributes system because it had fast bar code entry and built-in software for Pareto and defects analysis.

Before each product was shipped, operators identified the part number, job number, and defects if they found any. The charts produced by the DataMyte gave the feedback to make production corrections on the fly.

CASE 16-9 ENGINE PLANT

Plant Automates SPC on Crank Shaft Cam TIR

The engine plant of a major automotive company required control charting on critical crank shaft dimensions. By using DataMyte 761 fixed station data collectors, they improved measurement accuracy and could do trend analysis automatically.

Problem

The dimensions on the crank shifts included two readings per cam, a taper reading and a total indicator reading (TIR). The TIR reading involved an operator rotating the shaft in a fixture, and using his own judgement to determine the high and low spots. The operator would then have to write down both high and low readings and subtract the two to get the TIR.

Solution

See p. 19-10 for the 761 data collector.

Using a DataMyte 761 data collector, they could use a single instrument to record all readings. For the TIR reading, the operator had to simply rotate the part in the fixture once and the DataMyte would read and calculate TIR at all five points. See Figure 16.9.1. In addition, runs and trends were tracked, and the operator was alerted to them and asked for an assignable cause. Data collection time was reduced by a factor of ten.

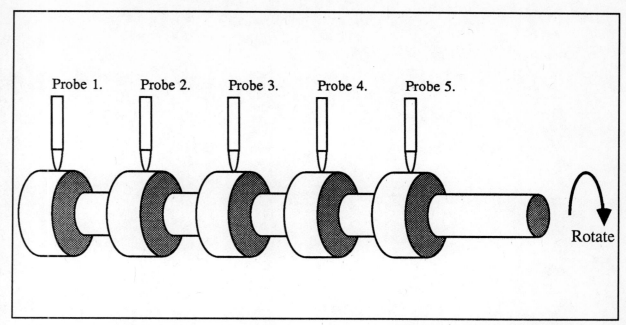

Fig. 16.9.1 Automatic TIR calculation.

PART III Products

17. COMPUTERIZED DATA COLLECTION FOR SPC

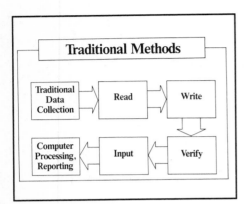

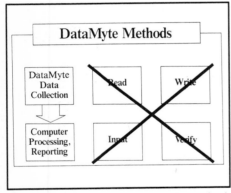

Fig.17.1.1 *Traditional handwritten methods of data collection compared to DataMyte methods.*

See p. 19-22 for the DataTruck.

17.1 INTRODUCTION

The DataMyte FAN® (Factory Area Network) system is a totally new concept for statistical quality control. It makes SPC work on a plant-wide scale. Data collection is rapid, accurate and automatic (see Figure 17.1.1). The analysis is in "real time," "on-line" rather than later. At each point in the process:

- Operators enter their own data.
- Readings are stored.
- Statistical computations are made.
- Analysis is immediate, giving operators the ability to control the process.

The FAN® system consists of:

- Gaging with electronic output (see Chapter 21)
- Data collectors (see Chapters 18 through 20)
- Special software for computers
- Training (see Chapter 22)

At each operator station, gages are used to measure process characteristics. Data collectors are used to record the data. The data collectors provide operators with control charts and other analysis to help them monitor the process. See Figure 17.1.2. Data that accumulates at each station is periodically picked up and transmitted to an IBM PC or other type of computer. At the computer, the data can be archived and analyzed further.

The FAN system enables communication between workers and management. The FAN system can be an "unwired" network, using a DataTruck® handheld data collector. Once a day, or as often as needed, the DataTruck can be taken on a route, stopping at each operator station in the plant (see Figure 17.1.3). The DataTruck plugs into and "harvests" the data from each data collector. It stores the data from each data collector in a special file. It can also program a data collector for a new operation. Up to twenty data collectors can be serviced in this way on a single route. At the end of the route, the DataTruck is connected to the IBM computer, and all the data is transmitted to the computer, where it can be stored on floppy disks.

This is one type of FAN system. The FAN system can include both fixed station data collectors dedicated to monitoring one part or process, and handheld data collectors for roving auditors who monitor many parts. A fixed station fully-wired FAN system can also be installed. See Chapters 18 through 20 for more information.

This chapter describes the FAN system, and introduces DataMyte products that are part of it. The system is quite simple. It shares these concepts with SPC itself:

- Fast and error-free data collection
- Worker responsibility for data collection
- Top-down and bottom-up communications

Fig. 17.1.2 DataMyte data collector and video monitor used at a machine tool.

Fig. 17.1.3 A QC supervisor uses a DataTruck data collector to "harvest" data from data collectors in the plant. At the end of his route, the DataTruck connects to a computer to store the data.

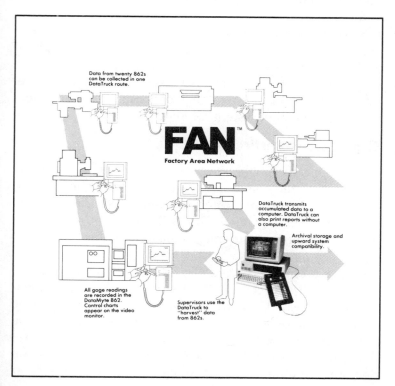

Data from twenty 862s can be collected in one DataTruck route.

FAN™
Factory Area Network

DataTruck transmits accumulated data to a computer. DataTruck can also print reports without a computer.

Archival storage and upward system compatibility.

All gage readings are recorded in the DataMyte 862. Control charts appear on the video monitor.

Supervisors use the DataTruck to "harvest" data from 862s.

17.2 GAGING FOR SPC

A New Generation of Gages

The gaging industry has been revolutionized by microelectronics. Mechanical dial indicators are being overrun by digital indicators. The hottest selling calipers and micrometers have computerized outputs, with LCD readouts, better resolution, functions such as maximum hold, inch/millimeter conversion, and displays within displays.

A trip to any trade show that exhibits metrology will convince you that technology sells. Handheld gages seem to be going the way of watches and pocket calculators. They have more functional ability, in smaller packages, for lower prices. Fixed station instruments have all taken on the ubiquitous "system" tag. They have many features, all of which cannot possibly be used in a given application.

Although it may not always be apparent, what gaging manufacturers are attempting to do with electronics is make measurement easier. By that we mean the act of measurement, or quantifying data, will be more reliable, repeatable, and efficient.

Since gaging is vital to statistical process control, let's consider what the new generation of gages has brought to SPC. Rather than reviewing specific gage features we should try to identify the characteristics a gage should have. We can then look at how a gage becomes part of an effective data collection system for SPC.

Gage Characteristics for SPC

The gage produces variables data — SPC requires resolution to detect minute changes in a process. The higher the resolution of a gage the better, as long as other criteria are served as well. In general, GO/NO-GO gages do not provide the kind of measurement data needed for SPC.

The gage promotes repeatability and reproducibility (R & R) — SPC requires precision. If a person uses the gage to measure the same part repeatedly, the person should obtain the same reading reliably. Similarly, another person should be able to reproduce those results. Error caused by R & R reduces confidence in the data and our ability to make good decisions. The human engineering put into a gage, such as how it must be held, how a part needs to be positioned in it for a measurement, and how much training is required to operate it, must evidence a concern for R & R.

The gage is efficient — SPC requires a lot of data. This calls for a gage that produces data fast and can be used over and over again reliably.

The gage supports the analytical effort — SPC is concerned with groups of data. A gage is essentially a serial device. It produces one piece of data at a time. A gage must support data collection by providing data in a uniform format, so it can be recorded and grouped.

Of course, not all of the enhancements of electronic gaging support data collection for SPC. In fact, there is one feature that makes many of the others nonessential. That feature is electronic data output.

A gage that does not have electronic data output still has to be read, no matter if it has LCD digits, columns of lights, mechanical pointers, a large display or a small display. If it has to be read by someone, it becomes subject to human error. Any gage can be read wrong and the reading interpreted wrong and written down wrong. Electronic output to a data collector solves this problem.

A gage with electronic output speeds both measurement and data processing. The person taking the measurement can concentrate on getting a good reading. The data collector can convert the serial stream of numbers coming from the gaging into groups of measurements for analysis.

Many of the new generation of gages now have electronic output. See Chapter 21 for details. The gages can be connected to a data collector which records and analyzes the data. Several somewhat standard gage outputs have been developed, such as serial BCD, serial ASCII, serial binary, and of course RS-232C. Some column gages and metrology displays have analog outputs at standard voltage levels.

Converting Gages Versus Replacing Gaging

Companies that use SPC need to have gaging for variables data. Gaging for SPC is an investment, either in the cost of converting existing gaging and fixturing or replacing

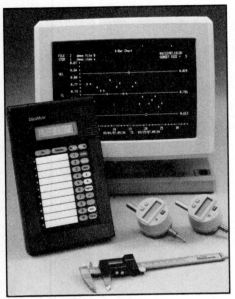

Fig. 17.2.1 *Digital indicators having electronic data output.*

Fig. 17.2.2 *Electronic column gages are an alternative to air gages.*

existing gages with newer ones. Since most of the cost of the original gaging is in the special fixturing needed for an application, there is a temptation to either convert existing gages and fixturing, or not do it at all. But in many cases, thanks to the new generation of gages, it can be less expensive to simply replace gages than convert them.

For example, a test fixture has sixteen mechanical dial indicators. To convert these dial indicators into ones with electronic output, an LVDT conversion package could be added. However, it may be less expensive to simply buy some of the popular LCD readout digital indicators which are designed to be a direct replacement (see Figure 17.2.1). The digital indicators already have digital output, use less power, and can be easily used in any similar fixtures in the plant.

Another example would be an air gage fixture. Air gages have high resolution, and air-actuated columns provide good readout, but they do not adapt well to electronic output. To convert air gages, a pressure transducer and custom circuitry is necessary. As alternatives, the air gages could be replaced with electronic columns or high resolution digital indicators such as the Federal Maxum indicator (see Figures 17.2.2 and 17.2.3).

Very often, if cost is the only criteria, completely replacing existing gaging with new gaging is cheaper in the long run. Whether converting or replacing, it is important to get a gaging system with good repeatability and reproducibility, with adaptability to fixturing, and with electronic output that is compatible with a data collector.

As a final example, several companies are using their CNC machines for measuring as well as machining. By mounting a digital indicator on the machine spindle and connecting it to a data collector, one company takes point measurements on aircraft wing molds. This eliminates the need for a separate coordinate measuring machine. The same program that machines the model can be used to check it, resulting in a considerable savings in equipment and labor.

A Gage Buying Philosophy

Several companies with active SPC programs have adopted gage buying philosophies that link gaging to data collection. The main points of one such philosophy are:

- Measurement that requires operator intervention should be handled with a gage that has electronic output, so the data can be sent automatically to a data collection device.

- Since it is also important that the operator see that measurement took place, the gages themselves should have digital readout.

- Ideally, gaging should be operator independent. An operator should simply place a part into a fixture, press a button or flip a switch, and remove the part. The data should be recorded automatically, although the operator should still have a readout to verify the activity.

- At a higher level, gaging should not even be done by an operator. If the application warrants it, 100 percent on-line gaging should be used. This can be done with fixturing that uses air actuated, spring return gages or with vision systems.

- At the highest level, gages should feed data back to the process in such a way that automatic adjustments to the process are possible. This would be a closed-loop control system, where data collection and analysis are integrated into the process itself. This type of system usually represents a large investment and a complete conversion to a computer integrated type of manufacturing.

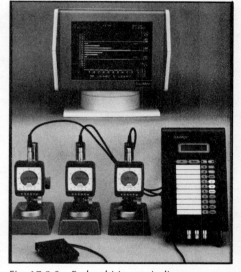

Fig. 17.2.3 *Federal Maxum indicators are an alternative to air gages.*

From Gage to Data Collector

Starting with SPC as an operator intensive activity, and moving to where gaging is more on-line, we find that data collection systems are broken up and defined as subsystems whereby:

- The gaging itself has minimal "smarts," but does have data output, and in some cases a digital readout.
- A data collection device is used at the source to record data in computer-readable form and provide feedback to the operator (or a process controller).
- Standard SPC software is used to perform analysis and produce reports.

Separating a data collection system into subsystems allows us to better identify what type of specialization belongs where. For example, it would be redundant to purchase multi-function gages that have statistical software built in, when several gages are needed at a test station, and the statistical software is already in the data collector. In this case, a gage with data output is all that is necessary.

DataMyte concentrates some "smarts" — that is, SPC software designed for machine operators — in the data collectors themselves. This provides several advantages.

- Operators need feedback at the process, more than what any one gage can give them. This includes prompting an operator for what gage to use and what dimensions to measure.
- A data collector can compare gage readings against reasonable limits and part specifications. It can compare subgroups against control limits and will alert the operator when something is wrong.
- A data collector can do math operations on several gage inputs at once, and plot charts on derived variables such as total indicator readings (TIR), minimums and maximums.

Where the process is largely unattended, data could be recorded at the process, but analysis could occur off-sight, typically on a desktop computer in the QC office. Or, a handheld data collector with SPC software could be used at the process on a periodic basis.

From Data Collector to Computer

A data collector is a vital link to the third subsystem, which is computer software for SPC. Although most of the actual SPC work is done at the process, most of the decisions that affect the process occur at the plant level. Good decisions require good data. A data collection system for SPC that can feed information to management serves to focus management resources on the big problems (the Pareto principle). SPC software on a computer provides a common language for discussing problems that occur at each process, and is a tool for estimating capability, scrap and rework, and cost of quality.

But the weakest link in all computer software is manual data entry. If the data has to be entered through the computer keyboard, most likely the SPC software will not be used. It becomes too labor intensive, and the data is subject to keypunching error. The answer, of course, is having a data collection system that can communicate directly with the computer and store data files which can be analyzed by SPC software.

DataMyte has desktop computer software as part of its data collection systems, and maintains an open architecture to its systems. Many commercial software vendors have developed interfaces to DataMyte equipment. This allows SPC users to choose among many valuable analysis and reporting features, and to maintain SPC data bases on a variety of different computers.

17.3   SPC AT THE OPERATOR LEVEL

SPC and Production Processes

The DataMyte 760 and 860 family of data collectors is for production-oriented SPC. The data collectors add these capabilities to gaging used at the process:

- Memory
- Statistical reporting
- Direct computer interface

At each station throughout a plant, operators can use gaging with electronic output to record measurements in the data collector (see Figure 17.3.1). Up to ten different characteristics can be monitored at any one time. Measurement is fast, resolution is maintained, and the data is automatically placed in subgroups logged with the time and date it was taken.

The data collector connects to many different gages, as described in Chapter 21. The keypad on the unit can also be used to enter data. The operator can enter process specifications and the data collector will alert the operator if any reading is outside of specifications. Process control limits and reasonable limits (gage range) can also be entered. If a subgroup of data is outside the process control limits, the operator is alerted immediately.

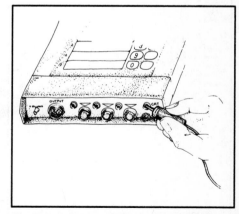

Fig. 17.3.1   Inputs and outputs of a DataMyte data collector.

Producing Control Charts

Control charts are created in a data collector by first specifying the size of the subgroup. The operator then records data in these subgroups, and each subgroup becomes a point on a chart. For an $\bar{x}$ & R chart, the data collector calculates the average and range of each subgroup and plots them. The data collector can calculate the control limits and include them on the chart after one subgroup of data has been collected. However, most statisticians recommend that a minimum of twenty subgroups be collected before calculating control limits.

The data collector displays control charts on a video monitor, or can print them on a printer. See Figures 17.3.2 and 17.3.3. The operator simply presses the $\bar{x}$ button for a $\bar{x}$ chart, or similarly the R button, sigma button and histogram button for those charts.

For process monitoring, a video monitor has two advantages over printed charts:

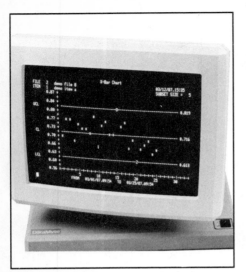

Fig. 17.3.2   Control chart on a video monitor.

- A monitor is more visible. It serves as a constant reminder. Paper charts can be soiled, torn, lost, or filed away and not used.
- A monitor can be updated instantly with each new subgroup.

Although paper control charts provide good permanent records, they are not the most efficient vehicle for data. Control chart data must be keypunched into a computer for the data to be made available for other types of analysis. The data collector stores data in computer readable format right at the process. So, in addition to producing control charts, the data collector provides for high speed communications.

The Smarts for Spotting Trends

A data collector won't tell you what went wrong, or why it went wrong, but it can help you spot it sooner. Once you establish the rules for spotting trends, some data collector models will display warning signals when a trend occurs. There are four rules you can set up:

- *2 points greater than 2.0 sigma*

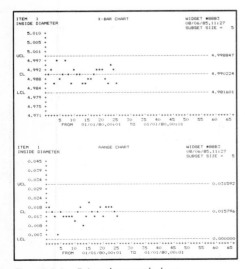

Fig. 17.3.3   Printed control chart.

- *7* points on one side of x-bar
- *3* points greater than *1.0* sigma
- *7* points ascending or descending

The italicized numbers can be changed to anything from 0 to 9 (0.0 to 9.9 for sigma). For example, changing the first rule to "3 points greater than 1.0 sigma," means that if there are already 2 subgroups on one side of 1 sigma away from a control chart centerline, a third subgroup will trigger a trend alert message. The operator is then asked for an assignable cause number before he or she can continue collecting data.

Math and TIR Capability

Some data collector models have math capabilities for addition, subtraction, multiplication, division and functions such as minimum, maximum and range of several inputs. The math can be performed on gage inputs, existing readings and constants. An example of use would be finding the weight of a liquid in a container. The container can be weighed dry and then filled. The data collector will subtract the two automatically and plot a control chart on the liquid weight.

The range function can be used to find the difference between the highest and lowest of several inputs. An example would be checking the flatness of a surface along several points by placing it in a fixture that has several column gage indicators. The data collector will read all of the column gage inputs, calculate the range, and plot the range on a control chart.

The total indicator reading (TIR) capability allows a data collector to calculate the minimum, maximum or difference from a series of readings from one gage. An example of use would be to check ovality with a bore gage, where the gage is inserted and rotated in the bore. The data collector finds and calculates the difference between the maximum and minimum and plots it on a control chart. Another example would be to check runout on a wheel, where the wheel is placed in a fixture and rotated against a digital indicator.

Auditing and Inspection Systems

DataMyte systems can automate two other areas where data collection and reporting are important:

- Auditing — where the person doing the data collection and analysis is not the machine operator
- Inspection — whether it is receiving, in-process, repair or final

Auditing is done usually with a handheld data collector such as the DataMyte 2000 series. Typical applications include roving SPC audits (Figure 17.3.4), torque verification (Figure 17.3.5), inspection, process capability studies, supplier certification and design of experiments. The data collector can audit low-volume processes, sequential operations, tasks on moving assembly lines and do periodic capability studies and quality checks. It can be assigned to a continuous daily route preprogrammed by a computer, or be used to randomly log defects and other attributes whenever they are found. The data collector summarizes and presents data on its own eight-line liquid crystal (LCD) display screen or an optional DataMyte video display terminal.

There are many qualities of a product or process that you can't easily measure, but you can plainly see. They include such things as scratches, dents, assembly errors, packaging errors, and whether or not the product works when it's shipped.

Almost all companies do visual inspection of one sort or another. Automating data collection for visual inspection does three things. It supports better problem solving on the part of production people and inspectors. It also provides documentation of inspection in readable form for management and customers. Lastly, it supports the transition to more prevention-based quality control methods by pointing the way to key areas requiring statistical process control. Data collection systems for inspection include both the 2000 series and the 769 data collector. Both collect and analyze attributes data. Bar codes are used to speed data collection. The DataMyte 769 data collector prints them for you. Every detail of an inspection routine can be coded and arranged on an inspection sheet, from the part description to the names of possible defects. During an inspection, the data collector prompts for each entry, ensuring that each record is accurate and complete.

Process control charts are displayed immediately on the video display terminal. See Figure 17.3.6. An interactive charting system makes the 769 data collector a powerful

Fig. 17.3.4 *Roving auditor uses a DataMyte handheld data collector to check part dimensions.*

Fig. 17.3.5 *Roving auditor checks torque*

17-12

Computerized Data Collection for SPC

problem-solving tool. You can sort and display attributes by any criteria, including lot number, machine operator, size, color, date and time. This allows you to quickly isolate a problem.

The DataMyte model 769 provides full traceability to an inspection. It automatically records the time and date of each record. Records can be transferred to a computer and combined with others to provide management analysis.

17.4 QUALITY MANAGEMENT SYSTEMS

The key characteristics of a computerized system for quality management include:

- The ability to interface sensors and gages on the factory floor.
- The ability to provide information to machine operators quickly to support problem solving and corrective action.
- The ability to document quality related activities and results.

Fig. 17.3.6 *A process control chart is displayed immediately on the monitor.*

Both quality and responsiveness are important to customers. By responsiveness we mean the ability to react quickly to new demands and changing conditions. Responsiveness includes the ability to analyze a problem and provide a solution after the product is in the hands of a customer. It requires having a large data base of information at hand, an ability to share the information with others, and a set of analytical tools.

DataMyte offers two basic quality management systems, OVERVIEW™ and FAN®. For more information about these systems, refer to Chapter 18 for OVERVIEW, and Chapters 19 and 20 for FAN.

18. DATAMYTE OVERVIEW™ SYSTEM

18.1 INTRODUCTION

The DataMyte OVERVIEW system is a quality management solution for today that addresses the emerging needs of organizations in the future. It provides an SPC and information reporting system to meet the special needs of such industries as aerospace, defense, and pharmaceutical, as well as commercial manufacturing.

With over five years in research and development, the genesis of the OVERVIEW system was a three-fold set of circumstances: an industry desire for computer-integrated quality management, flatter, leaner organizations, and the availability of context-based as well as event-based decision making tools.

Computer-Integrated Quality Management

Fact-based decision making is necessary in a complex manufacturing environment. Computer technology provides a way to make fact-based decision making easier and cost effective. Surveys show, however, that quality management is actually lagging behind other manufacturing disciplines in the application of computer technology.

In spite of this, customers are requiring documented proof of conformance to requirements from suppliers as a normal part of a business transaction. Major companies are demanding parts per million or less defect levels from their suppliers. Customers are also asking for greater responsiveness from quality departments. They are looking for advice about parts design, material selection and producibility. Increasing quality expectations are driven by world competition, just-in-time production methods, better manufacturing technology and a more sophisticated consumer.

In response to these new quality expectations, companies are stepping up their investments in computer-integrated quality management solutions.

Flatter, Leaner Organizations

The need to survive against world class competition has created organizations with fewer layers of management and less duplication of function. The result is that routine decision making is driven down to the lowest level possible in an organization, and managers find themselves with broader roles. Manufacturing and quality managers, for instance, find themselves dealing more with customers, and their roles becoming more that of customer service representatives. Internal quality issues are becoming more often a work team responsibility.

Companies see a strategic advantage in integrating computing and software resources within a company, and between key suppliers and customers. The advantage is to be better able to work together.

Context-Based Decision Making

In the last ten years we have seen the widespread application of event-based decision making tools such as SPC. Many companies now need to complement event-based decision making on the factory floor with context-based decision making at the supervisory level or above.

By context-based, what is meant is that data is not seen in isolation, or related to only one variable such as time. Rather, context-based data is fully integrated, such as in a relational data base, with all other relevant information. This offers a valid tool for making system-level improvements to complex manufacturing processes.

Given Dr. Deming's observation that 85% of all quality problems are system-related and not assignable to workers, it will take context-based decision making using computer-integrated quality management solutions to tackle most of the real work of quality improvement. The DataMyte OVERVIEW system is a quality management solution designed to meet this need.

18.2   BUILDING BLOCKS OF COMPUTER INTEGRATION

The DataMyte OVERVIEW system is a decision support tool consisting of the following basic building blocks:

- DataMyte 900 series data collectors — a new generation of data collectors for both operator-based SPC, inspection and lot control. See Figure 18.2.1.
- OVERVIEW quality management software — jointly developed by BBN Software Products Corporation and

Fig. 18.2.1   DataMyte 900 series data collector.

Fig. 18.2.2   OVERVIEW Software.

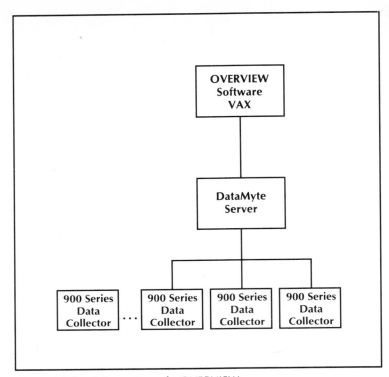

Fig. 18.2.3 Schematic of OVERVIEW network.

DataMyte, the software runs on VAX computers and is based on the powerful RS/1® data analysis software. See Figure 18.2.2.

- DataMyte Server — consisting of AppleTalk and Ethernet network components that link the data collectors to the VAX computer and manage communications. See Figure 18.2.3.

Top-Down Bottom-Up Integration

The OVERVIEW system integrates quality management at all levels of a plant organization. See Figure 18.2.4. At the machinery/process level, the OVERVIEW system ties to gaging, sensors and other metrology through the DataMyte 900 data collector. Quality-related data is quickly and accurately conveyed to the station level.

The DataMyte 900 data collector resides at the station level, where it provides local analysis and feedback to machine operators and inspectors. The data collector also serves as a terminal to enter notes on both quality and

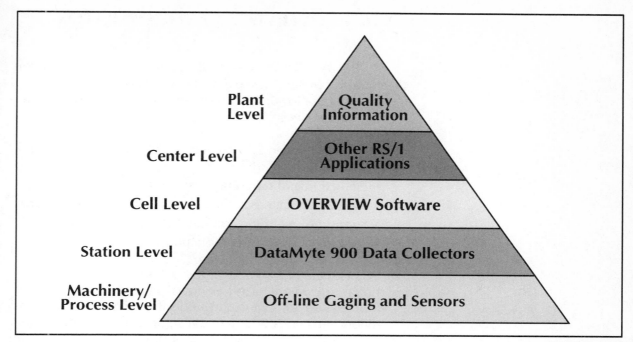

Plant Level	Quality Information
Center Level	Other RS/1 Applications
Cell Level	OVERVIEW Software
Station Level	DataMyte 900 Data Collectors
Machinery/ Process Level	Off-line Gaging and Sensors

Fig. 18.2.4 DataMyte OVERVIEW System as it relates to the Allen-Bradley Productivity Pyramid.

process performance. The data collectors are tied through a server network to a node on the VAX system or standalone MicroVAX at the cell level.

The OVERVIEW software is operated and managed primarily at the cell level, in support of station level activities. OVERVIEW system activities at the cell level include monitoring for exception events and managing parts files. In addition, statistical reports and capability analyses can be created and printed, and made available to the center level.

The center level is the gateway from OVERVIEW into other applications. Quality data can be made available to other RS/1 series programs such as RS/Discover® for designed experiments, RS/Decision® for creating expert systems and RS/Explore® for R&D.

All of the application software, including OVERVIEW, share the RS/1 data structure, offering communications capability across production, quality, engineering design, R&D and other departments in an organization. Such an integration of architectures and information serves the plant level, and its role in overall planning and execution.

See Chapter 23 for an explanation of the Productivity Pyramid.

See Chapter 24 for BBN.

18.3 OVERVIEW AT THE OPERATOR LEVEL

The OVERVIEW system offers a truly integrated solution to both operator-based SPC, inspection and lot control. At the operator level, the system serves these quality management objectives:

- Operator-based SPC on independently functioning work stations.
- Detection and reporting of exception events.
- Production data recording for traceability and lot control.
- Support of metrology for automatic data recording.
- Support of short-run production.

For operator-based SPC, the system provides the functional equivalent of other popular fixed station systems from DataMyte, such as the model 862 data collector. In addition, the series 900 data collector facilitates analog and digital gage input, variables and attributes data analysis, and short-run SPC in addition to more traditional SPC charting.

A Data Collector for Lot Control

Lot control means the management of information about a homogenous group (lot) of parts going through a manufacturing and quality control process in conformance with customer guidelines.

For lot control, the system provides a variety of formats for collecting and tracking production related information, such as piece counts, serial numbers, operational descriptions and notes on events. Figure 18.3.1 is a block diagram of how different types of information about a part interrelate. Although the system allows for a great deal of information types and relationships, it can be as simple as the application requires. For example, an application that asks for an operator to enter a part serial number, three measurements and any observed surface defects would look like Figure 18.3.2 in block diagram form.

Entering data — Each data collection station supports several measuring devices, such as calipers, weigh scales or column gages. The data collector keypad and a bar code wand serve as the other sources of data input. See Figure 18.3.3.

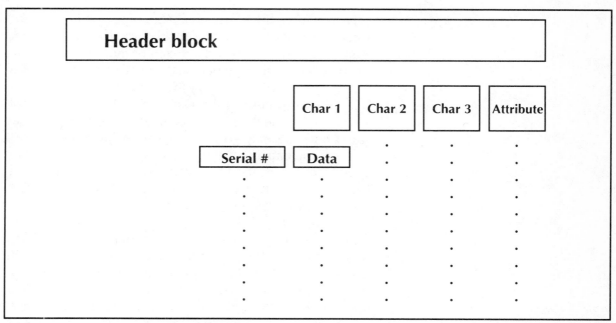

| Part Name: Very large left hand widget |
| Procedure Name: First stage rough cut |
| Revsion number: 73 |
| *Part Descriptors 1-5* |

12:52:42 PM 4/6/89

Descriptors entered in Setup.↑

Labels entered in Data Collect.↓

Char: Length	Char: Attributes
Process: Filling	Process: Backing
Char-Descriptors 1-5	*Char-Descriptors 1-5*

Assignable Cause	**Assignable Cause**
Machine	**Machine**
Char-Labels 1-3	*Char-Labels 1-3*

Lot #	Time Stamp	Unit I.D.	*Subgroup Labels 1-15*		**Serial #**	*Piece Labels 1-15*	11.37	Attr	Attr	Attr
					Serial #	*Piece Labels 1-15*	33.95			
					Serial #	*Piece Labels 1-15*	44.66	Attr		

Assignable Cause	**Assignable Cause**
Machine	**Machine**
Char-Labels 1-3	*Char-Labels 1-3*

Lot #	Time Stamp	Unit I.D.	*Subgroup Labels 1-15*		**Serial #**	*Piece Labels 1-15*	19.34	Attr	Attr	
					Serial #	*Piece Labels 1-15*	21.98	Attr	Attr	Attr
					Serial #	*Piece Labels 1-15*				

User defined optional labels and tags in Italic underline.   **System defined labels and tags in bold.**   System required labels and tags in outline.

Fig. 18.3.1   Block diagram of data types.

Header block

| Serial # | Data | Char 1 | Char 2 | Char 3 | Attribute |

Fig. 18.3.2   Diagram of data for simplified application.

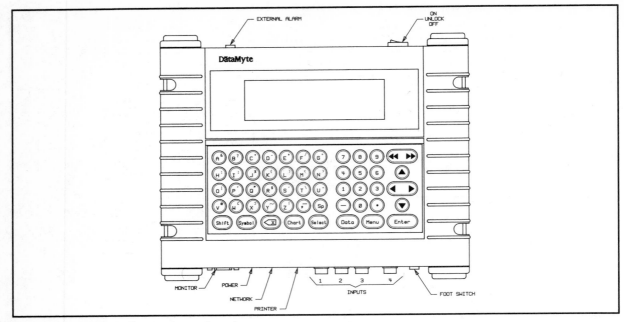

Fig. 18.3.3 Data collector keypad and connectors.

Each station, therefore, is capable of recording data on a variety of production and inspection applications, covering a dozen or more parts and a dozen or more characteristics per part. Although the type of data being collected from station to station differs, the method of collecting the data is standardized by use of the DataMyte 900 series data collector. The benefit of such standardization is easy to see when one part may go through a series of manufacturing centers, and data is collected about that part at each value-added step. With the OVERVIEW system, all the data collected for that part can be integrated by having the OVERVIEW software combine information from each step.

Math capability — A large selection of data manipulating functions complement the data entry capability of the OVERVIEW system. Math operations can be performed on variables data during gage and keyboard input. Basic math operations include add, subtract, multiply, divide, absolute value, square roots and powers. Derivative operations include standard deviations, range, minimum, maximum, sum and various TIR capabilities. These capabilities allow the calculation of values such as net weight, parallelism, perpendicularity, taper and others.

Local feedback — The data collector's built-in display, audible beeper and video monitor provide local feedback to the operator. The data collector display is used to prompt the user through the data collection process. The audible beeper signals when acceptable or unacceptable entries are made, as well as when a data subgroup does not meet certain statistical rules. The video monitor is used to display statistical information. All of the basic SPC charts are available to the user, including a short-run SPC chart. Table 18.3.1 lists the chart forms available and the earlier chapters of this book serve as a guide to selecting the form of analysis.

Exception reporting — A unique feature of the OVERVIEW system is its ability to detect and report exception events. An exception is defined as:

- A value beyond engineering limits.
- A completed characteristic subgroup beyond control limits.
- A completed characteristic subgroup that violates a trend rule.

Exception events are automatically logged by the data collector when they occur. This provides for rapid recognition and response to an event by both data collector users and supervisory people.

Data Collection Applications

Here are some typical data collection applications supported by the DataMyte 900 data collector. This is not an all inclusive list, but it does help indicate its capability to address real-world situations.

Multiple fixtures — An operator has a subgroup of four pieces to be measured on two separate fixtures. The first part will be placed in the first fixture and three characteristics recorded. The part is then placed in the second fixture and four more characteristics are recorded. The final characteristic is the difference of the average of three characteristics on the first fixture and the average of the first three characteristics of the second fixture. The procedure is repeated for the remaining three parts.

Reports
- **x-bar and Range**
- **x-bar and Sigma**
- **Short run x-bar and Range**
- **Histogram**
- **Pareto**
- **Summary statistics**

Table 18.3.1   Reports available from a 900 series data collector.

Multiple use fixture — An operator has a subgroup of four pieces to be measured in two separate operations on one fixture. The first part will be placed in the fixture and three characteristics recorded. The procedure is repeated for the remaining three parts. After all of the parts have been measured the fixture is broken down and set up for a second set of measurements. The first part is then placed in the fixture for the second time and four more characteristics are recorded. The final characteristic is the difference of the average of the three characteristics of the first setup and the average of the first three characteristics of the second setup. The procedure is repeated for the remaining three parts in order.

TIR — A part is placed in a fixture and rotated 360 degrees while the data collector samples the output of several gages. At the completion of rotation, the data collector stores several direct and several calculated characteristics. Note that all TIR readings are a dynamic measurement of gage travel and may be used to find minimum, maximum or difference values.

Manual TIR — The operator measures the outside diameter of a plastic bottle at five locations using a caliper. The readings are used to derive two characteristics, the min and max of the five. Note that standard TIR would not be appropriate because the gage is opened and closed between readings.

Analog gage two-point mastering — When the master values are known in advance, the operator enters the master values into the data collector and then aligns to the first one, and then the other master. Once the data collector has recognized both the master engineering values and the master analog values, the gage is mastered to the data collector.

Master values are not normally known in advance for gages that make destructive tests. For this type of gage the master procedure is much the same as the example above except that the analog bit values are determined before the master engineering values are entered.

For example, a maker of billiard balls needs to determine the force required to crush a ball. The gage used to make this measurement has a digital display that shows the peak force at crush and a real time analog output. The operator places a ball in the tester and performs the crush, the data collector captures the analog peak value, and the operator then enters the engineering value as displayed on the gage. The process is repeated for the second ball, selected to have a significantly different crush value.

Analog gage one-point mastering — One-point mastering is required when dealing with an analog gage where both the scale and the range are calibrated but there is an uncalibrated offset. If the master value is known in advance, such as with a torque wrench application, the operator sets the gage in a known value (normally zero) and performs the auto-zero function.

If the master value is not known in advance, as likely to be encountered in applications such as pressure and temperature where the gage is always loaded, the operator performs the auto-zero function at the data collector and then enters the displayed gage value as an offset or zero value.

Analog gage no-point mastering — No-point mastering is usable when dealing with an analog gage where the scale, range and offset are calibrated and known in advance and a calibration schedule is maintained. For example, where 1.000 volts equals 1.000 mm and 2.000 volts equals 2.000 mm.

Gage R&R study — In a short form study, the operators enter gage readings as prompted by the data collector. The test of gage R&R conforms to the General Motors Short Method analysis. The results are displayed as characteristic values named Gage Error (GRR) and GRR% or other names of the operator's choice.

In the long form study, the operators enter gage readings as prompted by the data collector. The test conforms to the GM Long Form analysis. The results are displayed as characteristic values named $\overline{R}$, $\overline{x}$ (diff.), A.V., R&R, %E.V., %A.V., %R&R, or other names of the operator's choice.

Several gage readings for two characteristics — The operator needs to measure the inside diameter in six different locations on each land on a bore that will be used to make an interrupted screw. See Figure 18.3.4. Only the maximum and minimum values are of concern. The operator aligns the gage to the center of land number one and the data collector takes a "snapshot" of the six gage values. This procedure is repeated for the other lands. After the four lands

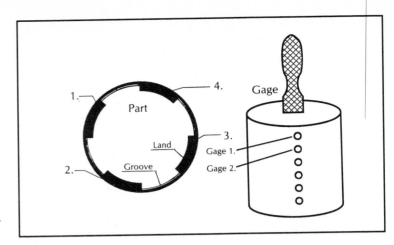

Fig. 18.3.4   Sample part in example of "snap-shot" measurement.

have been gaged, the data collector records just the highest and lowest of the 24 raw readings. Note that a TIR feature would not be appropriate in this application because of the grooves.

Short-run data collection — A machine is used to cut printed circuit board blanks with a production rate of 800 boards per day. All boards have a primary width and length and some have a secondary width and length as well. Batch sizes range from one to several hundred. A subgroup is collected for every batch and every hundred boards within a batch. If batch size is less than ten, then the subgroup size equals the batch size. If the batch size is greater than nine, then the subgroup size equals five. A given part is run at most once each day. Data is collected mainly for the purpose of creating charts and reports on each of the four processes (primary and secondary width, primary and secondary length) and also for batch and part history.

Piece attribute data collection — The operator enters the attribute and location for each defect discovered on a printed circuit board after wave soldering. The data will be used to track defects with this particular serialized piece as well as which location is most subject to difficulty. Control charts are appropriate for the board as a whole or for each location. It is not necessary to make an entry for each good location on each part for the analysis software to recognize that all locations are good on a piece that has no defects recorded.

18.4   SERVER AND NETWORK

The DataMyte Server is an independently functioning device serving as temporary data storage and communications manager for the 900 series data collectors. It supports these functions:

- Receive setup files from the OVERVIEW software on the VAX.
- Send setup files to the data collectors.
- Receive data from the data collectors on a periodic basis.
- Send data and setup file to the VAX.
- Receive exception log from the data collectors.
- Send the exception log to the VAX.
- Temporarily store setups, data and the exception log.
- Manage communications network.
- Print the exception log.

Network Configurations

Figures 18.4.1 and 18.4.2 are schematics of department and plant level network configurations respectively. Figure

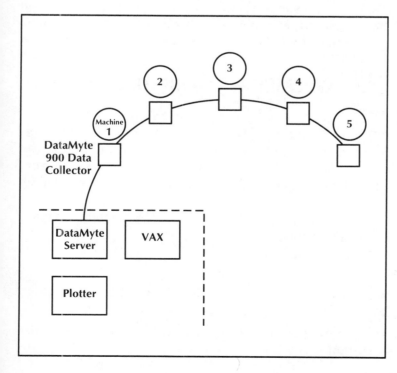

Fig. 18.4.1   Department level network.

See Chapter 25 for VAX equipment.

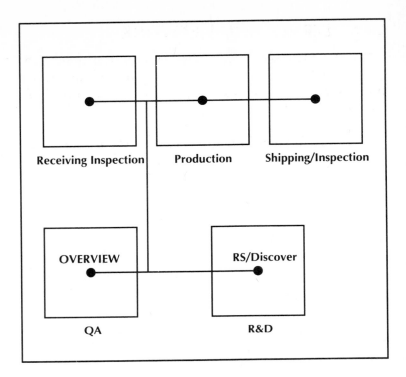

Receiving Inspection   Production   Shipping/Inspection

OVERVIEW   RS/Discover

QA   R&D

Fig. 18.4.2   Plant level network.

18.4.1 shows some of the many VAX peripherals that can be supported. Figure 18.4.2 shows how the OVERVIEW system fits into a plant level VAX network.

18.5   OVERVIEW AT THE SUPERVISORY LEVEL

The OVERVIEW system at the supervisory level centers around the OVERVIEW software system running on the VAX computer and the DataMyte Server. The supervisory objectives being served by the OVERVIEW system include:

- Support of operator data collection, by creating and managing parts files.
- Event monitoring.
- Editing and archiving data.
- Statistical analysis, both traditional SPC and short-run.
- Producing management reports.
- Maintenance and administration of the information system.

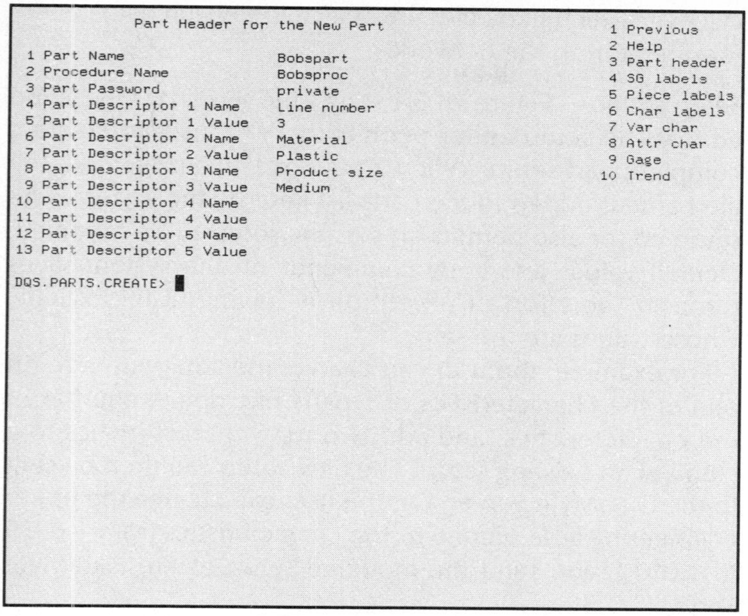

```
        Part Header for the New Part              1 Previous
                                                  2 Help
   1 Part Name                  Bobspart          3 Part header
   2 Procedure Name             Bobsproc          4 SG labels
   3 Part Password              private           5 Piece labels
   4 Part Descriptor 1 Name     Line number       6 Char labels
   5 Part Descriptor 1 Value    3                 7 Var char
   6 Part Descriptor 2 Name     Material          8 Attr char
   7 Part Descriptor 2 Value    Plastic           9 Gage
   8 Part Descriptor 3 Name     Product size     10 Trend
   9 Part Descriptor 3 Value    Medium
  10 Part Descriptor 4 Name
  11 Part Descriptor 4 Value
  12 Part Descriptor 5 Name
  13 Part Descriptor 5 Value

  DQS.PARTS.CREATE> █
```

Fig. 18.5.1   Setup Screen.

System Administration

The first major function of the OVERVIEW software system is to create the OVERVIEW operating environment. The environment consists of:

- Data collectors and gaging on the factory floor. The data collectors are identified and established through the software running on the DataMyte Server.
- VAX peripherals such as terminals, printers, plotters and the archive directory are identified for use through a customize menu on the OVERVIEW software system.

See Chapter 25 for Digital Equipment Corporation products.

Parts Manager

The OVERVIEW software system provides for the creation, storage and retrieval of part data and setup files. This takes advantage of the more complete keyboard and screen interface of the VAX, and facilitates more standardization of data file structure.

Through the parts manager, a parts file is created and made available in the DataMyte Server network. Once on the network, a data collector user can access the parts file. This allows a user to select the parts file appropriate to the work at hand, use it for data collection, and to select a new one when needed. The parts manager at the VAX has control over which parts files are available to each data collector operator on the network.

Setup editor — Figure 18.5.1 shows the screen for the setup editor. The setup editor permits the VAX user to create a complete part setup. When a new parts file is created, the file name is added to the parts list kept by the system. The setup editor also permits an existing setup to be modified. Default values and copy commands on the system saves the user the effort of repetitiously having to fill in fields whose values are the same.

For example, the user can change the subgroup size on one of the characteristics of a parts file, delete another of the characteristics, and add two new characteristics to a clone of an existing setup. The new setup can then be sent to the DataMyte Server. Or, the user can change the upper engineering limit on one of the characteristics from 1.0035 to 1.0032, and send the modified setup to the DataMyte Server.

Data editor — The data editor allows the VAX user to edit raw data associated with the current version of an existing part. The editing function is in the form of a smart table editor as shown in Figure 18.5.2. As a form of tracking data editing, the data editor keeps a record in the collected data structure of the date and time of the last data edit for each

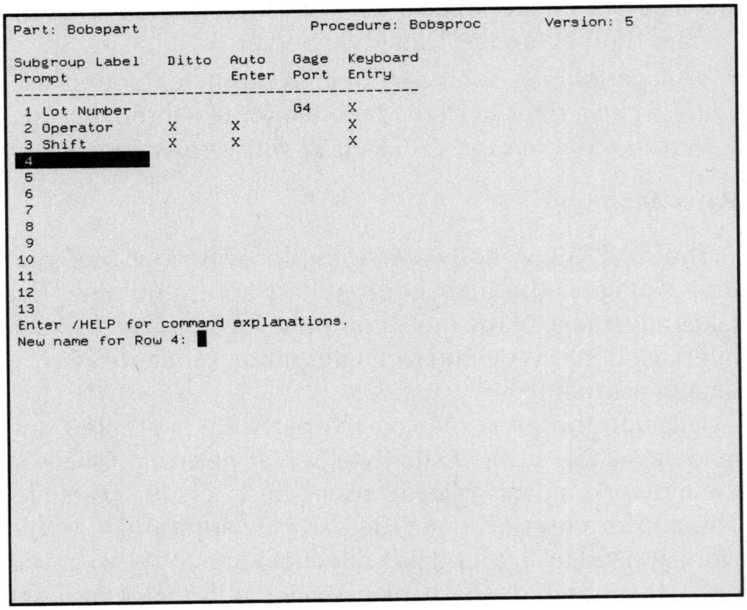

Fig. 18.5.2   Data editing screen.

```
Part: Bobspart                    Procedure: Bobsproc      Version: 5

Subgroup Label    Ditto  Auto  Gage Keyboard
Prompt                   Enter Port Entry
-------------------------------------------------
 1 Lot Number                    G4   X
 2 Operator         X      X          X
 3 Shift            X      X          X
 4
 5
 6
 7
 8
 9
10
11
12
13
Enter /HELP for command explanations.
New name for Row 4: █
```

characteristic for each subgroup. This information is displayed as part of the data report.

Parts directory — Figure 18.5.3 shows the parts directory. Each part entry consists of the part name, procedure name, version number and status (whether data exists or not).

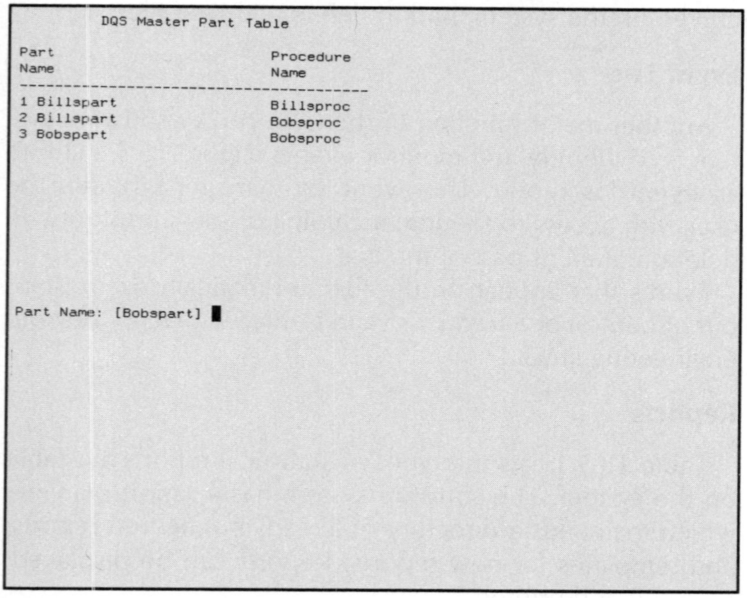

```
        DQS Master Part Table

Part                        Procedure
Name                        Name
-----------------------------------------------------
1 Billspart                 Billsproc
2 Billspart                 Bobsproc
3 Bobspart                  Bobsproc




Part Name: [Bobspart] █
```

Fig. 18.5.3   Parts directory.

Parts transfer, archiving and deletion — Three other features of the parts manager function are transfer, archiving and deletion. Parts transfer involves moving parts setup and data files to and from the DataMyte Server, thus making them available to data collector users.

The transfer function is used in a variety of ways. For example, prior to viewing a control chart, the user may want to transfer all recently collected subgroups for the part of interest from the DataMyte Server to the VAX. This may be a routine where once a day or week the user causes all recently collected subgroups for any part setup to be transferred.

In another example, an order arrives for a part that has not been manufactured for six months. Since the part setup has been removed from the data collectors and DataMyte Server, the setup along with 25 or so subgroups (to allow an ongoing control chart) can be sent to the DataMyte Server.

Parts archiving involves the permanent storage of parts files on the VAX. The parts files consist of both the setup and data for each part. For example, a user may archive all subgroups for a part setup or several part setups that are older than a certain date or are the "x" oldest. The user can de-archive a previously archived set of data and setups at any time.

Parts deletion provides a way to erase data not already archived on the system, and to delete archived files.

Event Log

Another major function of the OVERVIEW software system is to identify and monitor events. Figure 18.5.4 shows an event log report. The event log manager provides the user with access to the log, including display, printout and deletion of all or part of the log.

Events that appear on the log automatically are a point out of control or a trend, or values entered that are beyond engineering limits.

Reports

Table 18.5.1 lists the types of statistical reports available on the system. The software system has a report manager which maintains a directory of already configured reports, and templates for new reports. Reports can be displayed, archived and printed.

Figure 18.5.5 shows a sample $\bar{x}$ & R chart, and Figure 18.5.6 a histogram. Within each report type is a series of configuration choices available to the user, including the re-calculation of control limits and the editing of subgroups. These configuration choices can be saved in the form of templates to make successive report creation easier.

Reporting Scenarios

What follows are some typical scenarios that describe some of the supervisory reporting features of the OVERVIEW system.

Control charts — Some users will use OVERVIEW software at the VAX to produce control charts. In instances where daily or weekly reports need to be delivered to a manager for review, or placed in paper files, the software user can ask the system for control charts for the latest fifty or so subgroups for all characteristics for all active part numbers to be printed. The user can also ask for just the charts for

Unit	Part	Procedure	Char	Ver	Event	Time & Date
Bill's Unit	Large left handed widget	Flame cutting	Internal framistat fitting	90° 37	Stratification	10/19/89, 21:34
Mary's Unit	Very small right handed widget	Boring	External framistat fitting	90° 37	Exceeded RUCL	10/19/89, 21:52

Fig. 18.5.4   Event log.

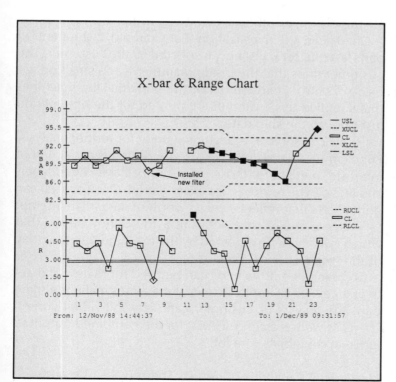

X-bar & Range Chart

From: 12/Nov/88 14:44:37   To: 1/Dec/89 09:31:57

Fig. 18.5.5   x-bar & range chart.

- **x-bar and Range**
- **x-bar and Sigma**
- **Short run x-bar and R**
- **Histogram**
- **Capability statistics**
- **c**
- **u**
- **Pareto**
- **Data**
- **Exception log**

Table 18.5.1   Graphs and reports.

18-19

DataMyte OVERVIEW System

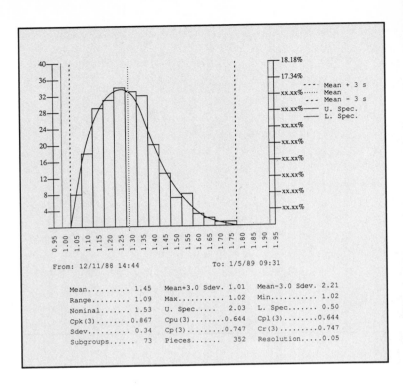

Fig. 18.5.6   Histogram.

for the plating or other selected departments.

Charts can aid investigation. For example, a shipment of parts last quarter is just now inspected by the customer. The customer finds that the inside diameter is too small on average. Control charts for all of the parts making machines that supplied this customer are printed for the time period just prior to the ship date.

Charts can serve as a beginning point for further analysis. For example, an analyst sees a series of points on a chart of interest. The data for those points and all associated labels and tags can be retrieved for examination. Or an analyst can print charts that reflect a year's worth of data in an attempt to observe cycles that occur at a frequency below that which is likely to be detected by the machine operator (day of the week, season, etc.).

Combined control charts can be used for review and acceptance. For example, if several machines make the same part, the analyst or customer can view individual control charts as well as a control chart for the combined output to judge acceptability of a lot.

Histograms, capability studies — These types of reports are used to assess process capability. For example a large sample of production from a new machine can be taken to determine the distribution of samples and calculate capability against engineering limits. This can be done after a major setup or overhaul.

Where several machines are making the same part, an analyst can view a frequency distribution for the combined output in an attempt to anticipate what the customer will see. The combined report can be sent to the customer with each shipment.

Where a machine that is not capable is making a certain part and pieces that are too large are reworked and pieces that are too small are scrapped, the analyst monitors this report to ensure the most economic balance between scrap and rework is maintained. If the costs are not balanced the production target size may need to be adjusted.

Pareto analysis — Pareto analysis can determine the most frequently occurring problems and their associated causes. An analyst can first view a Pareto chart of defects (possibly modified by severity) to find that the defect of greatest concern is burrs, for example. The analyst then views a Pareto chart for vendors where the defect on the finished piece was "burr." Finding a flat distribution, the analyst then views the same chart only for spindle instead of vendor and discovers that eighty percent of the burred product came from spindle number 13.

Data reports — Data reports can be characteristic or piece oriented. For example, after a capability study is run on a new machine the raw data for selected characteristics can be printed and sent to the machine vendor, along with the capability results. Or an aircraft engine has disintegrated in flight almost causing a crash. The analyst can retrieve all available data about the engine from several part setups using only the part number and serial number.

Product recalls are also supported by data reports. For example, it was determined that the last calibration of a pharmaceutical balance was performed incorrectly. In order to recall the potentially defective lots it will be necessary to print all lot numbers measured with that serialized balance and calibration date. This information was recorded as label entries by the data collector user.

Investigation — Using the event log, a user can print control charts or other reports for all characteristics from several different setups that are exceptions. As an example of selective investigation, a production manager knows that seven different parts were run last night for a contract. The manager retrieves data for all seven parts, runs control charts on all seven and prints histograms.

18.6   OVERVIEW AT THE PLANT LEVEL

At the plant level, the DataMyte OVERVIEW system becomes one of a series of RS/1 software applications available to technical professionals. Because it shares a common data structure with other RS/1 packages, data originating in OVERVIEW can be passed into other applications to support engineering, R&D, production control and other efforts. OVERVIEW thus collects and analyzes data for on-line, real time local quality applications but also consolidates data on the host VAX for off-line analysis for detecting longer-term trends.

Manufacturing Paradigm

To see the impact of the OVERVIEW system, we need to relate it to an organizational model for manufacturing quality (see Fig. 18.6.1). Each area of this manufacturing paradigm affects quality. From a flow standpoint, products gravitate from bottom to top, starting at design and going to development, planning, production, inspection and selectively to repair.

This paradigm also represents a good model of the impact each step has in the service of quality. Repair is the worst possible in terms of quality. Inspection is better. Working on quality during the production process has a better effect and ultimately, designing in quality has the biggest payoff.

In terms of how much quality costs, the paradigm should be inverted as shown in Figure 18.6.2. Repair is obviously the most expensive way to maintain quality. Design is the cheapest.

In terms of this model, the OVERVIEW system directly impacts the areas of production, inspection and repair. See Figure 18.6.3. The OVERVIEW system provides a real time quality management system and data base for directly attacking the areas of most potential cost to the organization

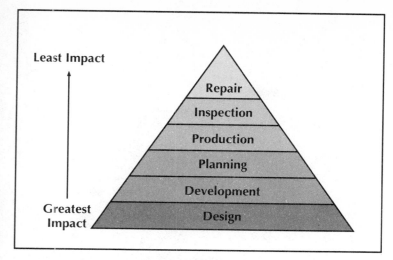

Least Impact

Repair
Inspection
Production
Planning
Development
Design

Greatest
Impact

Fig. 18.6.1 Paradigm of manufacturing quality.

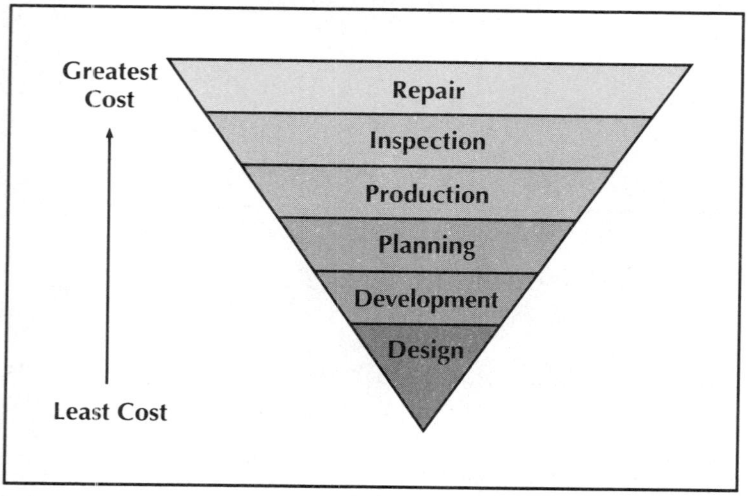

Greatest
Cost

Repair
Inspection
Production
Planning
Development
Design

Least Cost

Fig. 18.6.2 Paradigm of quality costs.

as it is currently operating. Coupled with RS/Explore, RS/Discover, and RS/QCA, OVERVIEW provides an integrated solution for better design, development and planning. These three areas offer the most potential savings to the organization in the future.

Data Input/File Interface

The OVERVIEW system interface with RS/1 offers a communications opportunity to a host of data bases, hardware and applications, as shown in Figure 18.6.4. This opens the door to virtually unlimited usages for quality data and support for quality management efforts. This is the result of the

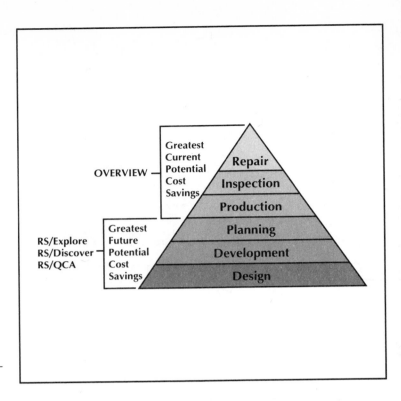

Fig. 18.6.3 Paradigm of Quality Management Applications

fundamental linkage that is part of all RS/1 series products (see Fig. 18.6.5) covering the following areas:

- Data analysis and graphics
- Statistical analysis
- Experimental design
- Statistical quality control
- Time-series analysis
- Clinical trials data management

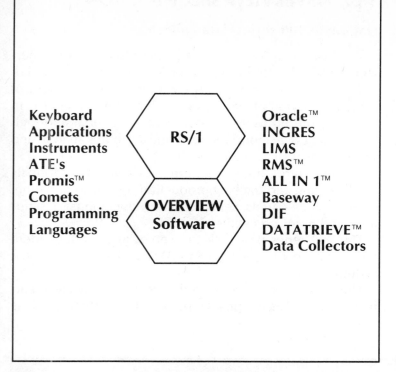

Keyboard
Applications
Instruments
ATE's
Promis™
Comets
Programming
Languages

RS/1

OVERVIEW
Software

Oracle™
INGRES
LIMS
RMS™
ALL IN 1™
Baseway
DIF
DATATRIEVE™
Data Collectors

RS/Discover

RS/Explore

RS/QCA

RS/1

OVERVIEW
Software

Fig. 18.6.4   *Potential data interfaces.*

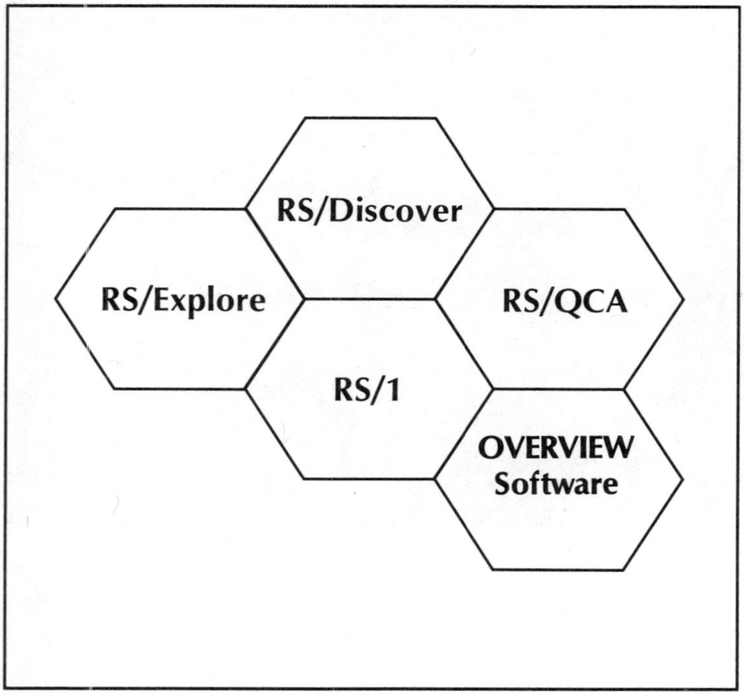

Fig. 18.6.5   *RS/1 series products.*

DataMyte

See p. 9-2 for an example of use.

18.7   OVERVIEW SPECIFICATIONS

DataMyte 900 series data collector

The DataMyte 900 series data collector is a new generation data collector designed for SPC and lot control. For SPC, the 900 series provides the functional equivalent of other DataMyte data collectors such as the model 862. In addition, the 900 series provides digital, serial and high level analog gage input, variables and attributes analysis and short run as well as standard SPC.

For lot control, the system provides a variety of formats for collecting and tracking production related information, such as piece counts, serial numbers, operational descriptions and notes on events.

The 900 series data collector serves as an independent work station on the OVERVIEW quality management system.

The data collector is powered by rechargeable nicad batteries or a plug-in power pack. The battery pack is removable.

Features

- Independent data entry and analysis
- Short-run SPC charts
- Events detection and reporting
- x̄, R, sigma, histogram and Pareto charts
- Summary statistics
- Variables data resolution to 0.0000001
- Piece attributes
- Part, characteristic, subgroup and piece labeling
- 512K memory capacity
- Math functions
- Password protection
- Menu driven with help screens and status displays
- Pull-down windows for selecting options
- Time/day/date labeling for all subgroups
- Weighs about 4.2 lb (0.9 kg)

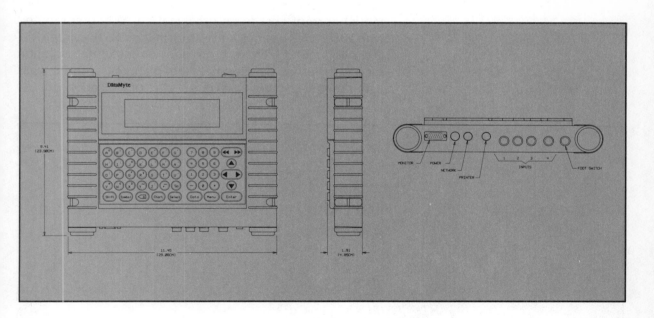

DātaMyte

OVERVIEW Quality Software

Jointly developed by BBN Software Products Corporation and DataMyte, the software is based on the powerful RS/1® data analysis software. OVERVIEW Quality software supports the activities of operators using DataMyte 900 data collectors.

The software is used to create and manage parts files. During data collection, the software is used for event monitoring, editing and archiving data, statistical analysis and producing management reports.

The software has extensive data management and report writing capabilities. It can combine data from different files, search for data by matching any number of pre-determined criteria, and analyze selected quality data.

The software provides information on the areas of production, inspection and repair — three areas of most potential current cost to an organization. When combined with other RS/1 series software, such as RS/Explore, RS/Discover, and RS/QCA, it provides an integrated solution for better design, development and planning — three areas with the most potential savings to the organization.

18-28

DataMyte OVERVIEW System

System Requirements

- VAX computer with Ethernet port
- 45,000 blocks of free disk space
- VMS version 4.0 or later

Charting

- x̄ & R chart
- x̄ & sigma chart
- Short-run x̄ & R chart
- Histogram
- Capability graph
- c-chart
- u-chart
- Pareto chart

Communicating with Data Collectors

- Communicates through DataMyte Server
- Loads and receives setups and data
- Receives exception log

Data Management

- Data stored in standard RS/1 tables
- Supports VAX peripherals such as printers, plotters and storage devices
- Smart spreadsheet editor for file creation and data editing
- Parts directory
- Communications capability to other RS/1 applications
- Enable and disable file security for data collectors

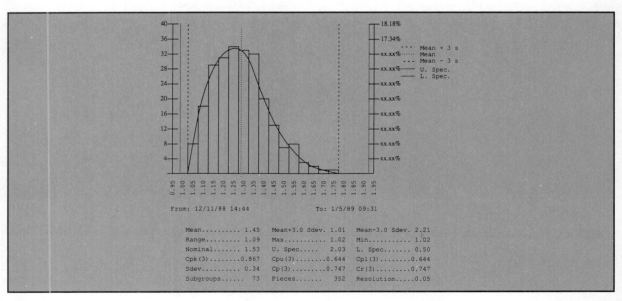

DataMyte

DataMyte Server

The DataMyte Server is a file server and network manager consisting of an Apple Macintosh computer and software. The server is used to configure communications between 900 series data collectors and the VAX computer running OVERVIEW software. The server then manages the flow of information going to and from the data collectors.

The server automatically harvests information from data collectors at selected intervals. Data collected by the server is then available to the software on the VAX.

The server provides a parts directory to the data collector user, allowing the user to load and change parts files when needed. New parts files created through the VAX software can be loaded on to the server and made accessible to data collector users.

Once configured, the DataMyte Server operates automatically between the two main functioning components of the OVERVIEW system, data collectors and OVERVIEW software running on the VAX.

System Requirements

- Macintosh SE30 or Macintosh IIcx
- At least 2 megabyte of RAM
- A 40 megabyte hard disk drive

Network

- AppleTalk network cabling
- AppleTalk to Ethernet interface
- Ethernet compatible cabling to VAX

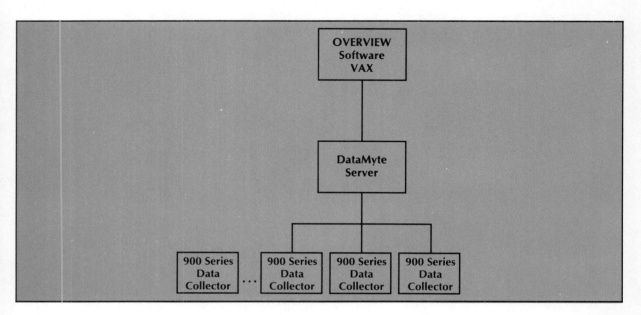

Accessories

Video Monitor

The DataMyte Video Monitor is used to display control charts and histograms. It provides visible feedback for a machine operator at the gaging station. The monitor connects by cable to the data collector.

Interface Cables

DataMyte has gage interface and communications cables for a large number of gaging and communications applications. An appropriate cable can be ordered from DataMyte to interface any of the gages featured in Chapter 21, as well as many other gages.

Battery Charger/Adaptor

The battery charger and adaptor is included with all DataMyte data collectors. There are both 110 and 220 volts AC models. The charger will bring a DataMyte up to charge in 14 to 16 hours. When connected to the DataMyte, the charger becomes an AC adaptor for fixed station operation. (The main battery must remain in the DataMyte during fixed station operation.)

19. DATAMYTE SYSTEMS FOR SPC OF PRODUCTION PROCESSES

19.1   INTRODUCTION

This chapter contains model-by-model descriptions of DataMyte data collectors and personal computer (PC) software for SPC of production processes.

Three other chapters in this Handbook are closely related to this chapter. Chapter 17 offers an introduction to automated data collection and explains the benefits of the data collector features listed in this chapter. Chapter 18 explains the DataMyte OVERVIEW™ system, a highly capable data collection and analysis system for both variables and attributes data collection. For model-by-model descriptions of DataMyte products for attributes data, inspection and roving auditing, see Chapter 20.

DataMyte fixed station data collectors are for real-time operator SPC at a stationary process. They make SPC easier, faster, error-free and more effective. Machine operators get better data, in more useable form, much faster than they could with paper and pencil.

DataMyte data collectors produce histograms, $\bar{x}$ & R charts, $\bar{x}$ & sigma charts and other control charts. Each time a subgroup (set of samples) is collected, the data collector automatically displays the desired control chart on the DataMyte monitor. The data collector automatically adds new points to a control chart as new subgroups are recorded. Each individual sample is also displayed as the operator makes the measurement.

In addition to showing control limits on the chart, fixed-station data collectors alert operators when a trend occurs, indicating the process may be out of statistical control. Once you have established the rules for treating successive data subgroups, the data collector displays a flashing warning sign and audible alarm when a trend occurs. Points out of control are clearly marked. During data collection, the data collector also alerts the operator to readings that are out of specification.

The data collector histogram displays include sigma, CP, CpK, CR and Zmin/3, both "actual" calculations and "estimated". "Estimated" statistics are based on the range of data ($\bar{R}/d_2$).

Fixed station data collectors are designed to connect to assorted electronic gages that people checking parts at a station would normally use. See Chapter 21 for some of the gages that can be used with these data collectors.

The 700/800 family, which comprises the fixed-station data collectors, can be used for multi-part or single part applications. Model 861 and 862 data collectors can track up to a total of 50 characteristics of over 15 parts. Model 761 and 762 data collectors are designed for processes or machines where 10 or fewer characteristics are being measured.

DataMyte 862 and 762 data collectors are for gages with digital output. 861 and 761 data collectors are for gages with high-level analog output, such as column gages. The DataMyte 753 data collector is for low-level analog gages (power for the gage is supplied by the data collector).

"Math" is an important feature of DataMyte data collectors. It enables you to add, subtract, multiply or divide gage readings, constants and stored data. Math can be used to compute and chart characteristics such as net weight, squareness, straightness, parallelism and total indicator reading (TIR).

Additional DataMyte products support SPC efforts at both the operator level and management level. The data collected by fixed station data collectors can be transmitted to an IBM PC by using a DataTruck® data collector. The DataTruck is a handheld device for picking up data, typically on a daily basis. Data collectors can also unload their data directly to a PC. TurboSPC™ or FAN® II software for the PC is used for data archiving, for management reports, and for further statistical analysis. The software options are explained at the end of this chapter.

DātaMyte

19.2 DIGITAL GAGE INPUT SYSTEMS

Model 862 Data Collector

The DataMyte 862 data collector is designed for any factory that has groups of automated or semi-automated machinery maintained by a single setup and inspection person. The data collector works especially well at gaging stations that handle a number of parts, or where there is a frequent changeover of parts.

The 862 data collector has three gage channels — record data from one channel at a time or from all three with a press of the footswitch. Many types of digital gages are compatible with the 862 data collector. See Chapter 21 for details. You can connect up to three DataMyte gage multiplexers to record readings from up to 24 digital indicators.

The DataMyte 862 data collector has several special features for statistical process control, including trend alert, assignable causes, and a choice of several different control charts in addition to $\bar{x}$ & R charts.

For examples of use, see pp. 9-12, 10-8, 11-5, 12-8, 13-3, 14-9, 15-14

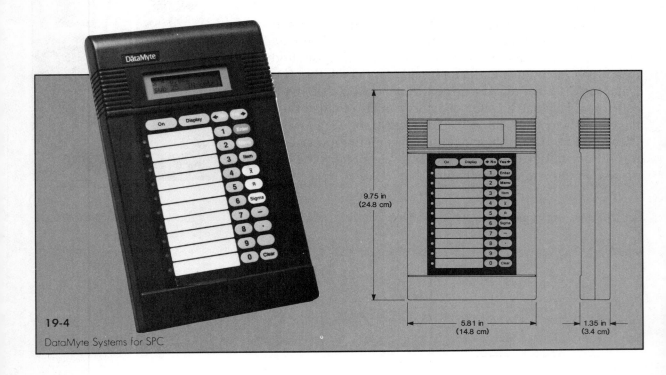

9.75 in (24.8 cm)

5.81 in (14.8 cm)

1.35 in (3.4 cm)

Special Features

- Selectable number of files and characteristics (up to 50)
- Records up to 10,000 readings
- Up to 24 gage connections possible with multiplexer
- Math: to add, subtract, multiply and divide readings, and for TIR
- Tracks more than one type of part at once, without changing setup
- Automatic scanning and recording of gage inputs
- Histogram, and $\bar{x}$ & R, $\bar{x}$ & sigma, moving $\bar{x}$ & R, and individuals with moving range charts
- Trend alert
- Assignable cause, machine and operator codes

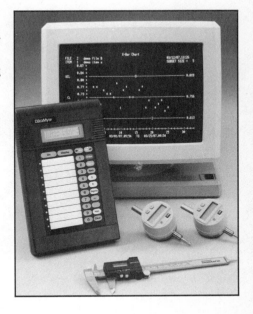

Other Features

- Interfaces with DataMyte software and DataTruck®
- Easy, menu-driven operation
- Automatic control charting
- Indicates files and memory available
- Password protection of setup and data
- Automatic time/date recording with each subgroup
- 24 sealed tactile membrane keys
- Two-line by 16-character LCD display
- Three digital/RS-232C gage ports
- Nicad battery operated, with AC charger/adaptor
- Turns off after eight minutes of non-use
- Rugged ABS plastic case, oil and water resistant
- Weighs about 2 lb. (0.9 kg)

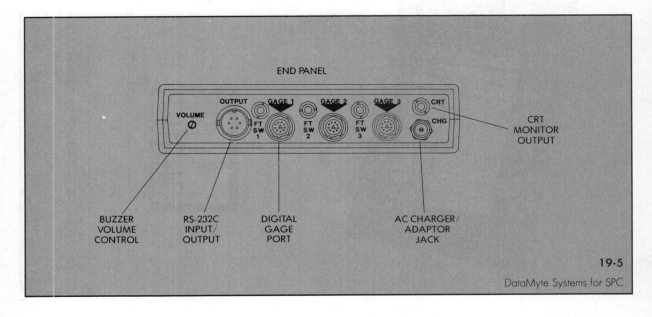

END PANEL

BUZZER VOLUME CONTROL

RS-232C INPUT/ OUTPUT

DIGITAL GAGE PORT

AC CHARGER/ ADAPTOR JACK

CRT MONITOR OUTPUT

DātaMyte

For examples of use, see pp. 9-7, 10-21, 12-9, 13-6, 14-4, 15-2

Model 762 Data Collector

Whereas the model 862 is for multiple part, quick-changeover applications, the model 762 data collector is for single-part applications. Up to ten characteristics may be charted.

Like the model 862, the model 762 connects to sophisticated measuring systems such as Mitutoyo, Sylvac and Trimos gages and many others. The data collector enables fast, easy and effective SPC in such applications as bore gaging, vertical measurement, and groove and recess gaging. The model 762 has a programmable trend-alert feature to let the operator know that the process may be out of statistical control.

Use the 762 data collector skip-counting feature to take readings from fully automated manufacturing lines. Set the 762 data collector to skip a certain number of readings before recording the next subgroup of data. This is quite frequently used in weigh-scale operations.

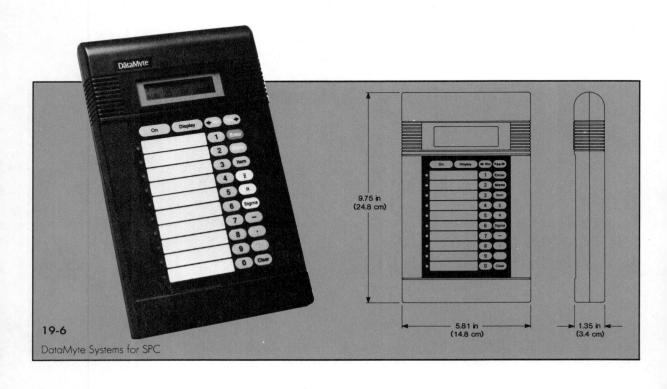

Special Features
- Selectable number of characteristics, up to 10
- 4000 reading memory size
- Expandable number of gage connections with multiplexer
- Math: to add, subtract, multiply and divide readings, and for TIR
- Automatic scanning and recording of gage inputs
- "Grouping," for flexibility in measuring characteristics
- Histogram, and $\bar{x}$ & R, $\bar{x}$ & sigma, moving $\bar{x}$ & R, and individuals with moving range charts
- Trend alerts
- Assignable cause, machine and operator codes

Other Features
- Interfaces with DataMyte software and DataTruck®
- Easy, menu-driven operation
- Automatically displays charts on CRT and printer
- Three digital/RS-232C gage ports
- Password protection of setup and data
- Automatic time/date recording with each subgroup
- 24 sealed tactile membrane keys
- Two-line by 16-character LCD display
- Nicad battery operated, with AC charger/adaptor
- Turns off after eight minutes of non-use
- Rugged ABS plastic case, oil and water resistant
- Weighs 2 lb. (0.9 kg)

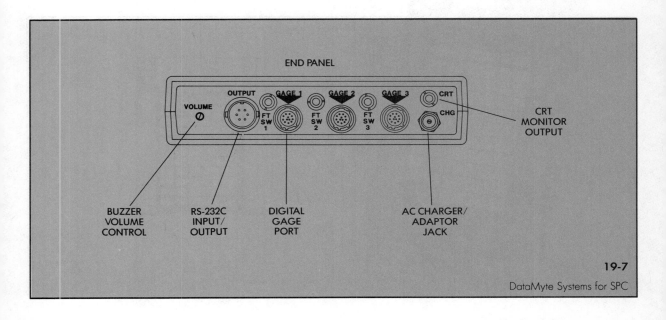

19-7

DātaMyte

19.3   ANALOG GAGE INPUT SYSTEMS

Model 861 Data Collector

Like the model 862 data collector, the model 861 data collector is designed for any factory that has groups of automated or semi-automated machinery maintained by a single setup and inspection person handling a large number of parts, or where there is a frequent changeover of parts.

The 861 data collector is used with up to 10 high-level analog output gages, typically column gages. Use the column gages in a test fixture to record several measurements of a single part at once. Readings for all gages connected to the data collector can be recorded at the same time by pressing a footswitch. The model 861 has the capacity for up to 50 characteristics in multiple parts files.

The 861 data collector has total indicator reading (TIR) to record the minimum, maximum or ranges of values of a part rotated in a fixture. It can also record the minimum, maximum or range of values from several gages for parallelism or other applications.

For examples of use, see pp. 10-7

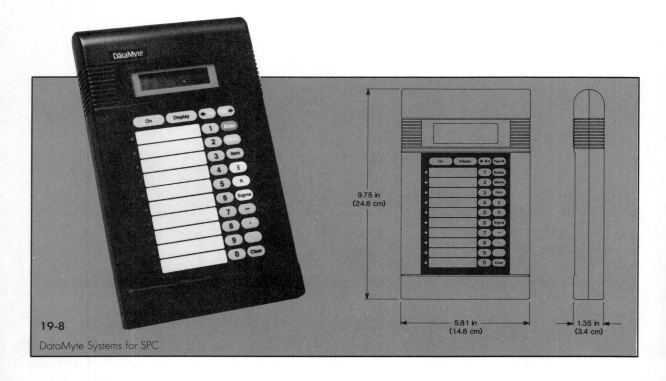

9.75 in
(24.8 cm)

5.81 in
(14.8 cm)

1.35 in
(3.4 cm)

Special Features

- Connects to up to 10 high-level analog gages (such as column gages)
- Selectable number of characteristics and files (up to 50)
- Records up to 10,000 readings
- Tracks more than one type of part at once, without changing the setup
- Math: to add, subtract, multiply and divide readings, and for TIR
- Automatic scanning and recording of gage inputs
- Histogram, and $\bar{x}$ & R, $\bar{x}$ & sigma, moving $\bar{x}$ & R, and individuals with moving range charts
- Trend alert
- Assignable cause, machine and operator codes

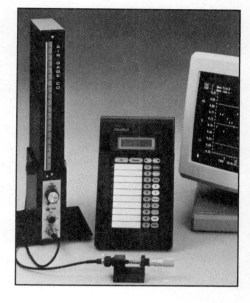

Other Features

- Interfaces with DataMyte software and DataTruck®
- Easy, menu-driven operation
- Automatic control charting
- Indicates files and memory available
- Password protection of setup and data
- Automatic time/date recording with each subgroup
- 24 sealed tactile membrane keys
- Two-line by 16-character LCD display
- Nicad battery operated, with AC charger/adaptor
- Turns off after eight minutes of non-use
- Rugged ABS plastic case, oil and water resistant
- Weighs 2 lb. (0.9 kg)

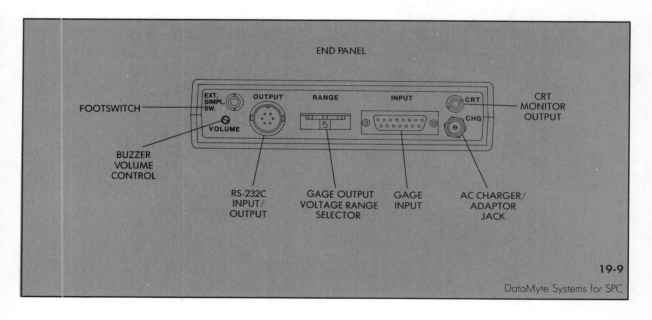

END PANEL

FOOTSWITCH

EXT. SIMPL. SW.

OUTPUT

RANGE

INPUT

CRT

CHG

CRT MONITOR OUTPUT

VOLUME

BUZZER VOLUME CONTROL

RS-232C INPUT/ OUTPUT

GAGE OUTPUT VOLTAGE RANGE SELECTOR

GAGE INPUT

AC CHARGER/ ADAPTOR JACK

DātaMyte

For examples of use, see pp. 9-17, 13-14, 16-16

Model 761 Data Collector

Whereas the model 861 is for mult-part, quick-change-over applications, the model 761 data collector is for single-part applications. Up to ten characteristics may be charted.

The DataMyte 761 data collector connects to column gages and analog displays. The 761 data collector adds memory, real-time statistical analysis and management reporting to bore gage, snap gage and other column gage applications.

Use the 761 data collector's skip-counting feature to take readings from fully automated manufacturing lines. Set the 761 data collector to skip a certain number of readings before recording the next subgroup of data.

Two models are available, for interfacing five or ten column gages. A junction box is included for interfacing. Readings are recorded with a footswitch. An autoscan feature lets the 761 data collector record up to ten readings at once. A math and total indicator reading operation allows plotting control charts on minimum, maximum, differences and other indirect variables derived from up to 10 points.

Like other DataMyte fixed-station data collectors, the 761 data collector displays complete histograms and control charts on a monitor or printer.

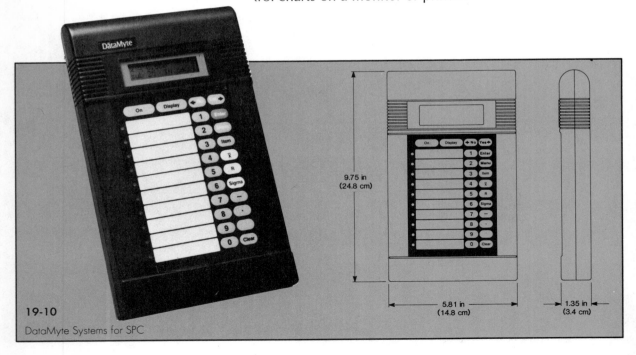

Special Features

- Connects to up to 10 analog gages (such as column gages)
- Selectable number of characteristics, up to 10
- 4000 reading memory size
- Adjustable voltage settings
- "Grouping," for flexibility in measuring characteristics
- Tracks more than one type of part at once, without changing setup
- Automatic scanning and recording of gage inputs
- Math: to add, subtract, multiply and divide readings, and for TIR
- Histogram, and $\bar{x}$ & R, $\bar{x}$ & sigma, moving $\bar{x}$ & R, and individuals with moving range charts
- Trend alert
- Assignable cause, machine and operator codes

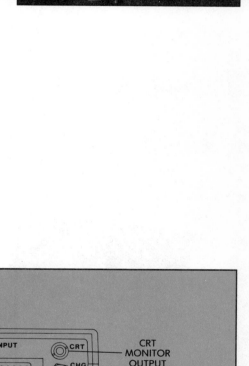

Other Features

- Interfaces with DataMyte software and DataTruck®
- Easy, menu-driven operation
- Automatic control charting
- Directory indicates files and memory available
- Password protection
- Automatic time/date recording with each subgroup
- 24 sealed tactile membrane keys
- Two-line by 16-character LCD display
- Nicad battery operated, with AC charger/adaptor
- Turns off after eight minutes of non-use
- Rugged ABS plastic case, oil and water resistant
- Weighs 2 lb. (0.9 kg)

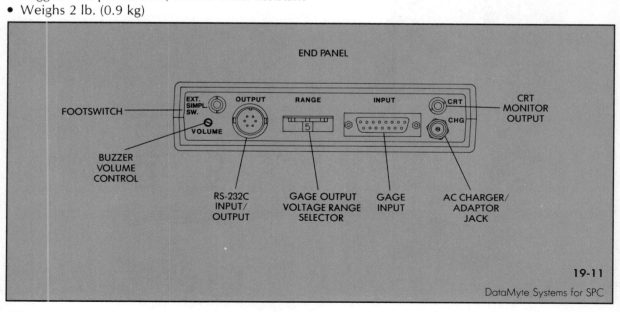

END PANEL

FOOTSWITCH

EXT. SIMPL. SW.

OUTPUT

RANGE

5

INPUT

CRT

CHG

CRT MONITOR OUTPUT

VOLUME

BUZZER VOLUME CONTROL

RS-232C INPUT/ OUTPUT

GAGE OUTPUT VOLTAGE RANGE SELECTOR

GAGE INPUT

AC CHARGER/ ADAPTOR JACK

DataMyte

Model 753 Data Collector

The DataMyte 753 data collector is a single-gage data collector for low-level analog gages (gages whose power is supplied by the data collector battery). Such gages include:

- Torque wrenches
- In-line torque transducers (for stall-type nutrunners)
- DataMyte 514 gap and 516 gap and flushness gages

Designed for production-line torque monitoring applications, the DataMyte 753 data collector has a skip function for efficient unattended statistical sampling. Special software allows reading the peak torque (or gap). To collect breakaway torque data, call DataMyte Customer Support for the appropriate DataMyte data collector. For sheet metal gap and flushness applications, a footswitch or gage sample switch lets you "back up" and retake a reading.

The 753 data collector may be used in mobile or fixed station applications. It displays complete $\bar{x}$ & R charts, $\bar{x}$ & sigma charts and histograms on a CRT monitor or printer. Like other 700/800 family data collectors, the model 753 communicates with the DataTruck, personal computers, and printers.

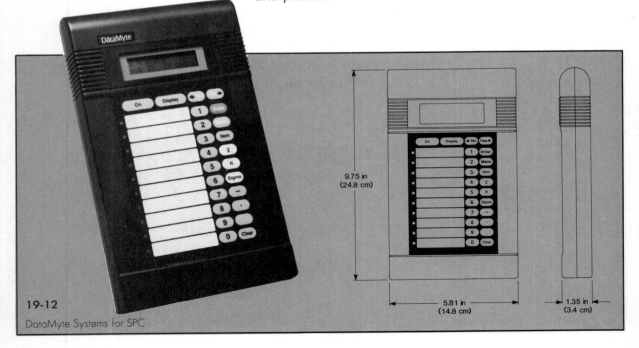

9.75 in
(24.8 cm)

5.81 in
(14.8 cm)

1.35 in
(3.4 cm)

Special Features

- Analog gage port, for 2mV or 0.8mV per 1V input
- Peak algorithm
- Skip function to bypass a certain number of readings before recording next subgroup
- Math: to add, subtract, multiply and divide readings
- 4000 reading memory size

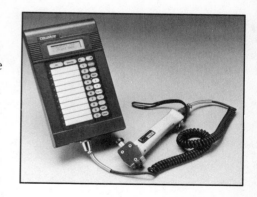

Other Features

- Interfaces with DataMyte software and DataTruck®
- Easy, menu-driven operation
- Displays charts on CRT and printer
- Ten items by 8 digits with plus and minus
- Automatic time/date recording with each subgroup
- 24 sealed tactile membrane keys
- Two-line by 16-character LCD display
- Nicad battery operated, with AC charger/adaptor
- Up to 16 hours continuous use with battery only
- Turns off after eight minutes of non-use
- Rugged ABS plastic case, oil and water resistant
- Weighs 2 lb. (0.9 kg)

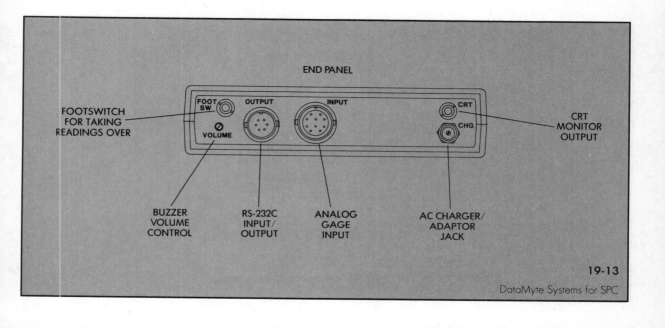

END PANEL

FOOT SW

OUTPUT

INPUT

CRT

VOLUME

CHG

FOOTSWITCH FOR TAKING READINGS OVER

CRT MONITOR OUTPUT

BUZZER VOLUME CONTROL

RS-232C INPUT/ OUTPUT

ANALOG GAGE INPUT

AC CHARGER/ ADAPTOR JACK

DataMyte

19.4 PERSONAL COMPUTER SOFTWARE

TurboSPC™ Quality Management Software

TurboSPC quality management software is a powerful SPC database package for the PC with many ease-of-use features, including pull-down menus and mouse support. TurboSPC software is used for:

- Reviewing control charts and capability performance
- Performing additional statistical analyses
- Performing "what if" type investigations
- Generating management reports
- Storing (archiving) SPC data from data collectors, or data that was collected manually
- Managing data collector setups
- Importing data from, and exporting data to, other programs

TurboSPC software is available in several versions, all with the same features except for data collector interfacing. The versions are:

- Keyboard input of variable or attribute data
- Model 769 attribute data collector interfacing
- Model 769, 2000 series, and 700/800 family data collector interfacing)

TurboSPC's powerful features include scatter plots, box and whiskers charts, and capability bar charting for comparing up to 12 characteristics at a time. The programs' ease-of-use and ease-of-learning features include:

- Pull-down menus
- Windows
- Mouse support
- On-line help
- Form-oriented data entry

System Requirements

- IBM PC/XT, PC-AT, PS/2 family or true compatible
- 512K memory
- Hard disk drive
- Graphics card (EGA, or better, recommended)
- Serial port and serial interface cable (available from DataMyte) if interfacing data collectors

TurboSPC supports the following equipment:

- Most popular mice, including IBM, Microsoft and Logitech
- IBM, Epson and Okidata serial or parallel dot matrix printers
- Any Postscript-compatible laser printer
- Hewlett-Packard LaserJet laser printers and compatibles
- Hewlett-Packard serial or parallel pen plotters
- CGA color graphics, EGA color graphics, VGA color or monochrome graphics or Hercules monochrome graphics

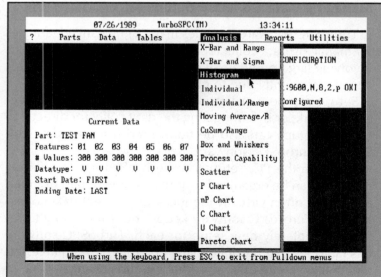

Pull-down menu showing chart analysis options.

Charting and Reporting

The charts and reports contain many statistics, as well as graphs of the data. TurboSPC software's charting and reporting capabilities for variables SPC include:

- Histogram with normal curve and normality tests
- $\bar{x}$ & R charts
- $\bar{x}$ & sigma charts
- Moving $\bar{x}$ & R charts
- Individuals with moving average charts
- Cumulative Sum (CuSum) charts
- Box and whiskers charts
- Capability bar charts
- Scatter plots

Data Analysis

- Mean
- Variance
- Standard deviation
- Normality tests, including Lillefors Test Statistic
- Kurtosis
- Min, max and median values
- Capability indices
- Other statistics, data list and frequency table reports

Managing Data Collector Setups

TurboSPC software provides many significant advantages for managing data collector setups, particularly in situations involving large numbers of data collectors or where setups change frequently due to changes in parts being run.

Setups can be transferred via the DataTruck data collector, network, or by connecting the data collector directly to the PC. Setups can also be transferred from the data collector to TurboSPC software.

Once on the PC, setups can be backed up on a floppy disk, and can be organized by job, machine or part number. Setups are often virtually identical for different data collectors, and TurboSPC software keeps you from having to re-key a completely new setup for each. TurboSPC software lets factories that produce many different parts change setups quickly and easily as production runs change.

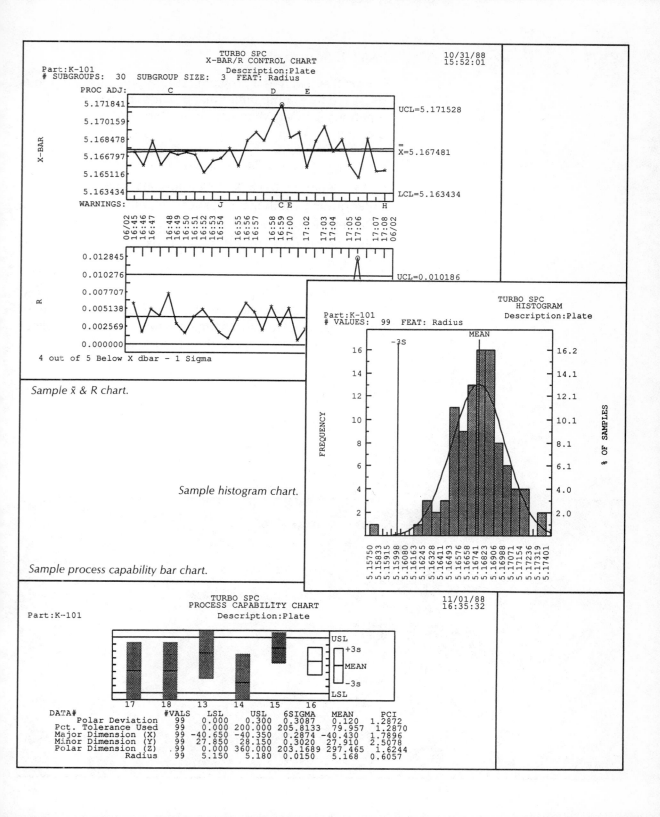

TURBO SPC
X-BAR/R CONTROL CHART   10/31/88
                          Description:Plate   15:52:01
Part:K-101
SUBGROUPS: 30 SUBGROUP SIZE: 3 FEAT: Radius

PROC ADJ: C   D   E

5.171841   UCL=5.171528
5.170159
5.168478   =
5.166797   X=5.167481
5.165116
5.163434   LCL=5.163434
WARNINGS: J   C E   H

06/02
16:45
16:46
16:47
16:48
16:49
16:50
16:51
16:52
16:53
16:54
16:55
16:56
16:57
16:58
16:59
17:00
17:02
17:03
17:04
17:05
17:06
17:07
17:08
06/02

0.012845
0.010276   UCL=0.010186
0.007707
0.005138
0.002569
0.000000

4 out of 5 Below X dbar - 1 Sigma

Sample x̄ & R chart.

TURBO SPC
HISTOGRAM
Part:K-101   Description:Plate
VALUES: 99   FEAT: Radius

-3S   MEAN

16   16.2
14   14.1
12   12.1
10   10.1
 8   8.1
 6   6.1
 4   4.0
 2   2.0

FREQUENCY   % OF SAMPLES

5.15750
5.15833
5.15915
5.15998
5.16080
5.16163
5.16245
5.16328
5.16411
5.16493
5.16576
5.16658
5.16741
5.16823
5.16906
5.16988
5.17071
5.17154
5.17236
5.17319
5.17401

Sample histogram chart.

Sample process capability bar chart.

TURBO SPC
PROCESS CAPABILITY CHART   11/01/88
                          Description:Plate   16:35:32
Part:K-101

                                                    USL
                                                    +3s
                                                    MEAN
                                                    -3s
                                                    LSL

      17   18   13   14   15   16

DATA#   #VALS   LSL   USL   6SIGMA   MEAN   PCI
  Polar Deviation   99   0.000   0.300   0.3087   0.120   1.2872
  Pct. Tolerance Used 99   0.000   200.000 205.8133   79.957   1.2870
  Major Dimension (X) 99 -40.650   -40.350   0.2874 -40.430   1.7896
  Minor Dimension (Y) 99   27.850   28.150   0.3020   27.910   2.5078
  Polar Dimension (Z) 99   0.000   360.000 203.1689 297.465   1.6244
          Radius   99   5.150   5.180   0.0150   5.168   0.6057

DātaMyte

FAN® II Software

FAN II software is a program for IBM personal computers (PCs) and true compatibles that provides management level functions for DataMyte data collectors. FAN II software is used for:

- Managing data collector setups
- Storing (archiving) SPC data from data collectors
- Performing additional statistical analyses
- Reviewing control charts
- Exporting data to dBase®, Lotus 1-2-3® and other programs

FAN II software is used for archiving SPC data collected by DataMyte data collectors. The program is also a convenient means for quickly changing setups of the data collectors. The program is menu-driven, for ease of use.

FAN II software has these capabilities:

- Graphs may contain up to 50 subgroup points, plotted by date, subgroup numbers, or other specified data ranges.
- Control limits can be entered or changed manually at any time, or calculated by the program upon request.
- Control and engineering limits may be plotted on the charts.
- Histogram reporting includes chi-square, skewness and kurtosis values, to help determine if data is truly normal.
- Capability reporting is available to help determine ability to meet spec, calculate percent parts defective, or to compare the capabilities of several items.

System Requirements
- IBM PC/XT, PC-AT, PS/2 family or true compatible
- 384K memory (RAM)
- MS-DOS or PC-DOS, version 2.0 or higher
- Two disk drives (hard disk recommended)
- Monitor (color optional)
- Asynchronous communications adaptor (serial port)
- Serial interface cable (available from DataMyte)

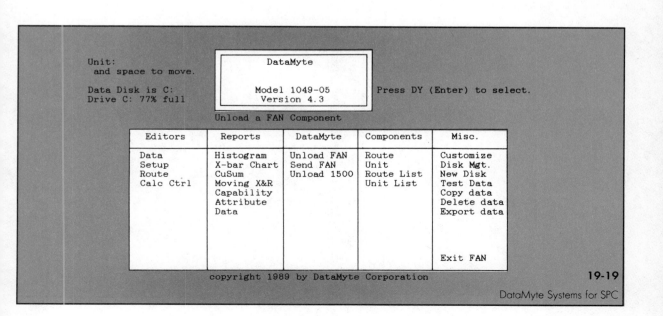

```
Unit:
  and space to move.
                              DataMyte
Data Disk is C:               Model 1049-05          Press DY (Enter) to select.
Drive C: 77% full             Version 4.3

                        Unload a FAN Component

      Editors       Reports       DataMyte      Components      Misc.

      Data          Histogram     Unload FAN    Route          Customize
      Setup         X-bar Chart   Send FAN      Unit           Disk Mgt.
      Route         CuSum         Unload 1500   Route List     New Disk
      Calc Ctrl     Moving X&R                  Unit List      Test Data
                    Capability                                 Copy data
                    Attribute                                  Delete data
                    Data                                       Export data




                                                               Exit FAN
```

Charting and Reporting

FAN II software can do the same control charting as DataMyte data collectors, plus some additional reporting functions for supplier certification, capability studies, or review of SPC data. The charts and reports contain many statistics, as well as graphs of the data. Most charts and reports contain a separate page listing assignable cause, machine and operator codes.

For each report, FAN II software lets the user select data by date or subgroup number and by characteristic. If the computer running FAN software has graphics capability, it displays charts on the screen — if desired — before printing them. When new data for a file is transferred to the computer, FAN II software adds the data to the end of the file. These merged data files help set up long-range studies of process performance. "What-if" analysis can be done with FAN II software by editing out data points and recalculating control limits.

Sample x̄ & R report, with assignable cause codes shown on the report.

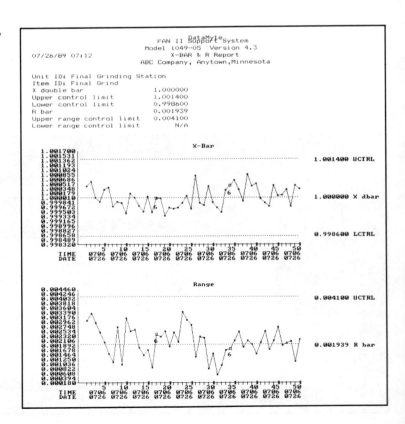

Managing Data Collector Setups

FAN II software provides many significant advantages for managing data collector setups, particularly in situations involving large numbers of data collectors or where setups change frequently due to changes in parts being run.

Setups can be transferred via the DataTruck data collector, network, or by connecting the data collector directly to the PC. Setups can also be transferred from the data collector to FAN II software.

Once on the PC, setups can be saved on a floppy disk, and can be organized by job, machine or part number. Setups are often virtually identical for different data collectors, and FAN II software keeps you from having to re-key a completely new setup for each. FAN II software lets factories that produce many different parts change setups quickly and easily as production runs change.

Sample histogram report.

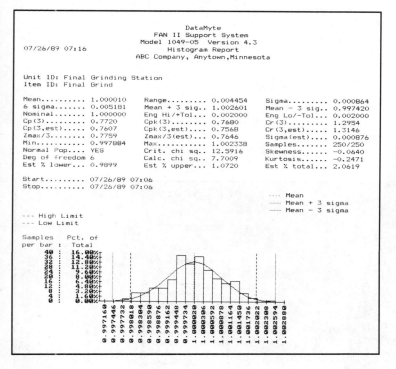

DātaMyte

For an example of use, see p. 15-13

19.5 ACCESSORIES

DataTruck® Data Collector

The DataTruck is used to harvest data from fixed-station data collectors located throughout a plant. It is the vital link in the unwired FAN® (Factory Area Network) system. The DataTruck data collector can transfer setup information to a data collector as well as receive its data and transmit it to an IBM PC. See Chapter 17 for a description of the FAN system and how the DataTruck works within it.

Transferring of a file to or from a data collector takes about ten seconds. Depending on the number of readings in the data collectors, a DataTruck can harvest data from up to 20 data collectors.

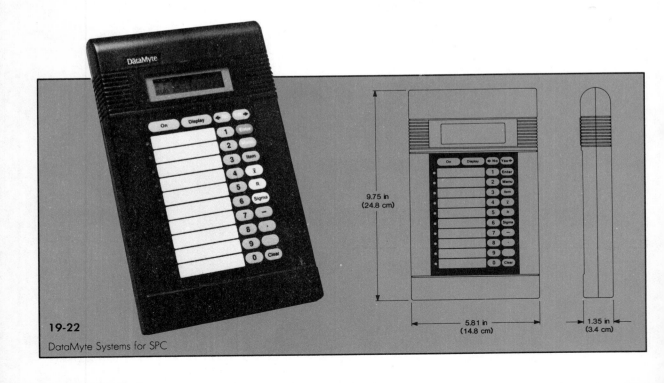

9.75 in
(24.8 cm)

5.81 in
(14.8 cm)

1.35 in
(3.4 cm)

Special Features
- High speed data transfer
- 50,000 reading capacity

Other Features
- Interfaces with DataMyte software
- Easy, menu-driven operation
- 24 sealed tactile membrane keys
- Two-line by 16 character LCD display
- Selectable RS-232C transmission rates
- Nicad battery operated, with AC charger/adaptor
- Up to 16 hours continuous use with battery only
- Unit turns off after eight minutes of non-use
- Rugged ABS plastic case, oil and water resistant
- Cabling for connection to PC
- Weighs 2 lb. (0.9 kg)

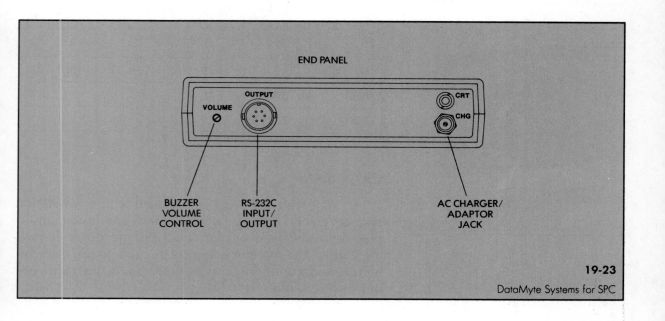

END PANEL

VOLUME

OUTPUT

CRT

CHG

BUZZER
VOLUME
CONTROL

RS-232C
INPUT/
OUTPUT

AC CHARGER/
ADAPTOR
JACK

DataMyte Gage Multiplexers

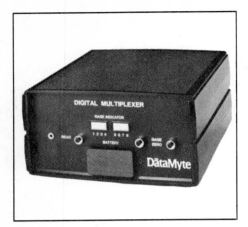

DataMyte gage multiplexers are ideal for multiple-gage fixtures, where many measurements of one part of assembly are taken at the same time. Once the part is in the fixture, and gages — typically digital indicators — are put into place, you record readings from *all* the gages in the data collector by pressing the footswitch once.

Each multiplexer can connect to eight gages. And each data collector can connect to up to three multiplexers, for a maximum of 24 gages connected to one data collector. Any one multiplexer supports the gages of only one manufacturer. Thus gages can be connected from up to three different manufacturers. The multiplexer extends gage capabilities of DataMyte data collectors to up to 24 gages.

See the multiplexer chart below for the appropriate multiplexer models for a particular DataMyte data collector. All necessary interface cables are available from DataMyte.

With model 762 and 862 data collectors, you can connect other gages (with the Mitutoyo Digimatic format output) to the multiplexer. For instance, you might have several Mitutoyo digital indicators connected to a multiplexer, as well as a height gage and a bore gage.

DataMyte Gage Multiplexer Model Numbers and Gage Compatibility

	Mitutoyo Digimatic 543	Fowler Ultra-Digit I or II	Federal Maxμm	Ono-Sokki
862/762	·529-15	529-16	529-17	529-18
2000	529-11	529-12		
1500	529-13	529-14		

CRT Monitor

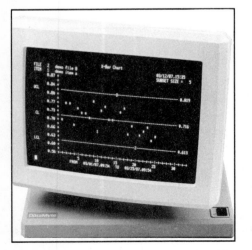

The DataMyte CRT Monitor is used with fixed-station data collectors to display control charts and histograms. It provides visible feedback for a machine operator at the gaging station. The monitor connects by cable to the data collector, and communicates at 9600 baud, so charts are quickly displayed.

The monitor has a 12-inch diagonal screen and uses the ANSI (American National Standards Institute) extended character set. The ANSI character set allows the monitor to display points that are part of a trend to blink continuously.

Printers

Printers are available from DataMyte for printing charts and data logs directly from the DataMyte data collector. All are configured for serial RS-232C communications. For the 700/800 family data collectors, one prints at 150 characters per second and the other at 240 characters per second. An interface cable is included with each printer.

Battery Charger/Adaptor

A battery charger and an AC adaptor are included with all DataMyte data collectors. There are both 110 and 220 volt AC models. The charger brings a DataMyte data collector up to charge in 14 to 16 hours. When connected to the DataMyte data collector, the charger becomes an AC adaptor for fixed station operation.

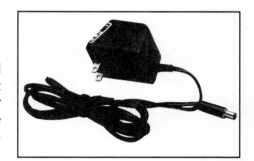

Interface Cables

DataMyte has gage interface and communications cables for a large number of gaging and communications applications. An appropriate cable can be ordered from DataMyte to interface with any of the gages featured in Chapter 21, as well as many other gages. Communications cables are available to interface with DataMyte data collectors to a number of popular computer systems.

Other Accessories

Footswitches are included for the 700/800 family of data collectors, for faster, more convenient data entry. Other accessories are also available.

20. DATAMYTE SYSTEMS FOR AUDITING AND INSPECTION

20.1 DATA COLLECTORS FOR INSPECTION

This chapter contains model by model descriptions of DataMyte data collectors used for auditing and inspection and the software that is used with them.

The DataMyte 769, 2000 Series (2003 and 2004), 1500 Series (1556 and 1558) and the 1005 and 1010 are used for auditing and inspection applications. These data collectors are used to record whatever you can see, count, grade, or note.

Traditional inspection systems often involve four steps: First the inspector collects data on paper check sheets. Then that information is entered into a personal computer. During the third step, the data is analyzed with a spread sheet or data base program. The process ends with reporting the results to management so they can make corrections.

The 769 data collector is used to record the results of a bottle inspection process.

With the inspection systems discussed in this chapter, in particular the DataMyte 769, this four-step process can be compressed into a single step.

The DataMyte 769 prints its own bar code data collection sheet that is used by the operator at the inspection station. The inspection sheet and bar code wand allow the inspector to quickly and accurately enter the data. The information is immediately analyzed by the data collector. And the outcome of the analysis can be displayed in pareto charts and data reports on the DataMyte monitor.

A traditional inspection process that currently takes hours or days is reduced to minutes.

The 769 data collector used to record inspection results.

DataMyte

For examples of use, see pp. 9-3, 9-8, 10-2, 10-24, 11-3, 11-11, 12-4, 12-19, 13-8, 14-18, 15-5, and 15-25.

769 Attributes Data Collector

The DataMyte 769 data collector lets operators quickly record defects with a bar code wand and look at Pareto and percent-defectives charts as they are working. The DataMyte 769 works well in electronics, packaging, automobile, defense and other industries where attributes data collection is used. The 769 data collector is set up by first describing the product, and then describing what you are looking for. You can use up to 250 fields to describe the product and list the attributes.

The 769 simplifies any application where a product is counted, inspected for appearance or defects, or is graded, classified or tested for proper operation. The 769 is effective for incoming inspection, in-process audits, surface inspection, pass-fail testing, packaging inspection, final inspection, and scrap reporting.

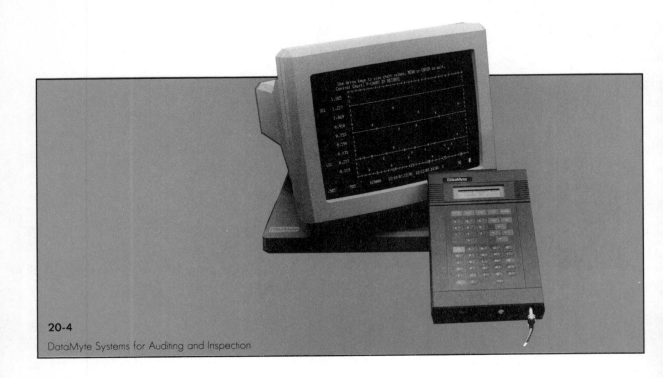

Special Features

- User definable subgroups
- Four data collection modes (counted units, batched units, single area and multiple areas)
- Code 39 bar code input
- Up to 30 charts can be defined and stored
- Print charts and data reports on optional printer
- Rework reports
- DataMyte monitor displays data collector screens and charts
- Accepts laser scanner input
- Support program for IBM PC or compatible computer

Other Features

- 512k bytes memory for data
- Menu-driven operation
- Date and time recorded with each record
- Two-line by 16-character LCD display, with adjustable contrast
- Complete alphanumeric keyboard
- Nicad battery operated, with AC charger/adaptor
- Memory retention: 6 to 8 weeks with fully charged battery, or continuous with AC adaptor
- Turns off after programmable number of minutes
- Rugged ABS plastic case, oil and water-resistant
- Weighs about 2 lb. (0.9 kg)
- Password protection

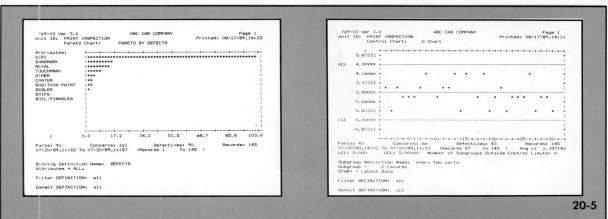

DataMyte Systems for Auditing and Inspection

20.2 DATA COLLECTORS FOR ROVING AUDITING

DataMyte® 2000 Series Data Collectors

The DataMyte 2000 Series is a comprehensive SPC system for quality professionals. It is used in roving applications for SPC, inspection and auditing. In a single handheld unit, a 2000 Series data collector provides:

- Direct input for both digital and analog gages
- Direct bar code input
- Variables and attributes auditing
- Complete alphanumeric keyboard
- Statistical charting capability on its own 40-character by eight-line display
- Memory capacity for auditing virtually an unlimited number of characteristics

For examples of use, see pp. 9-17, 10-8, 10-14, 10-17, 10-18, 10-24, 11-15, 12-5, 13-9, 14-2, 14-4, 14-23, 15-3, 15-7, 15-15, 15-19, 15-27, and 16-3.

The 2000 Series consists of the 2003 and the 2004 data collectors. The difference is this: the 2003 data collector can interface to Fowler Sylvac gages but not Federal Maxμm gages; the 2004 data collector can interface to Federal Maxμm but not Fowler Sylvac gages.

Total Quality Control With the 2000 Series

The DataMyte 2000 Series is made for real-world decision making. In the hands of a quality professional, 2000 Series data collectors are powerful aids to total quality control activities. Alphanumeric footnotes can be linked with data from a gage, or points on a control chart. Part numbers, production counts and percentage defective can be tied together so that all of the details — both observed and measured — are part of an analysis. Practically anything that can be done on paper can be done faster and with less error with 2000 Series data collectors.

2000 Series data collectors accomplish this with flexible memory and processing speed. Entries can be any length. Pre-defined operations can be executed with two key presses. Reports and graphs contain both data analysis and alphanumeric notes. Audit routes and studies can be set up, and data gathered and analyzed without tying up desktop computers and other resources.

The 2000 series data collector.

DataMyte 2000 Series data collectors can be used wherever process volumes do not require a fixed station data collector like the DataMyte 862 data collector. They can audit low volume processes, sequential operations, tasks on moving assembly lines, and do periodic capability studies and quality checks. They can be assigned to a continuous daily route preprogrammed by a computer, or be used to randomly log defects and other attributes wherever they are found.

DātaMyte

2003 Data Collector

The DataMyte 2003 data collector accepts direct input from both digital and analog gages including Fowler Sylvac gages.

The 2003 data collector has a 54-key keyboard that allows full alphanumeric input. The user can quickly enter footnotes to data, and execute user-defined function keys. The eight-line by 40-character display provides a complete status update during data collection, with statistical summaries and 80-column wide x̄ & R charts, x̄ & s charts and histograms. The data collector also connects to the DataMyte monitor to display graphs. Graphs can be output to a dot matrix printer, and data transmitted to a desktop computer for archiving.

The 2003 data collector is compatible with the DataMyte fixed-station data collectors and the DataMyte PC software packages. It has a terminal mode for sending and receiving large blocks of data. All functions of the 2003 can be programmed from a remote keyboard or computer program, allowing routes to be set up and downloaded on a daily basis.

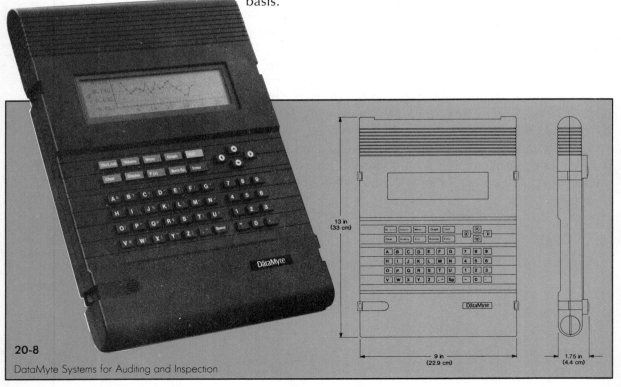

20-8

Special Features

- Fowler Sylvac compatibility
- Three input ports, for digital gages, analog gages and bar code wand
- Control charts and histograms on unit display
- Alphanumeric footnotes
- Math function for sums and differences of items
- 256K memory (512K numeric characters)
- Ten user-definable function keys
- Password protection of setup and data
- Displays charts on the DataMyte monitor

Other Features

- Interfaces with DataMyte software
- Menu driven with help screens and status displays
- Pull-down windows for selecting options
- User-definable headers for data collection
- Time/day/date labeling for subgroups
- Terminal mode
- RS-232C I/O port with fully selectable communications protocol
- Nicad battery powered, with removable battery pack
- Weighs 4.0 lb. (1.8 kg)

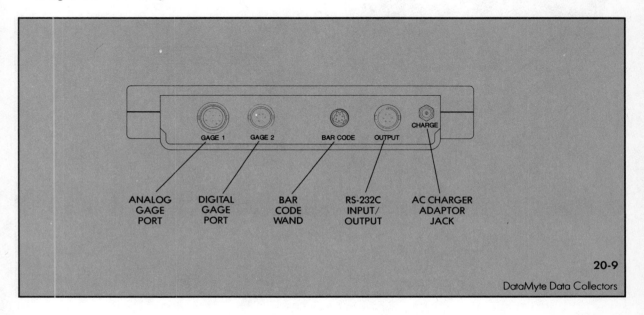

ANALOG GAGE PORT — GAGE 1
DIGITAL GAGE PORT — GAGE 2
BAR CODE WAND — BAR CODE
RS-232C INPUT/OUTPUT — OUTPUT
AC CHARGER ADAPTOR JACK — CHARGE

DātaMyte

2004 Data Collector

The DataMyte 2004 data collector accepts direct input from both digital and analog gages including the Federal Maxum.

The 2004 data collector has a 54-key keyboard that allows full alphanumeric input. The user can quickly enter footnotes to data, and execute user-defined function keys. The eight-line by 40-character display provides a complete status update during data collection, with statistical summaries and 80-column wide $\bar{x}$ & R charts, $\bar{x}$ & s charts and histograms. Graphs can be output to a dot matrix printer, and data transmitted to a desktop computer for archiving.

The 2004 data collector is compatible with the DataMyte fixed-station data collectors and the DataMyte PC software packages. It has a terminal mode for sending and receiving large blocks of data. All functions of the 2004 can be programmed from a remote keyboard or computer program, allowing routes to be set up and downloaded on a daily basis.

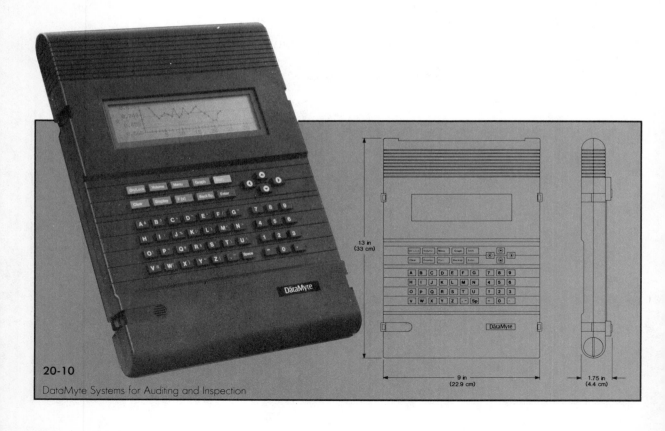

13 in (33 cm)

9 in (22.9 cm)

1.75 in (4.4 cm)

DataMyte Systems for Auditing and Inspection

Special Features

- Federal Maxum gage compatibility
- Three input ports, for digital gages, analog gages and bar code wand
- Control charts and histograms on unit display
- Alphanumeric footnotes
- Math function for sums and differences of items
- 256K memory (512K numeric characters)
- Ten user-definable function keys
- Password protection of setup and data
- Displays charts on the DataMyte monitor

Other Features

- Interfaces with DataMyte software
- Menu driven with help screens and status displays
- Pull-down windows for selecting options
- User-definable headers for data collection
- Time/day/date labeling for subgroups
- Terminal mode
- RS-232C I/O port with fully selectable communications protocol
- Nicad battery powered, with removable battery pack
- Weighs 4.0 lb. (1.8 kg)

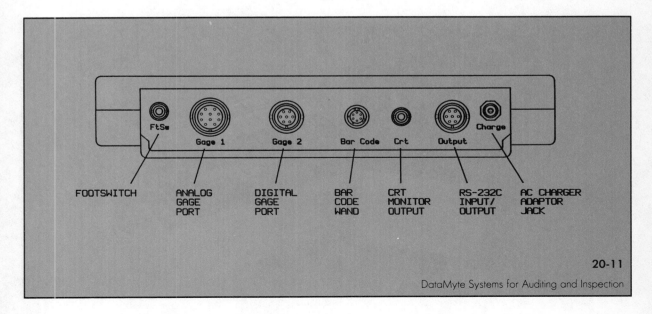

DataMyte

For an example of use, see pp. 10-12.

DataMyte® 1500 Series Data Collectors

1500 Series data collectors are rugged handheld data collection systems that interface to a variety of gages for direct input of data. They can print reports on a printer or communicate with a computer. The 1500 Series data collectors are made for a roving QC auditor — for applications where mobility and flexibility are needed.

The 1500 Series consists of the 1556 and 1558 data collectors. The 1556 data collector connects to analog gages such as torque wrenches, in-line torque transducers and gap and flushness gages. The 1558 data collector connects to digital gages such as micrometers, calipers, digital indicators and weigh scales. See Chapter 21 for gages.

Special Features of the 1556 Data Collector

- Accepts readings from torque wrenches, gap and flushness gages, force gages and other analog transducers
- Special algorithms for reading peak and torque breakaway
- Internal resolution of 0.004 of full scale

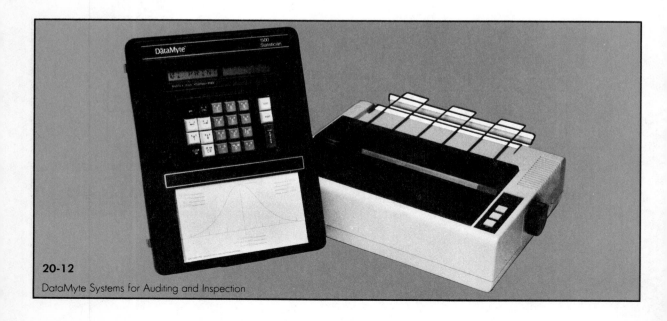

20-12
DataMyte Systems for Auditing and Inspection

Special Features of the 1558 Data Collector
- Accepts readings from calipers, micrometers, linear gages and other digital gages
- Resolution to 0.0001 in. (0.001 mm)

Other Features
- Calculates and prints $\bar{x}$ & R charts, $\bar{x}$ & s charts, histograms and capability studies
- Saves all $\bar{x}$ & R chart data
- Math feature for sum and difference of items and constants
- Alerts operator to item out of specifications and subgroups out of control
- Capacity for up to 40 ongoing control charts
- 64K character memory size
- Serial RS-232C I/O channel, with full parameter selection
- Remote control mode for download and upload of setups and data
- Fully selectable communications protocol
- Nicad battery operated, with 14 hours continuous use between chargings
- Weighs 4.25 lb.(2.2 kg)

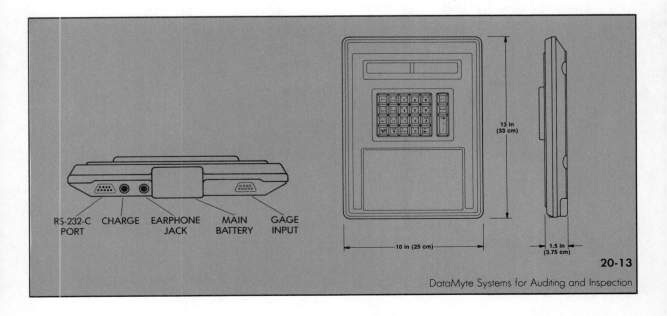

RS-232-C PORT   CHARGE   EARPHONE JACK   MAIN BATTERY   GAGE INPUT

13 in (33 cm)

10 in (25 cm)

1.5 in (3.75 cm)

DataMyte

1005 and 1010 Data Collectors

The model 1005 and 1010 can be used for attributes data collection and time study. The model 1005 and 1010 memory is flexible, allowing up to 32 alphanumeric characters per entry. It can be configured to fit a given application, whether paint auditing, inventory cycle counting, preventive maintenance data collection, or magnetic stripe and bar code reading.

The handheld data collector has a 16-character LCD display, 23-key keypad, RS-232C port, and audible buzzer. There are many features worth noting in particular. It has a programmable clock, letting you automatically enter a clock time with each data entry. It has prompting, including nested prompts — loops within loops of prompt messages — that sequence data collection in a variety of ways. The devices it can interface with include serial RS-232C gages or fixed station equipment such as electronic balances; also, an optional bar code wand, IBM magnetic hand scanner and others. The 1010 does a good job of processing time study data. Use your own coding system to enter data. You can simply plug the 1010 into its dedicated printer, and the 1010 calculates and prints the time study reports all by itself.

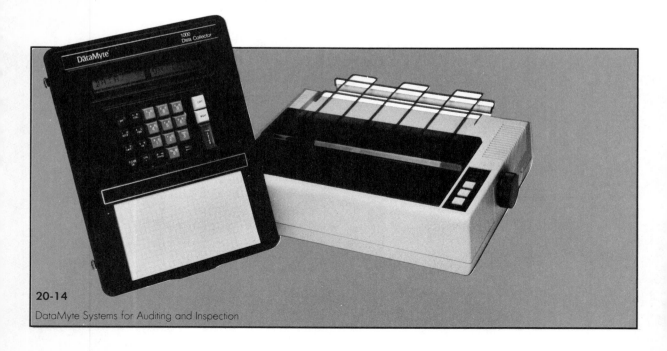

Distinguishing Features

- Full alphanumeric entry — Entries up to 32 characters wide
- 64K character memory size — Up to 8 separate records
- Programmable headers — Prompts, and parts list sections
- Terminal mode — Keyboard sign-on to a computer system
- Clock — Military time, tenths, hundredths of a minute and thousandths of an hour
- Serial — RS-232C data input capability
- Data collection — Data logging, auto entry, and limit checking
- Computer control — Downloading and uploading
- Reports formatter — Setting line lengths, etc.
- Weight — 4.25 lb. (2.2 kg)
- Operating Temp. — 32 to 122° F (0 to 50° C) for optimum battery performance
- Main Battery — Rechargeable, removable, 12 to 18 hours of operation
- Memory Retention — 15 hours without main battery, 200 hours with main battery
- Keyboard — Alphanumeric (0-9, A-Z, some symbols)
- Input/Output — ASCII serial RS-232C port or 20 mA current loop
- Transmission Rate — Selectable baud rates of 110, 150, 300, 600, 1200, 2400, or 4800
- External control — Stop transmission, resume transmission, transmit next line (ACK), transmit last line (NAK)
- Display — 16-digit, 0.315 inch (0.8 cm) LCD

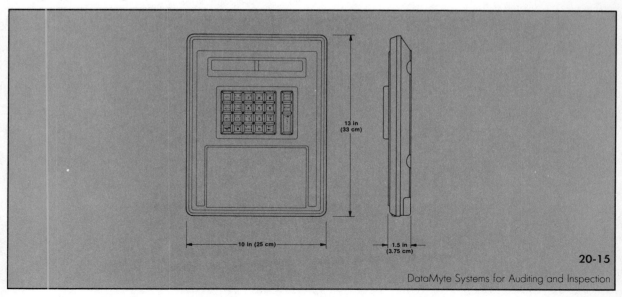

20-15

DataMyte

20.3 COMPUTER SOFTWARE FOR INSPECTION AND AUDITING

TurboSPC™ Quality Management Software

TurboSPC quality management software is a powerful SPC database package for the PC with many ease-of-use features, including pull-down menus and mouse support. TurboSPC software is used for:

- Reviewing control charts and capability performance
- Performing additional statistical analyses
- Performing ''what if'' type investigations
- Generating management reports
- Storing (archiving) SPC data from data collectors, or data that was collected manually
- Managing data collector setups
- Importing data from, and exporting data to, other programs

TurboSPC software is available in several versions, all with the same features except for data collector interfacing. The versions are:

- Keyboard input of variable or attribute data
- Model 769 attribute data collector interfacing
- Model 769, 2000 series and 700/800 family data collector interfacing

TurboSPC software has these charting capabilities for attributes data from the 769 data collector or 2000 series data collectors:
  Pareto, p, np, and u-charts.

The program's ease of use and ease of learning features include:

- Pull-down menus
- Windows
- Mouse support
- On-line help
- Form-oriented data entry

System Requirements
- IBM PC/XT, PC-AT, PS/2 family or true compatible
- 512K memory
- Hard disk drive
- Graphics card (EGA, or better, recommended)
- Serial port and serial interface cable (available from DataMyte) if interfacing data collectors

TurboSPC supports the following equipment:

- Most popular mice, including IBM, Microsoft and Logitech
- IBM, Epson and Okidata serial or parallel dot matrix printers
- Any Postscript-compatible laser printer
- Hewlett-Packard LaserJet laser printers and compatibles
- Hewlett-Packard serial or parallel pen plotters
- CGA color graphics, EGA color graphics, VGA color or monochrome graphics or Hercules monochrome graphics

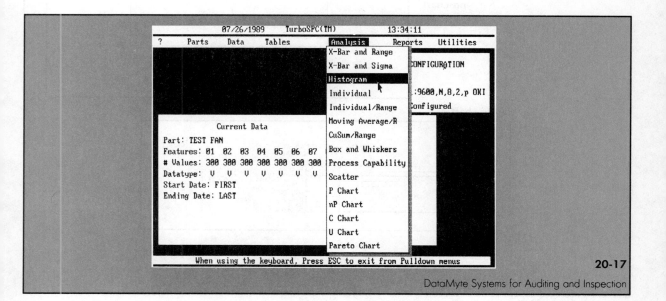

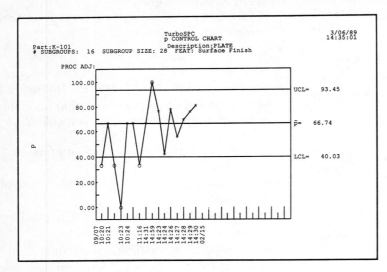

p-chart

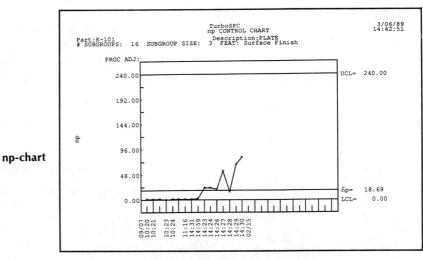

np-chart

Pareto chart

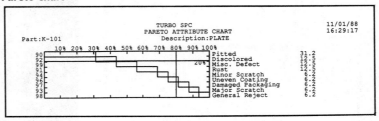

Managing Data Collector Setups

TurboSPC software provides many significant advantages for managing data collector setups, particularly in situations involving large numbers of data collectors or where setups change frequently due to changes in parts being run.

Setups can be transferred via the DataTruck data collector, network, or by connecting the data collector directly to the PC. Setups can also be transferred from the data collector to TurboSPC software.

Once on the PC, setups can be backed up on a floppy disk, and can be organized by job, machine or part number. Setups are often virtually identical for different data collectors, and TurboSPC software keeps you from having to re-key a completely new setup for each. TurboSPC software lets factories that produce many different parts change setups quickly and easily as production runs change.

DataMyte

DATAMYTE ATTRIBUTE SOFTWARE PROGRAM

The DataMyte Attribute Software program analyzes and stores SPC attribute data that has been collected by DataMyte 1000 series data collection systems. Data is changed into graphic form so better decisions can be made, rather than the QC manager having to rely just on numbers. This program lets the user send setup information to DataMyte 1005 and 1010 data collectors and unload data from these data collectors directly from the computer.

The user can edit and look at prompt loops used in the data collectors, edit and review collected data, format the data to be put into spreadsheets and data bases and back up data onto another disk. The program will also quickly and easily print bar code labels so data collection is more efficient, using a bar code wand.

20-20

System Requirements
- IBM PC or true compatible
- 256K memory
- Two disk drives
- Monitor (color optional)
- Asynchronous communications adapter
- Dot matrix parallel printer
- Serial interface cable (available form DataMyte)

Charting
- Pareto Charts
- p- and np- charts
- c-, c/100- and c/1000-charts
- u-charts
- Up to 50 subgroup points per chart
- Control limits are calculated by the program after 20 sets of data are collected

Communicating with the Data Collector
- Edit prompt loops for data collection
- Print bar code labels for data collection

File Management
- Data is stored on floppy or hard disks
- User specifies disk drive for data storage
- Data files are organized by Audit ID, date and time of data collection
- Spreadsheet-like editor for data files
- Save, copy, delete or edit data files

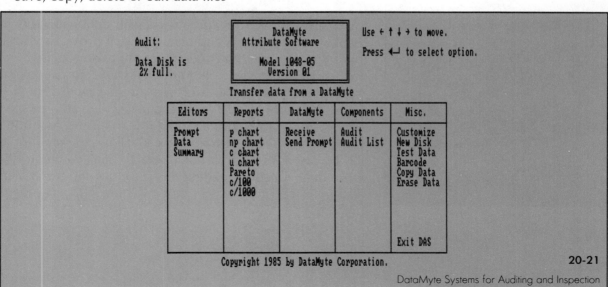

DataMyte Systems for Auditing and Inspection

20.4 ACCESSORIES FOR DATAMYTE DATA COLLECTORS

DataMyte® Gage Multiplexers

DataMyte gage multiplexers are ideal for multiple-gage fixtures, where many measurements of one part or assembly are taken at the same time. Once the part is in the fixture, and gages — typically digital indicators — are put into place, you record readings from *all* the gages in the data collector by pressing the footswitch once.

The 2000 series data collectors can be used with the DataMyte multiplexers for Mitutoyo and Fowler Ultradigit gages. Each multiplexer can connect to eight gages. Any one multiplexer supports the gages of only one manufacturer. The 2000 series can connect to more than one multiplexer. The 2000 series need to "daisychain" the multiplexers together. Daisychaining means that the multiplexers are connected to each other, and one of the multiplexers is connected to the data collector. All necessary interface cables are available from DataMyte.

DataMyte Monitor

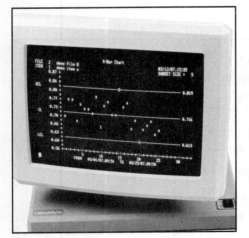

The DataMyte® Monitor is used with data collectors to display control charts and histograms. It provides visible feedback for a machine operator at the gaging station. The monitor connects by cable to the data collector, and communicates at 9600 baud, so charts are quickly displayed.

The monitor has a 12-inch diagonal screen and uses the ANSI (American National Standards Institute) extended character set. The ANSI character set allows the monitor to display points that are part of a trend to blink continuously, if the data collector sending the chart has the trend alert feature.

Printers

Printers are available from DataMyte for printing charts and data logs directly from the DataMyte data collector. A printer is required for obtaining graphic reports from the 1500 or 1000 Series. An interface cable is included with each printer.

Battery Charger/Adaptor

A battery charger and an AC adaptor are included with all DataMyte data collectors. There are both 110 and 220 volt AC models. The charger brings a DataMyte data collector up to charge in 14 to 16 hours. When connected to the DataMyte data collector, the charger becomes an AC adaptor for fixed station operation. (The main battery must remain in the DataMyte data collector during fixed station operation.)

Battery Charger for DataMyte 2000 Series

The 2000 battery charger accessory, included with all 2000 series data collectors, charges the 2000 battery pack while it is removed from the data collector. This lets you keep a spare battery pack charged all the time. The charger accessory plugs into the AC charger/adaptor.

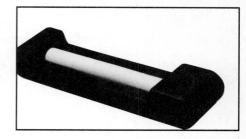

Interface Cables

DataMyte has gage interface and communications cables for a large number of gaging and communications applications. An appropriate cable can be ordered from DataMyte to interface with any of the gages featured in Chapter 21, as well as many other gages. Communications cables are available to interface with DataMyte data collectors to a number of popular computer systems.

Other Accessories

A footswitch is included with the 2004 data collector, for faster, more convenient data entry. A shoulder strap is included with each 2000 Series data collector. Other accessories are also available.

21. GAGES

21.1 INTRODUCTION

This chapter presents a sampling of current gaging that interfaces with DataMyte data collectors. Since DataMyte has engineered and tested interfaces to gages and instruments from over 200 manufacturers, those listed here are only a fraction of the gages to which DataMyte data collectors interface. This engineering and testing is necessary because no standard for digital output format for gages exists; gage manufacturers use many different output formats.

Brief specification information is listed for most of the gages. For more details, contact the manufacturer or distributor. DataMyte data collectors to which the gage interfaces are also listed.

Several specialized gap gages and torque wrenches offered exclusively by DataMyte are detailed at the end of this chapter.

21.2 HANDHELD GAGING

Following is a sampling of handheld gaging for SPC that interfaces to DataMyte data collectors — calipers, micrometers, contour and gap gages, force gages and torque sensing devices. Many other brands and types of handheld gages not listed here are also supported by DataMyte data collectors.

Mauser Caliper

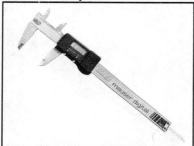

Chicago Dial Indicator Co.
1372 Redeker, Des Plaines, IL 60016
(312) 827-7186

- Resolution: .001 mm (.0005 in.)
- Various ranges from 150 to 1000 mm

Interfaces with DataMyte 762, 862, 2003, 900 Series

Max-Cal Caliper

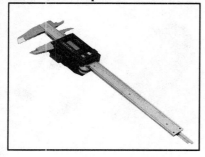

Fred V. Fowler Co., Inc.
66 Rave Street, Newton, MA 02166
(617) 332-7004

- Range: 150 mm (6 in.) or 200 mm (8 in.)
- Resolution: 0.01 mm (0.0005 in.)
- Accuracy: ± 0.03 mm (± 0.011 in.)
- Quantization Error: 1 digit

Interfaces with DataMyte 762, 862, 1558, 2000 Series, 900 Series

Mitutoyo Digimatic Caliper 500-321 and 500-322

MTI Corporation
18 Essex Road, Paramus, NJ 07652
(201) 368-0525

- Range: 6 in. (152 mm), 8 in. (203 mm)
- Resolution: .0005 in. (0.01 mm)
- Accuracy: ± .001 in.

Interfaces with DataMyte 762, 862, 1558, 2000 Series, 900 Series

Interfaces with DataMyte 762, 862, 2003, 900 Series

S-T Industries
301 Armstrong Boulevard, St. James, MN 56081
(507) 375-3211

- Range: 4 models from 6 to up to 24 in.
- Accuracy: ± .001 in. (.025 mm)
- Resolution: .0005 in. (.01 mm)

Scherr Tumico Group 10 Electronic Calipers

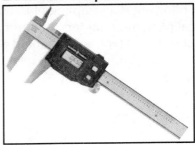

Interfaces with DataMyte 762, 862, 2000 Series, 900 Series

L.S. Starrett Company
121 Crescent Street, Athol, MA 01331
(508) 249-3551

- Accuracy ± .025 mm (± .001 in.)
- Range: 150 mm (0 to 6 in.)

Starrett 722 Caliper

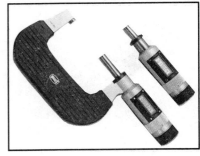

Interfaces with DataMyte 762, 862, 2003, 900 Series

Chicago Dial Indicator Company
1372 Redeker, Des Plaines, IL 60016
(312) 827-7186

- Range: 0-25 mm, 25-50 mm, 50-75 mm, 75-100 mm
- Resolution: .001 mm (.00005 in.)

Mauser Micrometer

Interfaces with DataMyte 762, 862, 1558, 2000 Series, 900 Series

Fred V. Fowler Co., Inc.
66 Rave Street, Newton, MA 02166
(612) 332-7004

- Range: Various models have ranges from 0-25 mm (1-1 in.) to 75-100 mm (3-4 in.)
- Resolution: .001 mm (.0001 in. or .00005 in.)
- Accuracy: ± .001 mm (.0001 in.)
- Quantization Error: 1 digit

Fowler Digitrix II Micrometer

ALL SPECIFICATIONS SUBJECT TO CHANGE WITHOUT NOTICE.

Mitutoyo Series 293 Micrometers

MTI Corporation
18 Essex Road, Parmus, NJ 07652
(201) 368-0525

- Range: Various models have ranges from 0-1 in. to 11-12 in.
- LCD Display: .00005 in. (.001 mm)

Interfaces with DataMyte 762, 862, 1558, 2000 Series, 900 Series

Demco Contour and Outline Transducer

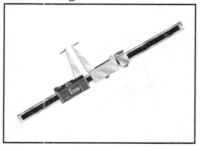

Demco Research and Development
4500 North Grand River Avenue, Lansing, MI 48906
(517) 321-4500

- Range and Resolution depends on gage and data collector used.

Interfaces with DataMyte 753, 1556, 2000 Series, 900 Series

Ultra-Cal II Groove and Recess Gage

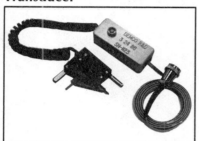

Fred V. Fowler Co., Inc.
66 Rave Street, Newton, MA 02166
(617) 332-7004

- Internal Range: 1.378-8.030 in. (35-204 mm)
- Outside Range: 9.433 in. (240 mm)
- Resolution: .0005 in. (.01 mm)

Interfaces with DataMyte 762, 862, 2003, 900 Series

J.S. Research Gap and Contour Gage

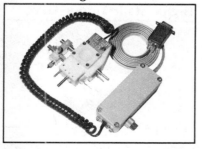

J.S. Research
81 Newman Road, Okemos, MI 48864
(517) 349-7674

- Range and resolution depends on gage and data collector used.

Interfaces with DataMyte 753, 1556, 2000 Series, 900 Series

Interfaces with DataMyte 762, 862, 2000 Series, 900 Series

AMETEK, Mansfield & Green Division
8600 Somerset Drive, Largo, FL 34643 (813) 536-7831

- Accuracy: ±.2% of rated capacity ± least significant digit
- Ranges: 7 models ranging in capacity from 250 x 0 g. to 500 x 0 lb (2220 x 0 N)
- Overload Range: 50% of rated full scale
- Weight: 1.7 lb

ACCUFORCE® III Force Gages

Interfaces with DataMyte 762, 862, 2000 Series, 900 Series

John Chatillon & Sons, Inc.
7609 Business Park Drive, Greensboro,NC 27409
(919) 668-0841

- Accuracy: ±.25% of full scale ± least significant digit
- Ranges: 7 models ranging in capacity from 8.0 x .005 oz. (2.5 x .001 N) to 500 x .2 lb (2220 x 1 N)

Chatillon DFGRS/DGGRS Series Digital Force Gages

Interfaces with DataMyte 762, 862, 2000 Series, 900 Series

Weigh-Tronix, Dillon Division
100 Armstrong Drive, Fairmont, MN 56031
(507) 238-4461

- Accuracy: ±.1% of rated capacity ± least significant digit.
- Ranges: 3 models:
2.2 x .002 lb (10 x .01 N)
22 x .02 lb (100 x.01 N)
44 x .02 lb (200 x .1 N)
- Weight: 11.5 oz.

Dillon Electronic Force Gage

Interfaces with DataMyte 762, 862, 2000 Series, 900 Series

AMETEK, Mansfield & Green Division
8600 Somerset Drive Largo, FL 34643 (813) 536-7831

- Accuracy: ±.5% full scale ± least significant digit
- Range: Eleven different models with ranges from 0-50 oz. in. to 0-6 Nm
- Weight: Including sensing head and adjustable chuck, 1.2 lb for 3 lowest ranges; 1.7 lb for all other models

ACCUFORCE TORQUE-CHEK® TC6000 Digital Static Torque Gage

ALL SPECIFICATIONS SUBJECT TO CHANGE WITHOUT NOTICE.

Industry Standard Rotary Transducers

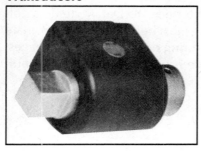

Crane Electronic Limited
4601 3rd Street, Moline, IL 61265
(309) 762-2285

Interfaces with DataMyte 753, 1556, 2000 Series, 900 Series

GSE Rotating Socket Wrench Torque Sensors

GSE Inc.
23640 Research Drive, Farmington Hills, MI 48024
(313) 476-7875

- Capacity: 9 models with ranges up to 100 lb in. (11.3 Nm) to up to 2000 lb ft (2411.6 Nm)
- Measures output torque of stall and clutch type nutrunners used in production fastening operations
- Not recommended for mechanical impact wrenches
- Measures CW or CCW direction

Interfaces with DataMyte 753, 1556, 2000 Series, 900 Series

Sensor I Model 125 Wrench

SPS Technologies
900 Newtown-Yardsley Road, Newton, PA 18940
(215) 860-3000

- Range: 170 Nm (125 ft. lbs.)
- Resolution: 1°, 1 Nm (1 ft. lb.)

Interfaces with DataMyte 1558, 2000 Series, 900 Series

21.3 FIXED STATION GAGING

The following is a sampling of fixed station gaging for SPC that interfaces to DataMyte data collectors. These gages include digital indicators, column gages, bore gages, height gages, coordinate measuring machines, bench micrometers, other dimensional measuring devices, thickness gages, digital protractors, spring testers, balances, hardness testers, surface roughness instruments, multimeters, pH meters, and various other measuring devices.

Interfaces with DataMyte 762, 862, 900 Series

Chicago Dial Indicator Company
1372 Redeker, Des Plaines, IL 60016
(312) 827-7186

- Various Ranges: (.25-4 in.)
- Resolution: .002 mm (.0001 in.) or .001 mm (.00005 in.)

Also marketed as the Fowler Ultra-Digit III.

Chicago Dial Indicator

Interfaces with DataMyte 762, 862, 2003, 900 Series

Chicago Dial Indicator Company
1372 Redeker, Des Plaines, IL 60016
(312) 827-7186

- Various Ranges
- Resolution: .01 mm (.0005 in.)

Mauser Indicator

Interfaces with DataMyte 762, 862, 2004, 900 Series

DataMyte Multiplexer: 529-17 (762, 862)

Federal Products Corporation
1144 Eddy Street, PO Box 9400, Providence, RI 02940
(401) 781-9300

- Five models with various ranges, resolutions and accuracies
- Ranges from ± .002 in. to ± 1.00 mm
- Resolutions from .0001 in. to .001 mm
- Accuracies from .05% to 1.0%
- A versatile gage used in many different fixtures

Federal® Maxμm™ Digital Indicator

ALL SPECIFICATIONS SUBJECT TO CHANGE WITHOUT NOTICE.

Fowler Ultra-Digit I and II Indicators

Fred V. Fowler Co., Inc.
66 Rave Street, Newton, MA 02166
(617) 332-7004

- Range: 25 mm (1 in.)
- Resolution: Ultra-Digit I: .01 mm (.0004 in.); Ultra-Digit II: .002 mm (.0001 in.)
- Accuracy: AGD Group 2, for 1 in. range
- Quantization Error: ± 1 digit
- Operating Temp: 0 to 40° C (32 to 104° F)

Interfaces with DataMyte 762, 862, 1558, 2000 Series, 900 Series

DataMyte Multiplexers: 529-16 (762, 862); 529-14 (1558); 529-12 (2000 Series)

Mitutoyo Digimatic Indicator (543-180 & 543-135)

MTI Corporation
18 Essex Road, Paramus, NJ 07652
(201) 368-0525

- Range: 12.7 mm (.5 in.)
- Resolution: .001 mm (.00005 in.) or .01 mm (.0005 in.)
- Accuracy: AGD Group 2

Interfaces with DataMyte 762, 862, 1558, 2000 Series, 900 Series

DataMyte Multiplexer: 529-15 (762, 862); 529-13 (1558); 529-11 (2000 Series)

Mitutoyo Digimatic Indicator (543-4xx)

MTI Corporationi
18 Essex Road, Paramus, NJ 07652
(201) 368-0525

- Ranges: 0-.4 in. to 0-2 in. Resolution: .001 in. or .0001 in.

Interfaces with DataMyte 762, 862, 1558, 2000 Series, 900 Series

DataMyte Multiplexer: 529-15 (762, 862); 529-13 (1558); 529-11 (2000 Series)

Ono Sokki EG-233 Digital Linear Gage

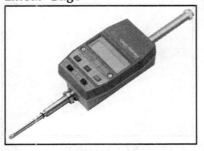

Shigma, Inc.
80 Martin Lane, Elk Grove Village, IL 60007
(312) 640-8640

- Range: 1.18 in. (30 mm)
- Resolution: .00004 in. (1 micron)
- Accuracy: .00008 in. (2 microns)

Interfaces with DataMyte 762, 862, 900 Series

DataMyte Multiplexer: 529-18 (762, 862)

Interfaces with DataMyte 761, 861, 2000 Series, 900 Series

Various Sources.

- Column gages available from Sheffield, Etamic, Edmunds Gage, G. T. E. Valenite, Dearborn and many others
- The DataMyte Junction Box allows up to 10 column gages, which are analog output devices, to be connected to the DataMyte 761 or 861 at once

Column Gages and Readouts

Interfaces with DataMyte 762, 862, 2000 Series, 900 Series

Brown & Sharpe Mfg. Co.
Precision Park, North Kingstown, RI 02852
(800) 426-1066

- Various Ranges: (.4-8.0 in.)

Micro-EBG Electronic Bore Gage

Interfaces with DataMyte 762, 862, 2004, 900 Series

Federal Products Corporation
144 Eddy Street, PO Box 9400, Providence, RI 02940
(401) 781-9300

- Used with Federal Maxμm indicator

Federal Electronic Plug (Bore) Gage

Interfaces with DataMyte 762, 862, 2000 Series, 900 Series

MTI Corporation
18 Essex Road, Paramus, NJ 07652
(201) 368-0525

- Various Ranges
- Resolution: .001 mm (.0001 in.)

Mitutoyo Digimatic Holtest Gage Series 468 with output

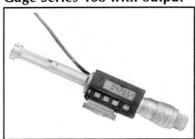

ALL SPECIFICATIONS SUBJECT TO CHANGE WITHOUT NOTICE.

Mauser Height Gage

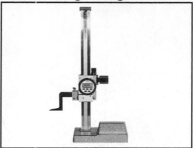

Chicago Dial Indicator Company
1372 Redeker, Des Plaines, IL 60016
(312) 827-7186

- Various Ranges
- Resolution: .01 mm (.0005 in.)

Interfaces with DataMyte 762, 862, 2003, 900 Series

Trimos Mini Vertical Height Gages

Fred V. Fowler Co., Inc.
66 Rave Street, Newton, MA 02166
(617) 332-7004

- Range: Two models: 300 mm (0-12 in.) and 620 mm (0-24.4 in.)
- Resolution: .01 mm (.0005 in.)

Interfaces with DataMyte 762, 862 900 Series

Trimos Vertical 3

Fred V. Fowler Co., Inc.
66 Rave Street, Newton, MA 02166
(617) 332-7004

- Can measure height, two coordinates, squareness, alignment of a part, etc.
- Measuring Range: 0-600 mm (0-23.6 in.)
- Application Range: 0-700 mm (0-27.5 in.), according to probe being used
- Resolution: .001, .01 mm (.0001, .001 in.)

Interfaces with DataMyte 762, 862, 2003, 900 Series

Mitutoyo Digimatic Height Gage, Series 192 Heavy-Duty Type

MTI Corporation
18 Essex Road, Paramus, NJ 07652
(201) 368-0525

- Ranges for 4 models: 12 in. (305 mm), 18 in. (457 mm), 24 in. (610 mm), 40 in. (1016 mm)
- Resolution: .0005 in. (.01 mm)

Interfaces with DataMyte 762, 862, 1558, 2000 Series, 900 Series

Interfaces with DataMyte 762, 862, 2000 Series, 900 Series

Numerex Corporation
7008 Northland Drive, Minneapolis, MN 55428
(612) 533-9990

- Resolution: 2µm (.0001 in.)

Numerex BRN-18 Height Gage

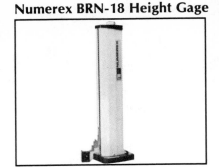

Interfaces with DataMyte 762, 862, 2003, 900 Series

Chicago Dial Indicator Company
1372 Redeker, Des Plaines, IL 60016
(312) 827-7186

- Various Ranges
- Resolution: .01 mm (.0005 in.)

Mauser Digital Linear Scale

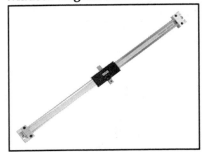

Interfaces with DataMyte 762, 862, 900 Series

Fred V. Fowler Co., Inc.
66 Rave Street, Newton, MA 02166
(617) 332-7004

- Range: Two models: 0-22 in. (550 mm) and 0-41.4 in.(1050 mm)
- Readout: .0001 in. or .01 mm and .001 mm

Trimos 54-196 Horizontal Setting/Measuring System

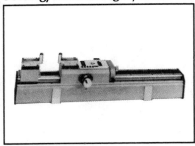

Interfaces with DataMyte 762, 862, 2000 Series, 900 Series

Anilam Electronics Corp.
5625 NW 79th Avenue, Miami, FL 33166
(305) 592-2727

- For coordinate measuring machines

Wizard Digital Readout

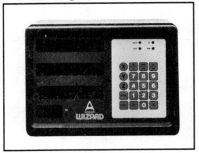

ALL SPECIFICATIONS SUBJECT TO CHANGE WITHOUT NOTICE.

Boeckler Microcode II Digital Readouts

Boeckler Instruments
31 E. Rillito Street, Tucson,
AZ 85705
(800) 552-2262

- Various Ranges

Interfaces with DataMyte 2000 Series, 900 Series

Federal 36-B I.D./O.D. Comparators

Federal Products Corporation
144 Eddy Street, PO Box
9400, Providence, RI 02940
(401) 781-9300

- Used with Federal Maxμm indicator

Interfaces with DataMyte 762, 862, 2004, 900 Series

Fowler Sylvac Measuring System

Fred V. Fowler Co., Inc.
66 Rave Street, Newton, MA
02166
(617) 332-7004

- Non-contact capacitive measuring probes
- Comparative and absolute measurements.

Interfaces with DataMyte 762, 862 900 Series

Quadra-Check III Metrology Display

Metronics Inc.
PO Box 5760, Manchester,
NH 03108
(603) 622-0212

- A digital readout for micrometer heads and optical comparators
- Calculates radii, angles and distances

Interfaces with DataMyte 762, 862, 1556, 1558, 2000 Series, 900 Series

Interfaces with DataMyte 762, 862, 1558, 2000 Series, 900 Series

Micro-Vu Corp.
7750 Bell Road, Windsor, CA 95492
(707) 838-6272

- Interfaces with optical comparators
- Provides digital readout, calculation of radii, center of radius and polar coordinate

Micro-Vu Q16 Metrology Computer

Interfaces with DataMyte 762, 862, 2000 Series, 900 Series

Numerex Corporation
7008 Northland Drive, Minneapolis, MN 55428
(612) 533-9990

- Resolution: 2μm (.0001 in.)

Numerex DMM 624 Coordinate Measuring Machine

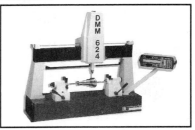

Interfaces with DataMyte 762, 862, 1558, 2000 Series, 900 Series

Sheffield Measurement Division
721 Springfield Street, PO Box 1127, Dayton, Oh 55428
(513) 254-5377

Cordax MPP-4 Coordinate Measuring Machine

Interfaces with DataMyte 762, 862, 1558, 2000 Series, 900 Series

G.C.A. Corporation
60 O'Connor Road, Fairport, NY 14450
(716) 377-3200

- Range: 203 mm (7.9 in.)
- Resolution: .01μm

Laseruler Vertical Bench Micrometer

ALL SPECIFICATIONS SUBJECT TO CHANGE WITHOUT NOTICE.

LaserMike Optical Micrometer

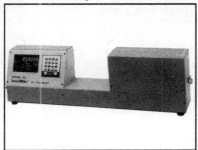

LaserMike Company
6060 Executive Boulevard,
Dayton, OH 45424
(513) 233-9935

- Resolution: up to .00001 in. (.0001 mm)
- Range: .003-1.5 in. (.075-15 mm)

Interfaces with DataMyte 762, 862, 1558, 2000 Series, 900 Series

Zygo 1201 Series Laser Telemetric System

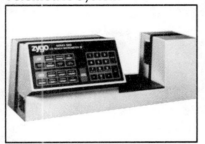

Zygo Corp.
Laurel Brook Road, PO Box 448, Middlefield, CT 06455
(203) 347-8506

- High-speed, non-contact diameter measurement
- Resolution: .00001 in. (.0001 mm)
- Range: .01-2.00 in. (.25-50 mm)
- Requires RS232 option

Interfaces with DataMyte 762, 862, 1558, 2000 Series, 900 Series

Nanoscope Ultrasonic Flaw Detector (with RS 232 upgrade)

Erdman Instruments, Inc.
1179 Romney Drive,
Pasedena, CA 91105
(818) 792-5184

- Range: from .012 to 1 in.

Interfaces with DataMyte 762, 862, 1558, 2000 Series, 900 Series

PERMASCOPE® Coating Thickness Gage

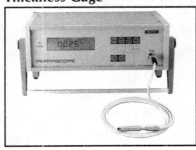

Fischer Technology, Inc.
750 Marshall Phelps Road
Windsor, CT 06095
(203) 683-0781

- PERMASCOPE E111 measures non-conducting coatings such as paint, plastics and oxides on non-ferrous metals
- PERMASCOPE M11 measures non-magnetic coatings such as tin, paint, and plastics on iron and steel

Interfaces with DataMyte 762, 862, 2000 Series, 900 Series

Interfaces with DataMyte 762, 862, 900 Series

Krautkramer Branson
PO Box 350, Lewistown, PA 17044
(717) 242-0331

- Range: delay mode: .005-1.000 in. (.15-25 mm); contact mode: .06-15 in. (1.5-380 mm)
- Resolution: delay mode: .0001 or .001 in. (.001 or .01 mm); contact mode: .001 in. (.01 or .1 mm)
- Portable operation (4 lbs)

CL304 Ultrasonic Thickness Gage

Interfaces with DataMyte 762, 862, 2000 Series, 900 Series

MTI Corporation
18 Essex Road, Paramus, NJ 07652
(201) 368-0525

- Various Ranges
- Resolution: .0005 in.

Mitutoyo MU-Gage (Ultrasonic Thickness)

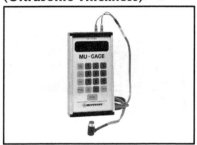

Interfaces with DataMyte 762, 862, 2000 Series, 900 Series

UPA Technology, Inc.
60 Oak Drive, Syosset, NY 11791
(516) 364-1080

- Resistance Range: 0-20000 µohms

Caviderm CD-8 Thickness Gage

Interfaces with DataMyte 762, 862, 2000 Series, 900 Series

ELECTROMATIC Equipment Co., Inc.
600 Oakland Avenue, Cedarherth, NY 11516
(516) 295-4300

- Range: 0-1000 µohms
- Accuracy: ± 1%

Check Line Coating Thickness Tester Model DAC-40S

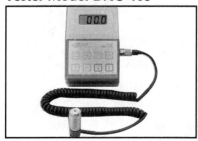

ALL SPECIFICATIONS SUBJECT TO CHANGE WITHOUT NOTICE.

Mintest 3000 Coating Thickness Tester

Elektro-Physik
778 W. Algonquin, Arlington Hts, IL 60005
(800) 782-1506

Interfaces with DataMyte 762, 862, 2000 Series, 900 Series

Fowler Sylvac Electronic Digital Protractor

Fred V. Fowler Co., Inc.
66 Rave Street, Newton, MA 02166
(617) 332-7004

Interfaces with DataMyte 762, 862, 2000 Series, 900 Series

- With a push button, different ranges can be selected: 1 x 360°, 2 x 180°, or 4 x 90°.
- Accuracy: ± 1 min of arc or ± .01 degrees of arc
- Read-out: 1 minute of arc or .01 degrees of arc

AngleStar® Model DP-60 Digital Protractor

Schaevitz Sensing Systems Inc.
21640 N. 14th Avenue, Phoenix, AZ 85027
(602) 582-3741

Interfaces with DataMyte 762, 862, 2000 Series, 900 Series

- Linearity: ± .05° at 0 to 20°; ± .1° at 20 to 60°
- Resolution: .01° (± 20°); .1° (20° to 60°)
- Range: ± 60 °
- Reading modes: degrees, mils, inches/ft., % grade, mm/m

Carlson Spring Tester

The Carlson Company
PO Box 71, Clinton, AR 72031
(501) 745-4811

Interfaces with DataMyte 762, 862, 2000 Series, 900 Series

- Measures loads and deflections
- For compression and extension springs
- Accuracy: .25 percent of load, within 1 digit
- Up to 600 tests per hour can be made

21-17

Gages

Interfaces with DataMyte 762, 862, 2000 Series, 900 Series

Larson Systems, Inc.
5205 Lakeland Avenue, Crystal, MN 55429
(612) 535-6299

● Force Range: 0-1 lb. to 0-10,000 lbs.

Larson Spring Tester

Interfaces with DataMyte 762, 862, 2000 Series, 900 Series

Torque Specialities Div., A.K.O., Inc.
110 Broad Brook Road
Hanfield, CT 06082
(203) 749-7441

● Torque Capacities: To 2,000 in. lbs.

Bottle Cap Torque/Force Tester

Interfaces with DataMyte 761, 861, 2000 Series, 900 Series

Dresser Industries Instrument Division
153 S. Main Street, Newton, CT 06482
(203) 426-3115

● Torque Capacities: To 2,000 in. lbs.

Heise Series 620 Pressure Transducer

Interfaces with DataMyte 762, 862, 2000 Series, 900 Series

Mettler Instruments Corporation
PO Box 71, Princeton Heights, NJ 02850
(800) 638-8537

Mettler Electronic Balances

ALL SPECIFICATIONS SUBJECT TO CHANGE WITHOUT NOTICE.

Ohaus Scales

Ohaus Scale Corporation
29 Hanover Road, Florham
Park, NJ 07932
(201) 377-9000

Interfaces with DataMyte 762, 862, 2000 Series, 900 Series

Uson Leak Tester

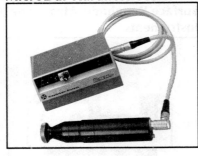

Uson Corporation
5215 Hollister, Houston, TX
77040
(713) 460-1700

Interfaces with DataMyte 762, 862, 1558, 2000 Series, 900 Series

MicroDur Hardness Tester

Krautkramer Branson
PO Box 350, Lewistown, PA
17044
(717) 242-0331

Interfaces with DataMyte 762, 862, 900 Series

**Mitutoyo Hardness Tester
(Series 940 with output)**

MTI Corporation
18 Essex Road, Paramus, NJ
07652
(201) 368-0525

Interfaces with DataMyte 762, 862, 2000 Series, 900 Series

- Test Loads: 60, 100 and 150 kg
- 24 in. high, 8 in. wide, 17.5 in. deep
- Weight: 135 lb.
- Penetrator travel: 9 in.

Interfaces with DataMyte 762, 862, 1558, 2000 Series, 900 Series

Page-Wilson Corporation
6 Emma Street, Binghamton, NY 13905
(607)770-4500

Series 500 Digital Hardness Tester

Interfaces with DataMyte 762, 862, 2000 Series, 900 Series

Vector
675 44th Street, Marion, IA 52302
(319) 377-8263

Schleuniger-4M Pharmaceutical Hardness Tester

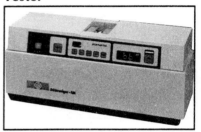

Interfaces with DataMyte 762, 862, 2000 Series, 900 Series

Feinpruef Corp.
(Mahr/Perthen)
8848A Red Oak Boulevard, Charlotte, NC 28217
(704) 525-7128

Perthometer M3P, M4P Surface Roughness Measuring Instruments

- Surface Parameters: R_a, R_z, R_{max}, R_{pm}, P_c (R_{3z} optional)
- Operating Range: 2800 or 6000 μin.)
- 3 tracing lengths; 3 cutoff lengths
- Weight: 1.3 lb
- Battery-operable

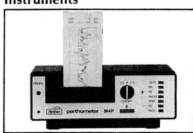

Interfaces with DataMyte 762, 862, 2000 Series, 900 Series

Precision Devices, Inc.
PO Box 220, Milan, MI 48160
(313) 439-2462

- Various Ranges

Surfometer Digital Surface Roughness Measuring System

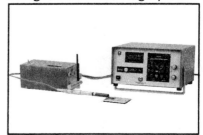

Sheffield Profilometer Surface Roughness System

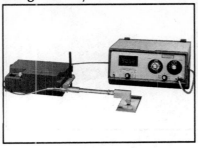

Sheffield Measurement Division
721 Springfield Street, PO Box 1127, Dayton, OH 45401
(513) 254-5377

- Average roughness (R_a) measured in µin. or µm

Interfaces with DataMyte 762, 862, 2000 Series, 900 Series

Fryer Electronic Score Residual Gage

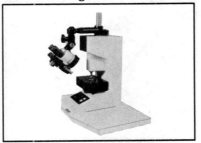

Frank E. Fryer Company, Inc.
36 East Main Street, Carpentersville, IL 60110
(312) 426-6731

- Range: 50 mm (2 in.)
- Resolution: .00015 in.

Interfaces with DataMyte 762, 862, 2000 Series, 900 Series

WACO Digital Enamel Rater

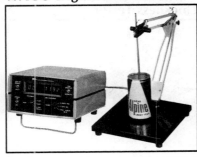

Wilkens-Anderson Company
4525 W. Division Street, Chicago, IL 60651
(312) 384-4433

- Range: Low range 0-249.9 ma; High range 0-500 ma

Interfaces with DataMyte 762, 862, 2000 Series, 900 Series

Portable Distinctness of Reflected Image Meter

ATI Systems, Inc.
PO Box 71460, Madison Heights, MI 48071
(313) 589-1580

Interfaces with DataMyte 762, 862, 1558, 2000 Series, 900 Series

Interfaces with DataMyte 762, 862, 2000 Series, 900 Series

Macbeth Process Measurements
PO Box 230, Newburgh, NY 12550
(914)565-4440

- For density only on pre-1987 production

RD 918 Macbeth Densitometer

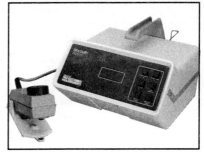

Interfaces with DataMyte 762, 862, 2000 Series, 900 Series

X-Rite, Inc.
3100 44th Street SW, Grandville, MI 49418
(616) 534-7663

- Models for color or black and white printing
- All models measure density
- Various models also measure dot area, trap, print contrast, hue error grayness, and cast brightness
- Battery-powered

X-Rite 400 Series Reflection Densitometers

Interfaces with DataMyte 762, 862, 2000 Series

CMI International
2301 Arthur Avenue, Elk Grove Village, IL 60007
(312) 439-4404

Model MR300 Copper Coating Thickness Tester

Interfaces with DataMyte 762, 862, 2000 Series, 900 Series

MetraByte Corporation
440 Miles Standish Boulevard, Catauon, MA 02780
(508) 880-3000

MBC Digital Panel Meter

ALL SPECIFICATIONS SUBJECT TO CHANGE WITHOUT NOTICE.

Omega M2110-2 Multi Meter

Omega
1 Omega Drive, PO Box
4047, Stamford, CT 06907
(203) 359-1660

Interfaces with DataMyte 762, 862,
2000 Series, 900 Series

**Corning Model 155 pH/ion
Meter**

Corning Glass Works
MP-21-5, Corning, NY
14831
(607) 974-7996

- Range: -2 to 14 pH Resolution: .001
  pH

Interfaces with DataMyte 762, 862,
2000 Series, 900 Series

**Orion Model EA 940
Expandable Ion Analyzer**

Orion Research Incorporated
529 Main Street, Boston, MA
02129
(800) 225-1480

- pH Range: -2,000 to 19,999
- pH Relative Accuracy: ± .002 pH
  units

Interfaces with DataMyte 2000 Series,
900 Series

**Brookfield Viscometer
(Analog)**

*Brookfield Engineering
Laboratories, Inc.*
240 Cushing Street,
Stoughton, MA 02072
(617) 344-4310

- Measures coefficient of viscosity
  (cp), a characteristic of liquid flow
  that also affects coating and spray-
  ing characteristics of a liquid

Interfaces with DataMyte 761, 861,
2000 Series, 900 Series

21.4 GAGING SOLD AND SUPPORTED BY DATAMYTE

514 Gap Gage Available from DataMyte

The 514 Gap Gage is used to quickly measure and instantly record gap measurements between adjacent parts, parts in ring or margin fixtures, and any application where the amount of gap is critical to either the appearance or function of a product. This gage was designed for versatility. By substituting various easily fabricated fingers and attachments, the gage can be used to check depth, diameters, clearance in hard-to-reach areas, and other special types of measurements.

Several models are available, in different resolutions and with either gap fingers or a depth checking attachment.

To order, select the required range from the specifications table and specify the corresponding model number.

Interfaces with DataMyte: 753, 1556, 2000 Series, 900 Series

Specifications

- Operating Temp: 0 to 50° C (32 to 122° F)
- Storage Temp: -30 to 70° C (-22 to 158° F)
- Battery: Powered from the DataMyte data collector battery
- Battery Life: 12 hours minimum at 25° C
- Interface cable
- Built-in re-try switch
- Will measure in inches or millimeters

* Call DataMyte Customer Support for availability.

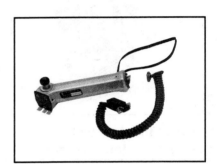

ALL SPECIFICATIONS SUBJECT TO CHANGE WITHOUT NOTICE.

516 Gap and Flushness Gage

Available from DataMyte

This gage enables fast dimensional audits between mating components. The .25 in. (6.35 mm) wide fingers allow measurement of small radii gaps in hard to reach places. A single gage can be used instead of two because both gap and flushness readings can be made one after another.

Gap measurements are made by placing the upper edge of the fingers against one inner surface, and pulling the trigger until the movable finger contacts the opposite surface of the gap. The data collector automatically records the maximum separation. Flushness is measured using a movable rod and foot. The gage does not need to be perfectly perpendicular to take a reading because of the matching radius of the rod and foot.

Specifications

- Model: 516
- Range: Gap = 2.54 to 14.48 mm (.100 to .570 in.)
  Flushness = +6.35 mm to -5.72 mm (+.250 to -.225 in.)
- Resolution: .050 mm (.002 in.)
- Inaccuracy: ± .5%
- Non-repeatability: ± .20%
- Quantization Error: ± .025 mm (± .001 in.)
- Operating Temp: 0 to 50° C (32 to 122° F)
- Storage Temp: -30 to 70° C (-22 to 158° F)
- Weight: 596 g (21 oz.)
- Battery: Powered by the DataMyte data collector battery
- Battery Life: 12 hours minimum at 25° C
- Wrist strap
- Built-in re-try switch
- Will measure in inches or millimeters

Interfaces with DataMyte: 753, 1556, 2000 Series, 900 Series

Torque Wrenches and Drivers

Available from DataMyte

DataMyte Torque Tools have been developed to meet the requirements of the industrial user. They are rugged but accurate tools suitable for assembly line auditing for critical fasteners subject to vibration or stress, such as heavy machinery and automobiles. When connected to DataMyte data collectors, the torque tools can measure breakaway torque — the point at which a fastener begins to turn — with great accuracy.

The torque tools can be used with the DataMyte 1500 Series and 2000 Series. DataMyte 2000 Series data collectors can be used to audit fasteners with either clockwise or counter clockwise rotation without the need to change setup parameters or cabling. To order, select required range from specifications table and specify the corresponding model number.

NOTE: Torque wrenches and drivers are not ratcheting tools. They must be used within their specified range.

Specifications for Torque Wrenches

- Inaccuracy: ± .25%
- Non-repeatability: .1%
- Operating Temp: 0 to 50° C (32 to 122° F)
- Storage Temp: -30 to 70° C (-22 to 158° F)
- Battery: Powered from the Data-Myte data collector battery
- Battery Life: 12 hours minimum at 25° C
- Interface cable
- Certificate of calibration
- User's Guide

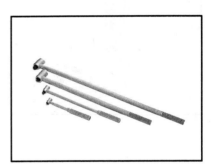

ALL SPECIFICATIONS SUBJECT TO CHANGE WITHOUT NOTICE.

22. TRAINING AND CUSTOMER SUPPORT

22.1 TRAINING CONSIDERATIONS FOR SPC AND AUTOMATED DATA COLLECTION

The addition of new technology to a manufacturing process is almost always a carefully planned process. Decisions on the type of equipment, its placement in the process and its expected benefits are based on an analysis of fact: time and material savings, greater capacity and new capabilities. These considerations are certainly true when purchasing an automated data collection system for SPC. An additional consideration that is especially significant in the implementation of a new SPC data collection system is training for all members of the production team. DataMyte systems provide people with the process information they need to make better decisions on process control. If this process information cannot be interpreted and acted on by the person controlling the process the entire system breaks down.

Figure 22.1.1 shows the elements and interactions of an effective automated SPC system. Missing elements or a failure between elements stops the process. People are the

Fig. 22.1.1   The elements and interaction of an effective automated SPC system.

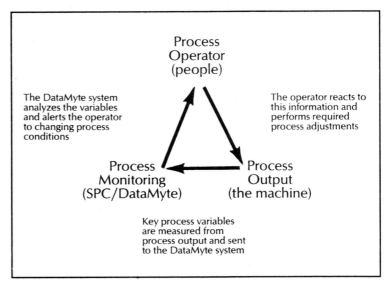

Process
Operator
(people)

The DataMyte system analyzes the variables and alerts the operator to changing process conditions

The operator reacts to this information and performs required process adjustments

Process
Monitoring
(SPC/DataMyte)

Process
Output
(the machine)

Key process variables are measured from process output and sent to the DataMyte system key to system success. They must be able to use a system effectively to gather process data, interpret SPC charts and indices and be able to adjust the operating levels of the process.

Unique Training Requirements

Each data collection system is a unique combination of standard components: data collectors, gages, cables, computers and software. A careful analysis of data collection needs is completed and a system configuration is proposed. The same is true for the training requirements of each system. An analysis of the knowledge and skill needs of all people involved in the project results in a "training prescription." The training prescription may include SPC training, data collector training and software training. It is essential that this analysis of training needs be completed. In some cases it may reveal that many of the required skills are already in place and the system may be implemented with only a minimum of "on the job training." And in other cases, it may show that a series of structured training courses are critical for the success of the system. In either case, the training decision is based on fact. Intelligent training decisions prevent the omission of required training programs and the delivery of unneeded training programs. The actual process of assessing training needs parallels the complexity of the data collection system: a simple system with one or two data collectors and one or two operators and a project coordinator will have easily defined training requirements. A more complex system would consist of multiple data collectors of different models, many operators who may be spread across more than one shift, and a centralized management system for collected information. A system like this requires a structured assessment of training requirements that is given the same care and consideration as the initial data collection system proposal.

22.2 DEFINING TRAINING NEEDS

The questions that must be answered when assessing the training needs for a new data collection system are:

- What do the work stations consist of? What process

equipment? What data collector? What kind of gaging and measurement systems?

- Who are the team members? How do they match up to each work station? Who are the project coordinators? Who is the in-house system expert?
- What are the job performance requirements of each team member? Who backs up each team member?
- Which of these job performance requirements can be completed by team members right now?
- What are the expectations for the system? How fast is it to be brought to full operation? What are the specific results expected by management?

The answers to these questions provide a clear definition of training requirements. The worksheet shown in Figure 22.2.1 is used to identify the components and structure of a DataMyte system. The worksheet shown in Figure 22.2.2 is used to identify the production team members for this system and the specific training requirements for each member. Completing worksheets like these helps answer the key questions relating to training needs: What are the system configuration and goals? Who are the people that are going to make it happen? What knowledge and skill do they require? And what knowledge and skills do they currently possess?

22.3 DATAMYTE TRAINING RESOURCES

The training components of a successful data collection system are very similar to the hardware components in one respect: both are made up of standard components combined in a custom system to meet the customer's specific needs. To meet these needs DataMyte has developed several different training resources: product reference manuals, regional training courses, self instructional video courses and custom courses presented at the customer's site. This range of options provides an effective and affordable training solution for all of our customers.

The product reference manual — Every data collection system comes with a complete set of reference manuals. These manuals are the best single source of information on all of

Station and System Configuration

Instructions: Use this worksheet to identify the elements of each work station in your data collection system and the management system used for the total system.

Workstation ID:

Process/Station Description:

Data Collector(s):
- ☐ 761   ☐ 861   ☐ 769   ☐ 900
- ☐ 762   ☐ 862   ☐ 2003 ☐ _____

Gaging

SPC Requirements: Capability
  Control Charts: ☐ Cp
- ☐ $\bar{x}$ ☐ p   ☐ Cr
- ☐ R   ☐ c   ☐ CpK
- ☐ Short-run   ☐ u
- ☐ Moving $\bar{x}$ ☐ np   ☐ Pareto

Workstation ID:

Process/Station Description:

Data Collector(s):
- ☐ 761   ☐ 861   ☐ 769   ☐ 900
- ☐ 762   ☐ 862   ☐ 2003 ☐ _____

Gaging

SPC Requirements: Capability
  Control Charts: ☐ Cp
- ☐ $\bar{x}$ ☐ p   ☐ Cr
- ☐ R   ☐ c   ☐ CpK
- ☐ Short-run   ☐ u
- ☐ Moving $\bar{x}$ ☐ np   ☐ Pareto

Data Management System

Computer System:	Analysis Software:	Network System:
☐ IBM PC/AT/PS2	☐ DataMyte FAN II	☐ DataTruck
☐ VAX	☐ TurboSPC	☐ Vista Lan I
	☐ DataMyte/BBN (OVERVIEW)	☐ OVERVIEW
		☐ N/A

Workstation ID:

Process/Station Description:

Data Collector(s):
- ☐ 761   ☐ 861   ☐ 769   ☐ 900
- ☐ 762   ☐ 862   ☐ 2003 ☐ _____

Gaging

SPC Requirements: Capability
  Control Charts: ☐ Cp
- ☐ $\bar{x}$ ☐ p   ☐ Cr
- ☐ R   ☐ c   ☐ CpK
- ☐ Short-run   ☐ u
- ☐ Moving $\bar{x}$ ☐ np   ☐ Pareto

Workstation ID:

Process/Station Description:

Data Collector(s):
- ☐ 761   ☐ 861   ☐ 769   ☐ 900
- ☐ 762   ☐ 862   ☐ 2003 ☐ _____

Gaging

SPC Requirements: Capability
  Control Charts: ☐ Cp
- ☐ $\bar{x}$ ☐ p   ☐ Cr
- ☐ R   ☐ c   ☐ CpK
- ☐ Short-run   ☐ u
- ☐ Moving $\bar{x}$ ☐ np   ☐ Pareto

Fig. 22.2.1   Station and System Configuration Worksheet.

Training Plan for Work Station(s): _____

Instructions:

1. Complete this sheet for each work station. If 2 or more stations are the same, a single sheet may be used.

2. List the job titles/positions for this station.

3. Identify the job performance requirements for each position by blackening the triangle in the upper left corner:

4. Match a training resource to each training requirement by noting the appropriate choice:

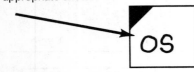

Job Performance Requirements

SPC Level 1: Be able to identify and intrepret the SPC charts and indexes presented by the DataMyte data collector.

SPC Level 2: Be able to identify and describe the role of SPC in quality and productivity improvement. Identify and complete the steps necessary to implement an SPC program.

Data Setup: Be able to set up the DataMyte data collector for the parts and processes at the work station.

Data Collect: Be able to collect SPC data with the data collector and specified gaging.

Data Communications: Be able to communicate collected data back to a host personal computer for analysis and storage.

Data Management: Be able to analyze data from multiple stations and manage data collector activity at each station.

Training Resources

OK: Not required; already trained

IT: Internal training by project staff

OS: Use on-site training by DataMyte

RG: Attend a DataMyte regional training course

VT: Use a DataMyte video training package

Station Team Members	# to be Trained	SPC Level 1	SPC Level 2	Data Setup	Data Collect	Data Management	Data Communication
Operators							
Supervisors							
DataMyte Project Coordinator							
Additional Job Title at This Station:							
Additional Job Title at This Station:							

Fig. 22.2.2   Training Worksheet.

the features and operating characteristics of each data collector. Reference manual training works best in instances where only one or two individuals require training and they have had success in the past using reference manuals to learn about computers and software.

Regional training courses — In many cases DataMyte customers do not have the time or inclination to devote time to studying the reference manual. These individuals want personalized attention and answers to their specific questions. To accommodate these people DataMyte has set up a series of training courses around the country. Each course is offered a minimum of twice each month in different Allen-Bradley training centers. A typical course lasts one day and focuses on a single model data collector. The course is an intensive, hands-on session on setting up, operating and managing data from a data collector. The knowledge and skill gained in each course allows participants to return to their own applications and implement their data collection system.

On-site training courses — In instances where five or more people must be trained, the best alternative is to present a custom training course at the customer's site. This is the most economical alternative and the most effective training solution. An on-site training course is structured to meet the customer's specific needs. It can include provisions for SPC training, operator training, data collector specialist training and trainer training. Following the structured training class the DataMyte trainer works with course participants down on the plant floor to make sure that they are able to apply what they have learned. By the time the trainer leaves the site, the data collection system is up and running and the customer's expectations are being met.

Video based training programs — To support all of our training options DataMyte has developed a series of video tape programs on the setup and operation of each data collector. These programs can be used as a self instructional training alternative or as a part of a formal in-house training session. The video tapes are a valuable asset when new operators come on-line or when refresher training is required.

By analyzing training needs and then matching those needs to DataMyte's training resources an effective and affordable training solution can be found.

Results and Benefits

Companies who give equal and careful consideration to people and technology when implementing an automated data collection system for SPC will attain these benefits:

- Immediate return on investment. System payback is dependent on full implementation. A comprehensive training strategy is the best guarantee to getting hardware and software out of the box and on to the plant floor.

- Acceptance on the plant floor. Apprehension and resistance to new technology results from fear, uncertainty and doubt on the part of operators. They are often unsure if they can master the new technology and they do not see the potential benefits of it. A comprehensive training program demystifies technology so it can be understood by all. The delivery of required training also demonstrates commitment to employees and to their future in the organization.

- The system stays on-line. The majority of support and service calls handled by DataMyte relate to system operation failures not equipment failures. A trained operator is knowledgeable about system operation and able to resolve operational questions without slowing or stopping the process.

Commitments are made to statistical process control programs and prevention based quality because of the belief in "doing things right the first time." When designing and implementing your data collection system for SPC, let us help you "do it right the first time" through a careful analysis of your technological needs and the training needs of your people.

22.4 CUSTOMER SUPPORT CONSIDERATIONS FOR SPC AND AUTOMATED DATA COLLECTION

Customer Support is as much a part of the product as a power cable or an instruction manual. It should be considered up front with any equipment purchase. After all, what you are buying is a complete system and customer support is a part of the package.

Manufacturers should treat the customer as a business partner. Both parties have an investment in the long term success of the system; the manufacturer can count on continued business and the customer can count on a successful SPC and automated data collection program. Therefore, the manufacturer's primary goal is to provide total customer satisfaction after the sale.

Strong customer support programs are designed to make it easier for the customer to do business with the manufacturer. Customer support is the manufacturer's customer advocacy board and should be used by the customer for any problems he encounters. Manufacturers, in order to facilitate this, need to provide a support program that meets the following requirements:

- On-site technical assistance
- Phone support
- Application evaluations
- Implementation assistance
- Cost effective maintenance
- Protection against technical obsolescence

22.5 DATAMYTE CUSTOMER SUPPORT RESOURCES

DataMyte's customer support fits the needs of a changing industry with a varied client base that requires several kinds of support.

On-site technical assistance — On-site technical assistance is one of the ways manufacturers show that they are dedicated to keeping customers running at full operation. At the customer's request, DataMyte can schedule a visit to the customer's site to assist in surveying applications, supporting implementation, evaluating gage interfaces and troubleshooting problems.

Phone support — A customer's first call for help may be by phone. Manufacturers should provide timely and adequate support. DataMyte's customer support group has dedicated troubleshooters and customer service representatives for phone support. With the customer support direct dial number, customers can generally reach a customer support specialist within three rings.

Application evaluations — Field support includes on-site troubleshooting as well as pre-sales assistance. DataMyte's customer support staff is also the first line of support for its sales force. Customer support specialists conduct application evaluations prior to implementing a system at a customer site to ensure that the customer is getting the most appropriate system for the application.

Implementation assistance — Implementation involves more than equipment installation. The customer relies on the expertise of the manufacturer to assist in configuring the equipment and getting it up and running to best fit the customer's needs. DataMyte customer support assists in implementation by providing either on-site support or phone support.

Cost effective maintenance — Equipment that has a low purchase price but high maintenance costs is not an element of a support program that treats the customer as a business partner. The program must be flexible enough so that the customer can afford to maintain his equipment to continue its effectiveness. This is beneficial to both the manufacturer and the customer. DataMyte's program allows customers to purchase a support agreement that covers most maintenance costs. The SUSA (System Update and Support Agreement) is the customer's worry-free maintenance contract.

Protection against technical obsolescence — DataMyte's SUSA agreement also is the customer's protection against technical obsolescence. In fast changing industries like automated data collection, customers must be assured that the vendor will be there to continue to support the product after technological advances. DataMyte is a long term vendor in the industry and the SUSA policy offers customers the opportunity to trade in equipment for equipment with the latest advances.

Meeting the requirements of a good support program is an important part of DataMyte's business philosophy. It is essential to treat customers as business partners to enhance both DataMyte's position and customers' positions.

Results and Benefits

Again, DataMyte's customer support demonstrates the philosophy that companies who give equal and careful consideration to people and technology when implementing an automated data collection system for SPC will attain many benefits both for the manufacturer and the customer. The benefits the customer gains include:

- Reduced startup times when implementing a system because of the pre-sales, applications, and installation support.
- Continued successful operation because of the timely and effective phone and field support.
- Continued long term product use because of proactive maintenance and upgrade policies.

DataMyte develops partnerships with customers. An investment has been made in the customer's future and the commitment has been made to support the customer.

23. ALLEN-BRADLEY QUALITY MANAGEMENT

Fig. 23.1.1 Allen-Bradley trademark.

23.1 INTRODUCTION

No other industrial automation supplier is better known for quality than Allen-Bradley. The word *Quality* at the base of the Allen-Bradley trademark summarizes a longstanding commitment. (Figure 23.1.1)

From its beginning in 1903, as a pioneer in the field of industrial motor control, Allen-Bradley earned its reputation for quality from years of experience on the plant floor. Its founders, Lynde and Harry Bradley had a consuming dedication to quality, and not just in regard to products, but in every part of the enterprise.

Today, Allen-Bradley is a Rockwell International Company, and employs over 13,000 people in plants and sales offices in North America, Latin America, Europe, Australia and Asia. Its business is predominantly data acquisition, control and communications devices for industrial automation.

To help ensure the highest quality for all its customers in every one of its businesses throughout the world, Allen-Bradley has an ongoing quality program called the *Total Quality Management System* (TQMS).

Key to the TQMS philosophy is that quality achievement is a never-ending process and primarily a management responsibility. Through the application of specific principles to real factory situations, Allen-Bradley has found that quality influences productivity, and can be measured objectively using scientific sampling and analysis.

As a result of TQMS, Allen-Bradley saved $125 million since 1982 and received a $13 return for every dollar spent. In addition, the company's market share increased due to cost reductions and productivity improvements.

Allen-Bradley product offerings follow the TQMS philosophy. An integral part of Allen-Bradley industrial control and communication products is the concept of preventive quality control — the avoidance of nonconformities. In addition, these products produce data for statistical analysis and traceability.

Today, including the acquisition of DataMyte, Allen-Bradley has the broadest, deepest line of products and systems for quality management. It includes SPC modules for

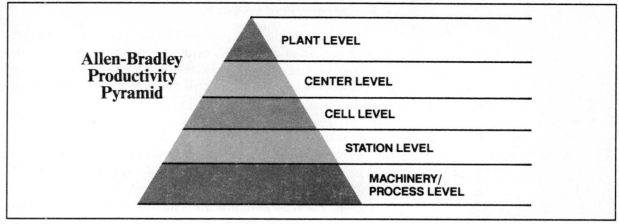

Allen-Bradley Productivity Pyramid

PLANT LEVEL

CENTER LEVEL

CELL LEVEL

STATION LEVEL

MACHINERY/ PROCESS LEVEL

Fig. 23.2.1   Productivity Pyramid.

programmable controllers, industrial PCs and minicomputer systems for data analysis, and automated quality systems for threaded fasteners, and programmable vision systems.

Quality management products are part of the Allen-Bradley Productivity Pyramid, which is a blue-print for a fully integrated plant.

23.2   THE PRODUCTIVITY PYRAMID

The Allen-Bradley Productivity Pyramid is a master plan for increasing productivity and quality through the application of computer, control, and communications technology. It provides the basis for plant-wide communications, permitting minute-by-minute management of information and the plant itself.

The Allen-Bradley Productivity Pyramid views a plant as five separate and distinct levels, each representing a different set of tasks requiring a different set of controls. See Figure 23.2.1. It's a systems approach to plant automation in which each control level builds on the information provided by the level below.

The ultimate goal is to weld each of these levels into a single, seamless automation system. Communications networks link all the plant levels from top to bottom, from computer mainframe to the most elementary production element. The result is a company controlled from the top down and informed from the bottom up.

Machinery/Process Level

The machinery/process level is the basic interface with production and process equipment on the plant floor. Here, sensing and control devices respond to upper level commands. At this level, the Allen-Bradley product line includes:

- Sensors such as limit switches, pressure and temperature controls, proximity switches, photoelectric controls, automatic identification including RF systems, bar codes, and encoders
- Logic controls and indicators including push buttons, selector switches, relays, low-cost programmable controllers, and intelligent panel systems
- Power products such as motor starters, contactors, and motor protection devices

Station Level

The station level converts input from lower levels to output commands, based on direction from above. Allen-Bradley real-time control products at this level include an integrated family of programmable controllers, vision systems, intelligent motion control, intelligent I/O modules, and drive systems.

Cell or Supervisory Level

The cell level bridges the gap between data processing and plant floor control. It coordinates production flow among various stations and integrates them into an automated system. Products for this level include Pyramid Integrator and other cell level control systems, advanced programmable controllers, color graphics systems, PC management systems, and distributed numerical control (DNC) systems, all linked through Allen-Bradley subnetworks. Cell level products also include the Allen-Bradley fastening systems which ensure manufacturers of consistent bolt, tension, and joint quality.

Center Level

The center level integrates industrial computers and management computers. It schedules production and provides management with information by monitoring and supervising lower levels, managing programs and parameters, collecting information from cell controllers, analyzing data, and reporting to higher-level computers.

Plant Level

The plant level provides overall planning and execution. This requires two-way communication between the mainframe computer and lower levels. Utilizing internationally accepted communications standards, including MAP, Allen-Bradley VistaMAP products provide for the integration of activities at upper pyramid levels.

A Standardization of Communication Links

Plantwide automation requires standardized communications links. The success of tomorrow's industrial plant will depend on its ability to gather, share and use data for effective planning and control.

The Productivity Pyramid provides a top to bottom blueprint for automating the gathering and sharing of production information. It creates a closed-loop real-time feedback system, which provides a wealth of information at low cost.

For several years, Allen-Bradley has been in the forefront of a cooperative effort to establish a standard, industry-wide data architecture. This effort has involved major companies and international standards organizations, and has resulted in the Manufacturing Automation Protocol (MAP) family of standards. MAP makes it possible to tie an entire plant together through standardized communications. It has been adopted by many companies representing many industries.

Allen-Bradley now offers a family of MAP-compatible communications networks and gateways to enable the linking of multi-vendor control equipment at every level of the pyramid.

Today's typical industrial plant contains large numbers of relatively low-cost programmable devices and operator interfaces, directly controlling machines and processes at the lowest levels of the Productivity Pyramid. These must communicate with each other, and report upward through the pyramid. Relatively small amounts of data are involved, communications are usually event-driven, and response times must be very fast. Up to now, the only way to accomplish this type of communication has been to employ proprietary communications subnetworks, which are then linked into the overall broadband LAN.

Allen-Bradley now offers products which link various proprietary subnetworks into the MAP broadband LAN, allowing the integration of existing equipment from different manufacturers into the overall system.

How Quality fits in the Productivity Pyramid

Quality management means more to Allen-Bradley than the attainment of consistent quality. That is aiming too low. Allen-Bradley industrial automation is geared toward:

- Progressively higher quality
- Progressively lower costs
- Making quality a competitive edge

The Productivity Pyramid symbolizes a highly flexible, highly responsive production system where quality is prevention-based rather than inspection-driven. Data collection at the machinery/process and station levels is attuned to gathering the data most relevant to process quality. Data is provided in real-time to the appropriate decision-making levels. Analysis is designed to highlight trends and patterns that give advance notice of a process problem. In addition, it provides data archiving for historical tracking.

The Allen-Bradley quality management offering fits into the pyramid as follows:

Level 1 — Machinery/Process Level
- Sensors to PLCs
- Automatic test equipment
- Torque sensing (instrumented tools)

Level 2 — Station Level
- PLCs
- DataMyte data collectors
- Fastening system controllers
- Programmable Vision System

Level 3 — Cell Level
- Phoenix/VX Productivity System
- Allen-Bradley Industrial Computers

- Fastener remote modules
- DataMyte software for desktop computers

Level 4 — Center Level
- Central data collection system (host computer)
- DataMyte OVERVIEW System

See Chapter 18 for the DataMyte OVERVIEW system.

Earlier chapters in this book acquaint the reader with DataMyte product offerings. The rest of this chapter covers these Allen-Bradley systems:

- Pyramid Integrator
- Configurable Vision Input Module (CVIM)
- Phoenix/VX Productivity System
- Pro-Spec 6000 Threaded Fastening System
- EXPERT Programmable Vision System

23.3   PYRAMID INTEGRATOR®

The Pyramid Integrator offers a new approach to linking control and information that results in a highly integrated system solution. With the Pyramid Integrator, there is a seamless flow of information from sensors through plant computers, as well as station-level device integration on the backplane.

The Pyramid Integrator provides the first direct link between plant-floor control devices and plant-wide computers by linking a MicroVAX Information Processor to the Allen-Bradley PLC-5/250 programmable controller, CVIM vision processor, and universal I/O system. A complex network is not needed because the information processor, programmable controller, and visions processor modules share a common 32-bit backplane. See Figure 23.3.1.

Benefits

One of the greatest benefits of the Pyramid Integrator is that you can enhance your quality management capability. The Pyramid Integrator enables you to integrate your production systems, planning systems, and management information systems. More people can access the data they need, and when they need it. There is better decision-mak-

Fig. 23.3.1 Allen-Bradley Pyramid Integrator.

ing and tighter control of every operation. In addition, the Pyramid Integrator offers these benefits:

Reduce data acquisition time — The MicroVAX Information Processor reduces data acquisition time by its tight coupling with other Pyramid Integrator modules via the high-speed backplane. The information processor communicates with the PLC-5/250 and CVIM modules over the backplane, providing virtually instant access to plant-floor data and instant delivery to management.

Install on the plant floor like a traditional PLC — Each module is tested to Allen-Bradley standards for shock, temperature, and vibrations, so you can install the Pyramid Integrator on the plant floor.

Select the modules you need — Various functional modules let you perform programmable control, vision inspection, and information processing. You can select the modules you need for your present and future applications.

Access the data table from other modules — Each module has a data table that can be accessed by the other modules. For example, you can send statistical information from the vision module to the programmable controller modules to modify a process or a control point.

Control processes more efficiently — Use Sequential Function Chart (SFC) programming to perform true parallel processing. Use Independent Background Programs (IBP) to run programs independently and asychronously of the main PLC program.

Configure the system for specific applications — The modular design of the PLC-5/250 programmable controller lets you configure the amount of I/O, program memory size, and data table memory size depending on your application.

Process plant-floor information in a VMS environment — Allen-Bradley offers a module that processes information in a VMS environment. This allows you to distribute your VMS computing to the plant floor.

Support numerous VMS software packages — The

MicroVAX Information Processor is a true Digital MicroVAX computer. Because of this, the information processor supports a wide range of existing Digital and third-party application programs.

Applications

The Pyramid Integrator can be used in any application where programmable control, artificial vision, and information processing are used. The modular design and control environment packaging mean easier integration and simplified maintenance, ideal conditions for the following applications:

- Process industries
- Flow-through manufacturing
- Verification and inspection throughout processes

Process Industries — Process industries that are automating or integrating their facilities can use the Pyramid Integrator to:

- Allow information to flow from the plant floor to the business system
- Improve configuration flexibility
- Eliminate communication bottlenecks via tight coupling of control and information processors
- Reduce application development costs

Flow-through manufacturing — In a flow-through manufacturing environment, use the Pyramid Integrator to:

- Collect needed data
- Turn collected data into useful information

Verification and inspection — In various process industries, use the vision system's features to:

- Perform in-process inspection
- Configure the system using menus and icons
- Update the vision system as the plant process changes
- Perform high-speed inspection and verification
- Integrate the vision system with control and information

Fig. 23.4.1 *The CVIM module mounted in the Pyramid Integrator chassis.*

23.4 CONFIGURABLE VISION INPUT MODULE (CVIM)

The Pyramid Integrator offers non-contact inspection with the Configurable Vision Input Module (CVIM) and its associated components. The CVIM acquires images from up to two cameras and provides results that can be used by other Pyramid Integrator modules. CVIM can be used with other modules or as a standalone product. See Figure 23.4.1.

The CVIM module is designed for sophisticated inspection applications, and offers advanced image processing for high speed applications. But, even though it can handle sophisticated applications, it is easy to configure through a simple icon and menu-based user interface that lets you set up your application using a monitor and a light pen. You can use one or two cameras with the CVIM module, and the module can communicate over the Remote I/O link with other control devices. The CVIM module also offers area-of-interest scanning, which lets you inspect a specific area on a part.

Applications that are ideal for CVIM include assembly verification, package inspection, and product measurement.

CVIM offers you low-cost, high-performance vision and its icon-based configuration saves you setup time. One of the biggest benefits of the CVIM module is its ability to compensate for the manufacturing environment, including poor lighting conditions, X-Y part positioning, and part rotation.

Either as a standalone system or a module in the Pyramid Integrator, the CVIM module is a vision solution for sophisticated process control applications.

23.5 PHOENIX/VX PRODUCTIVITY SYSTEM

The Phoenix/VX productivity system is a cell level on-line SPC system. It organizes and manages data in a way which shows where improvement efforts must be focused. Plant and test data are collected and processed by the Phoenix/VX productivity system on-line and presented to plant personnel in summary form. Operators, technicians, supervisors, engineers and managers can take action in time to effect appropriate changes in process or machine operation in order to improve quality and productivity.

The powerful advantage of the Phoenix/VX productivity system is that it can handle thousands of parameters simultaneously. It allows a user to get a much quicker representation of what is happening on the plant floor.

System Architecture

The Phoenix/VX productivity system would typically be at the cell level of the productivity pyramid. It obtains data from devices at the station level via the Data Highway. The system can configure up to seven Data Highway networks.

The Phoenix/VX productivity system gathers information in raw data form. It then converts the data into a format that is meaningful for SPC functions, and useable for the operator. In the case of multiple production lines, it is possible to have systems at the cell level lock into a Phoenix/VX productivity system at the center level of the pyramid.

The system consists of the following hardware:

Main Phoenix/VX productivity system processor — This is a DEC MicroVAX, VAX workstation, or A-B Pyramid Integrator featuring a hard disk for dynamic data storage and a tape drive for system back-up. See Figure 23.5.1. The processor pulls up data from the Data Highway and generates pictures of machine, raw material and production data as well as quality performance over time.

Fig. 23.5.1   *Main Phoenix/VX Productivity System processor.*

Allen-Bradley Quality Management

Data Highway local area network — The network links data sources such as Allen-Bradley programmable controllers with the processor as well as peripheral operators' terminals, displays, and printers.

Operator color graphic display terminals and printers — The operator's color graphic display terminal generates graphs, such as $\bar{x}$ & R charts, histograms, Pareto charts and scatter plot diagrams, and also provides tables of data. The terminals can graphically depict plant floor operation. The operator has a picture of the machines and process, with colors indicating machine status.

The operator's display terminal also provides quality management information, such as current status and historical information, operating system messages and alarm messages. Downtime can be logged manually at the terminal. The processor merges the manual inputs with data that is automatically recorded. A printer provides hard copies of reports or any of the screens.

Configurer's terminal — System configuration is done at the configurer's terminal. The terminal is a black and white terminal for customizing the Phoenix/VX productivity system to a particular application. Reconfiguration can be done on-line.

Applications for the Phoenix/VX Productivity System

The Phoenix/VX productivity system can provide on-line SPC data to operators and management. It can help analyze the cause of a defect in a product coming off a production line. If an operator knows what parameters would cause a defect, the system can measure all needed data and apply SPC to establish a correlation to the highest probability of failure.

For the startup of new machinery, the Phoenix/VX productivity system can be used to develop accurate models of how the machines will actually function. A user can determine if the new equipment will be fast enough to meet desired production levels. The modeling function in some cases has led to paybacks on the system in just a few months.

One of the most important characteristics of the Phoenix/VX productivity system is its simplicity of use and operation. It's easily configurable by non-computer people who are experienced in the plant floor operations. A series of menus prompt the configurer who simply answers the questions. The only information the configurer must know includes the output, the appropriate report or display format, and the control devices that are involved.

The simplicity of the system greatly reduces the customer's cost of implementation. On-line, as the process is taking place, the system evaluates data and displays it in usable form for the operator. The system can also track raw materials and parts through the process on the plant floor. It keeps a data history so that as productivity improves, current operation of the plant can be compared to prior periods.

23.6 PRO-SPEC™ 6000 THREADED FASTENING SYSTEM

The Pro-Spec 6000 system is a microprocessor based threaded fastening system for the automotive, off-highway and aerospace industries. The primary function of the Pro-Spec 6000 system is to improve the quality of threaded fastener operations.

The Pro-Spec 6000 system has the capability to gather torque and angle data (see Figure 23.6.1). The system can play an active role in achieving desired specifications by controlling the fastening operation, or it can take a passive role by monitoring and reporting on the fastening operation. The system provides:

Fig. 23.6.1   Computer-based fastening system in auto plant.

- Accurate monitoring of the fastening cycle
- Accurate and flexible control of fastening tools
- Feedback on the performance of the fastening operation in the form of reports and charts

The Pro-Spec 6000 system monitors the state of the fastener at the end of the tightening cycle. The system compares this state to a range of engineering specifications, providing a basis for quality control. The Pro-Spec 6000 system can also decrease clamp load variations via the control of tension. By controlling tension, the Pro-Spec 6000 system improves the integrity of a joint, thereby optimizing the strength of a fastener.

Pro-Spec Fastening Concept

Conventional fastening systems measure only torque. Determination of a quality joint, however, requires measurement and precise control of not only torque, but also clamp load. Clamp load is the bolt tension required to hold the pieces of the joint together. The primary job of the fastener is to exert clamp load.

Torque measurements alone are not reliable indicators of the quality of a fastened joint. A typical nut and bolt are assigned to work together as a clamp load machine and apply clamp force to a joint, but threaded fasteners are not very efficient clamp load machines. A relatively small amount of clamp load is generated proportional to the work, or torque applied to the fasteners.

In producing clamp load the torque applied to the fastener must overcome several sources of friction:

- Friction on the fastener itself
- Friction under the bearing surface
- Friction in the mating threads

When torque is applied, friction in the threads consumes approximately half the work put into the fastener. Friction on the bearing surface consumes another forty percent of the work. Only ten percent of the torque applied to the fastener generates clamp load.

The actual clamp load generated will vary inversely with the coefficient of friction. A small increase in bearing face friction, three percent for example, reduces the clamp load by thirty percent. Such reductions in clamp force can mean the difference between a quality joint and a joint that has poor durability. Reductions in friction can also cause problems. For example, oil contamination of the fastener threads would reduce the friction, thus increasing the clamp load produced on the joint. Increases in clamp load can result in deformation of a bearing or overload of a joint component.

Many factors affect friction including fastener finishes, coatings, nicks, burrs, rough machined surfaces, sealants, and even color identification. Product designers must take friction variables into consideration when calculating the clamp load needed to hold an assembly together. For critical joints, fastening strategies other than torque monitoring must be considered, as the fundamental concern is clamp load, not applied torque.

Torque Graphing

An excellent tool for evaluating predicted clamp load and for modeling fastener mechanics is a graph indicating torque versus clamp load. (See Figure 23.6.2.) The torque output of a power tool is plotted along the vertical axis. The tension, measured by strain gages, load cells, and other devices is plotted along the horizontal axis.

The typical torque curve has three segments:

Prevailing torque — This is the "free" rundown segment. At this stage, the fastener has not yet begun to actually clamp parts together. Using the prevailing torque feature of the Pro-Spec 6000 system, limits can be assigned to the free rundown segment to identify defects, bends, cross threading and stripped threads.

Mid-point pause — This segment of the torque curve is the area between prevailing torque and a mid-point pause. The mid-point pause is a point at which the system analyzes torque/angle data and checks the performance of the instrumented fastening tool. The tool is subject to overrun, characterized by the brief span of time between the shut off command and the point at which the tool actually stops. During the mid-point stop, the Pro-Spec 6000 system can build a history of the tool's performance and measure the tool's overrun characteristics.

Torque value — The third segment of the torque curve represents the area between the mid-point pause and the desired torque value. Based on previous measurements of the tool's overrun, the system shuts off the tool at a calculated distance prior to the desired torque value. When the tool stops, the Pro-Spec 6000 system decides whether to accept or reject the fastened joint.

The Pro-Spec 6000 system continually compares the mid-point tool overrun measurements with the tool's final

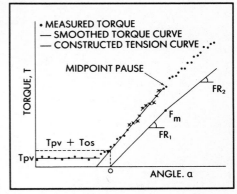

Fig. 23.6.2   *Torque graph.*

Fig. 23.6.3 Multi-spindle application.

positions. If significant deviations are noted, the system signals marginal tool performance.

Torque Synchronization for Multi-Tool Applications

The Pro-Spec 6000 system has special features for multiple joint fastening applications, such as automotive wheel hubs. (See Figure 23.6.3.) The pre-torque pause feature, in conjunction with spindle synchronization, evens the clamp load of a multi-fastener.

When multiple tools are used for multiple joint fastening applications, all of the tools get an initial command to tighten to a low level of torque. This low, pre-torque level is usually five to ten percent of the ultimate desired value. As each tool reaches the pre-torque level, the Pro-Spec 6000 system waits until all tools are at the same level. This tool synchronization eliminates uneven clamping at the beginning of the rundown cycle. When all of the tools in the application have reached the pre-torque level, they are simultaneously powered to advance to the mid-point pause. The system again waits for all tools before commanding them to proceed to the point of shut off for the desired torque value.

Origin of Tension

Origin of tension is a patented concept that allows the Allen-Bradley Pro-Spec 6000 system to consistently achieve a specified clamp load, independent of the friction variables in the joint. The origin of tension is the point of reference used to calculate clamp load (see Figure 23.6.4). The Pro-Spec 6000 system uses the Logarithmic Rate Method (LRM) to find the origin of tension. The algorithms for tension are derived from logarithmic rate formulas.

Using the LRM approach, the Pro-Spec 6000 System minimizes clamp load variations. Controlling tension increases the integrity of the joint, maximizes the strength of the fastener and improves the quality of the assembly.

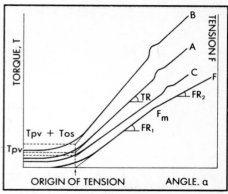

Fig. 23.6.4   Origin of Tension.

Pro-Spec 6000 System Architecture

The Allen-Bradley Pro-Spec 6000 system is a station-level real-time controller. It is designed to be integrated into the A-B architecture via Data Highway Plus area networks and remote I/O. Additional communication networks will be made available in the future.

The primary components of the Pro-Spec 6000 system are a station controller, terminal, optional printer and fastening modules which interface with the machine/process-level fastening tools. (See Figure 23.6.5.)

The heart of the system is the station controller. The station controller has computer functionality and connects to printers or CRT displays to generate reports for statistical analysis and quality management. The station controller supports parameter storage and report generation for each fastening module operation.

A remote terminal, such as the Allen-Bradley T-30 plant floor terminal, is required for system configuration and report display. Instructions to the system are made through menu selections and function keys. New tools, fastening strategies and channels can be added to a configuration, or existing assignments can be modified. Quality and productivity reports include shift end reports, trace reports, histograms and $\bar{x}$ & R charts. An optional printer may be connected to the station controller for hard copy reports.

The station controller supports up to 32 fastening modules. Each fastening module controls a single air or electric instrumented fastening tool. Any fastening module can be

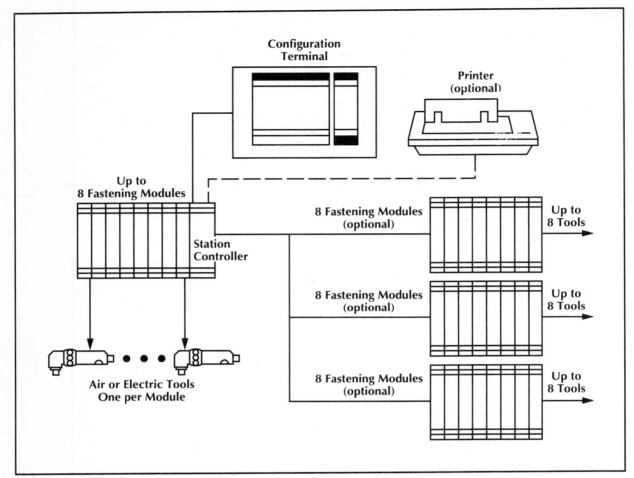

Fig. 23.6.5 Pro-Spec 6000 Threaded Fastening System Architecture.

inserted or removed from the rack under power. The rest of the system continues to operate if a module is removed. The fastening modules support a comprehensive set of algorithms for multiple fastening strategies. These fastening strategies offer application flexibility to suit virtually any fastening requirement.

Each application requires a specific level of tool instrumentation, such as a torque transducer, an angle encoder and a shut off mechanism. Throughout operation, the fastening module continuously responds to signals from the torque transducer and angle encoders and makes in-process adjustments. The fastening modules also send information to the station controller where it is logged for use in reports and analysis.

23.7 EXPERT PROGRAMMABLE VISION SYSTEM

The EXPERT programmable vision system allows inspection of 100% of products at production line speeds. This system allows identification, location and sorting of objects. The marriage of video imagery and computer technology offers the manufacturer a means of reducing product waste, costly inspection errors, and human fatigue.

In cases of product nonconformance, corrective action can be taken to reduce shipped waste. Overall product quality increases are possible, while product warranty losses can be decreased. A constant production rate can be maintained 24 hours a day, seven days a week, without endangering product integrity. And A-B's communications capability means the system is compatible with existing Allen-Bradley products.

For more information about Allen-Bradley Quality Management, contact Allen-Bradley, 1201 South Second Street, Milwaukee, Wisconsin, 53204.

24. BBN SOFTWARE PRODUCTS AND TOTAL QUALITY IMPROVEMENT

24.1 BBN SOFTWARE PRODUCTS CORPORATION

See Chapter 18 for the OVERVIEW system.

DataMyte — the leader in SPC data collection, and BBN Software Products — the leader in high-level data analysis software, have combined their strengths in the jointly-developed DataMyte OVERVIEW™ system. The DataMyte OVERVIEW system puts the power of BBN's RS Series data analysis software into a factory data collection network. This network is designed specifically for manufacturing environments and provides a tailored, user-friendly interface to the software's powerful features, reducing training time and making it easier to use. Chapter 18 explains the DataMyte OVERVIEW system more fully. This chapter looks specifically at BBN Software Products, its approach to quality, and its products.

BBN Software Products Corporation is a leading developer of data analysis software for manufacturing, engineering, research and development applications. BBN software is installed in 24 of the Fortune 25 and in 75 of the Fortune 100 industrial companies. BBN Software Products is a wholly-owned subsidiary of Bolt Beranek and Newman Inc., a pioneering technology firm since 1948.

Typical applications of BBN software are for:

- Statistical quality control
- Design of experiments
- Engineering data analysis
- Process optimization
- Decision support
- Research and development analysis

BBN's line of software products is used across the spectrum of manufacturing industries, from packaged goods to aerospace. These products operate on a wide range of Digital Equipment, IBM, Sun and Hewlett-Packard computer hardware. BBN also provides consulting services along with its high-level software, to develop integrated custom solutions and to help its customers gain the highest return from their hardware and software investments.

BBN Software Products' software is in use in over 30 countries around the world. Corporate headquarters and research and development are located in Cambridge, Massachusetts. Training facilities are based in London, Paris, Munich and Tokyo, as well as several U.S. locations.

24.2 TOTAL QUALITY IMPROVEMENT (TQI)

The mission of BBN Software Products is to give manufacturers statistical software and decision support tools that help them be more competitive by:

- Cutting development lead time
- Improving existing manufacturing processes
- Creating proprietary processes to reduce cost, improve quality, and increase yield
- Designing products for manufacturability
- Better deploying engineering knowledge and manufacturing expertise in the plant

To maximize the productivity of manufacturing technologies, BBN Software Products Corporation supports Total Quality Improvement (TQI), a four-pronged statistical approach to quality. The four "prongs" of the TQI approach are:

- Statistical Process Control (SPC)
- Design of experiments
- Analysis of historical data
- Expert systems to leverage manufacturing knowledge

TQI is a cost effective approach, and is often a requirement for implementing JIT, factory automation and other technologies to improve competitiveness. Moreover, while many of these other technologies require a large capital investment in new equipment, TQI maximizes existing resources.

This strategy of using statistical software tools to improve productivity and quality is one that is shared by many manufacturers, according to the results of a recent Dataquest survey. Dataquest polled manufacturers about what new technologies and new software solutions they anticipated implementing during the upcoming 18 months. Quality control received the highest response with 34 percent of those surveyed ranking it the highest. JIT manufacturing came in second with a 20 percent response.

By Total Quality Improvement, BBN Software Products means *statistical process control and beyond* — in other words, implementing statistical process control (SPC), analyzing historical data, and then advancing new techniques such as design of experiments to improve processes, and expert systems to capture and leverage manufacturing knowledge.

Statistical Process Control (SPC)

SPC is largely the subject of this handbook, but a brief summary of how SPC works — and what it achieves — is still appropriate here. In SPC, a production process is monitored, measuring key variables. SPC is based on the principle that the average values of a measured variable vary according to a random pattern that follows a normal distribution. Thus, when the average of measured variables does not follow a normal distribution, one can conclude that variations are not due to normal randomness of the product or process, but due rather to some special cause or causes. SPC identifies when these special causes occur and provides clues as to what the causes might be.

Statistical process control is a powerful tool for monitoring a manufacturing process to ensure that it is producing product within control limits, to raise a flag when the process goes out of control, and to define the problems or the causes and ultimately to help fix them.

Pareto analysis is another technique under the umbrella of SPC that is frequently used. Pareto analysis is a powerful tool for prioritizing efforts while striving to improve quality. The Pareto approach incorporates what is sometimes called the 80/20 rule — 20% of the sources of error cause about 80% of the problems. For example, a printed circuit board manufacturer may discover a large number of defects at post-wave inspection. Pareto analysis might reveal that 80% of the defects are blow-through holes and solder voids. By determining the causes and resolving these two defects alone, the manufacturer can significantly reduce defects.

The benefits of statistical process control are many: reduced scrap and rework, higher yield, higher quality products, and lower inspection costs. Today, more and more customers are demanding that their suppliers ship control charts or process capability studies with their products as assurance that the product is within specifications. These supplier-provided documents eliminate 100% inspections at the customer's receiving dock. See Chapter 8 for information on supplier certification. See Part II of this handbook for numerous examples of the benefits of SPC.

Design of Experiments

Statistical process control — as powerful as it is in helping to improve quality — is conducted mid-process or after the fact. It doesn't engineer quality into a product or process. *Design of experiments,* however, can be used to improve quality by improving the process or helping to design a product for manufacturability. Design of experiments is an emerging component of total quality that is producing outstanding results.

Today, most process improvement attempts are based on best guesses and troubleshooting experiments with the process. To debug a process that is encountering problems, engineers often experiment with the process by varying one variable at a time until they find the best level for that variable. Then, once that variable's optimum setting is found, the engineer varies another variable until that variable is optimized. The engineer continues varying process variables until each of the process variables has been optimized.

There are two problems with this traditional "one variable at a time" approach:

- *It is inefficient* — Because the number of runs needed expands exponentially as the number of variables increases, many experimental runs can be required, often far more than are feasible.

- *It does not detect the interaction between process variables* — For example, there may be interaction between temperature and pressure so that optimizing the individual variables produces results that are far from optimal for the total process.

Design of experiments is a method for setting up experiments to obtain the exact information needed in the most efficient way possible. By determining the information needed about a process or product in advance, the engineer can set up experiments that use fewer runs, require simpler analysis techniques and yield more information than traditional, ad hoc approaches to experimentation. Designed experimentation, moreover, can detect variable interaction.

The objective of designed experiments varies. The objective may be to increase yields, find lower cost materials, reduce defects, or improve product performance. A common objective is to develop a more "robust" process, insensitive to changes in the input of that process such as operator, machine or vendor's materials.

Designed experiments are extremely effective in the development of a new product. A good rule of thumb often used is that 80% of a product's cost is fixed in the design stage. For example, once a design engineer decides that a product will be made out of steel instead of plastic and will consist of four individual sub-components instead of two, then manufacturing can only reduce the cost of the product by reducing the remaining 20% (largely labor costs), and by making small modifications to the product manufacturing process. Therefore, it is extremely important for the design engineer to work with manufacturing to develop designs that manufacturing can produce easily and cost effectively. In design of experiments, the engineer first identifies the factors or inputs to the process. A "process" can be either a product or manufacturing process. For example, factors in a wave soldering application would include the conveyer speed, the temperature of a solder, the mix of the flux, the amount of the flux, and the pressure of the air knife. The engineer must also identify the "responses," or the results to be optimized. Responses in the wave solder application might be the number of solder defects.

Depending on the objectives of the study, the engineers select the type of design to use. Screening experiments are used to review a number of production factors and determine the key factors. If the key factors have been identified,

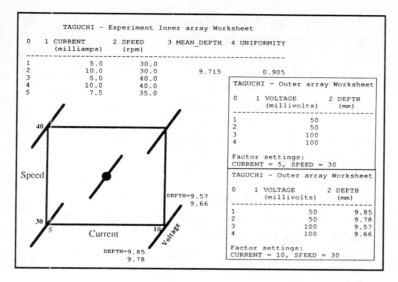

```
      TAGUCHI - Experiment Inner array Worksheet

  0    1 CURRENT      2 SPEED     3 MEAN_DEPTH   4 UNIFORMITY
        (milliamps)    (rpm)
  -------------------------------------------------------------
  1         5.0         30.0
  2        10.0         30.0          9.715
  3         5.0         40.0
  4        10.0         40.0
  5         7.5         35.0                       0.905
```

```
         TAGUCHI - Outer array Worksheet

  0        1 VOLTAGE        2 DEPTH
            (millivolts)      (mm)
  -----------------------------------------
  1            50
  2            50
  3           100
  4           100

  Factor settings:
  CURRENT = 5,  SPEED = 30
```

```
         TAGUCHI - Outer array Worksheet

  0        1 VOLTAGE        2 DEPTH
            (millivolts)      (mm)
  -----------------------------------------
  1            50             9.85
  2            50             9.78
  3           100             9.57
  4           100             9.66

  Factor settings:
  CURRENT = 10,  SPEED = 30
```

Fig. 24.2.1 *Experiment worksheets for one type of designed experiment, Taguchi methods. In this example, current, speed and voltage are the factors (variables). The responses are mean depth and uniformity.*

the engineers would select another type of designed experiment, such as response surface modeling or Taguchi designs for optimizing processes or products.

In any designed experiment, each of the input factors is varied in a systematic fashion. The experimental design determines the settings of each of the different runs within the experiment such that maximum information can be generated from a minimum number of experimental runs (an example is shown in Figure 24.2.1). A worksheet is developed that lists each of the experimental runs within the experiment and the setting for each of the process factors. The experimenter then performs each experimental run and records the results on the worksheet.

Once all the runs of the experiment have been completed the results are analyzed. In a screening experiment, the analysis identifies the input factors having the largest impact on the process — the "trivial many" from the "vital few." Optimization experiments such as response surface modeling produce mathematical models that can then be used to determine optimal settings.

Taguchi designs can be used to help identify the process setting regions that are most sensitive to inherent process variation. Taguchi experiments can also be used to develop a robust process that is insensitive to the variations and factors that are not easily controlled within the manufacturing process such as operator, humidity or time of day.

Design of experiments can be used to help increase

yields, and reduce scrap and rework. More importantly, design of experiments is a critical tool in helping to reduce the development time for new products and processes and to develop products and processes that are designed for manufacturability.

Results of design of experiments include:

- Determining optimum equipment/process settings
- Decreasing throughput time
- Improving yields
- Reducing scrap and rework
- Identifying which key variables drive processes
- Reducing process variability
- Decreasing development time for new products and processes
- Establishing more cost-effective operating levels for non-critical variables

Design of Experiments: A Typical Scenario

This example shows how BBN's RS/Discover® Software for design of experiments can be a powerful tool for designing products for manufacturability. A composites manufacturer recently purchased several new ovens for drying and curing of composites. These new pieces of equipment were intended to increase product consistency and quality. However, after installation and setup, the new ovens barely exceeded the performance of the old equipment.

Designed experiments were used to try to improve the performance of the new equipment. Each of the ovens was identified as a factor in the experiment. Some of the ovens were assigned to be drying ovens, while other ovens were assigned to the curing step in the composite process. The temperature of each of the ovens was allowed to vary within certain ranges. The rate of the conveyor speed through the oven was identified as another factor. The strength and consistency of the composites were identified as the response parameters to be optimized. After performing several runs, each at different factor settings, the results of the experiment showed that yield could be improved by slowing the conveyor speed while at the same time lowering the temperature of the ovens. With designed experiments the composites manufacturer was able to improve process performance far beyond the process equipment's specifications.

Analysis of Historical Data

Analysis of historical data is another approach that can be used to improve the quality of a process or product. Frequently, there is a *great deal of data* residing in factories but *little useful information*. The analysis of such data can provide many insights. It can aid in understanding the relationship between variables in a process. It can also be used to help identify key variables that drive processes — much like design of experiments. In analysis of data, however, historical data is used instead of data that is the result of a designed experiment. Typical users of analysis of data techniques are quality engineers and analysts, process engineers, development engineers and statisticians.

Historical data is usually collected on a number of process variables. The intent is to provide useful information, preferably displayed visually, for decision making. The key to analysis of historical data is to find the source of the problem rather than simply identifying that a problem exists (as is often the case in statistical process control). To do this, analysis of variance and analysis of residuals techniques may be used, for example.

Historical Data: Case Study

The benefits of analysis of historical data can be shown in an example of a glass manufacturer who was having a problem with poor quality and a high defect rate (more than 50%). The plant was operating at a loss. The feeling within the plant was that they were not monitoring the process sufficiently. They used exploratory data analysis on historical data.

A team of engineers and production personnel reviewed the process data and used multiple regression capabilities of BBN's RS/Explore® software to screen the many factors. They found that the glass left at the bottom of each tank was the biggest contributor to their quality problems. Once they changed their process to clean out the glass after each run, profitability was restored. The plant is now back to full production, and they claim an annual savings of $8 million and a dramatic reduction in scrap.

Expert Systems to Leverage Manufacturing Knowledge

Leveraging manufacturing knowledge is the fourth prong of the TQI approach to improving quality. Expert systems in manufacturing companies leverage scarce knowledgeable resources such as experienced engineers or shop floor operators who know the equipment or process better than others. These people are not always available on all shifts or at all locations.

Expert systems put this manufacturing knowledge to work and distribute it to others by placing the expertise *where and when it is needed most.* By capturing the expertise of the shop floor operator or engineer who best understands the process in an expert system, the expert's skills and experience can be placed at the fingertips of other operators to help them perform at the same high quality levels. For example, when SPC identifies that a process has drifted out of control, an expert system can begin a "conversation" with the operator. By asking such questions such as, "when was the equipment last maintained?" or "from which vendor did the raw materials come?" the expert system can recommend corrective action based on the set of answers. Expert systems are most appropriate for troubleshooting and problem resolution.

Expert Systems to Leverage Manufacturing Knowledge: Case Study

A chemical manufacturer was having a problem with over-correction of a process. SPC was being used to monitor the characteristics of the process, using both on-line instruments and extensive lab results.

A chemical engineer who had spent twenty years supporting the process had developed extensive expertise about the process. By comparing the results of on-line SPC charts with the results of off-line lab analysis, he was able to diagnose problems such as sensor failures, lab failures and process problems. When the engineer went home at night and the operators were on their own, their decisions were in error 25% of the time. Using expert system tools, the chemical engineer was able to build a simple expert system that captured his expertise about the process. He was able to automate his decision making process and make that expertise available to the second and third shift operators.

There have been no errors in decision making in over two years since that expert system was installed. Expert systems like this have been used extensively at this company, which now reports over $100 million dollars in annual savings company-wide from the use of the RS/Decision™ expert system tool kit.

24.3   BBN SOFTWARE PRODUCTS FOR TQI

BBN's software products are tools for total quality improvement (TQI): advanced statistical process control, design of experiments, analysis of historical data and expert systems.

OVERVIEW Software

Jointly developed by BBN and DataMyte, this software provides a link between DataMyte 900 Series data collectors and BBN's RS Series of data analysis software. Once data has been loaded into OVERVIEW software, it can be accessed and used in the RS Series products. OVERVIEW software is part of the DataMyte OVERVIEW data collection and analysis system, explained more fully in Chapter 18.

See Chapter 18 for OVERVIEW software.

RS/1® Software

RS/1 data analysis software is a comprehensive data management, data analysis and decision support tool specifically designed for scientific, engineering and manufacturing professionals. This fully integrated system includes data management, statistics, graphics (see Figure 24.3.1), curve fitting, modeling and report-generating capabilities. These features are complemented by a powerful, built-in programming language, RPL, for customized applications.

The additional BBN software products that follow are options to RS/1 software.

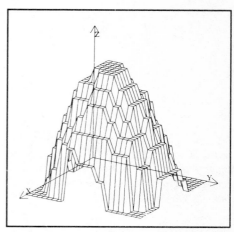

Fig. 24.3.1   A three-dimensional graph produced by RS/1® software showing the results (Z axis) as two factors (X and Y) are varied.

RS/QCA II™ Software

RS/QCA II (Quality Control Analysis) is on-line and off-line SPC software, including control charts, CuSum charts, process capability studies, Pareto analysis, trend analysis, inspection sampling plans, and reliability analysis.

Both process and discrete manufacturers use RS/QCA II to monitor, control and understand manufacturing processes, thus providing the basis for implementing changes in production, reducing product variability, tightening specification limits and adjusting other parameters that affect production yields and product quality.

RS/Discover® Software

RS/Discover software is a problem-solving tool that helps manufacturing and development engineers optimize processes by taking advantage of designed experimentation techniques. Created especially for use in industrial research, development, engineering and manufacturing environments, the RS/Discover system is the first software package for the design of experiments that is fully integrated with a complete system of data management, graphics and analysis tools.

RS/Discover software enables users to define experimental conditions, select an appropriate design, enter collected data, and perform a complete analysis and interpretation of experimental results (see Figures 24.3.2 and 24.3.3).

RS/Explore® Software

RS/Explore software is an innovative system for graphical data exploration and statistical analysis. It guides non-sta-

Fig. 24.3.2 The process is represented in terms of factors (inputs) and responses (outputs). RS/Discover® software helps the user understand the behavior of the process (black box).

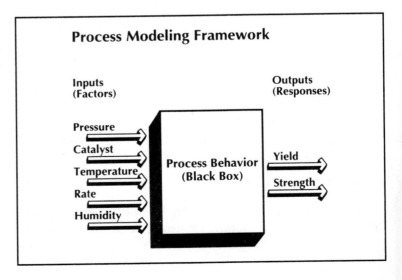

Process Modeling Framework

Inputs (Factors)

Outputs (Responses)

Pressure
Catalyst
Temperature
Rate
Humidity

Process Behavior (Black Box)

Yield
Strength

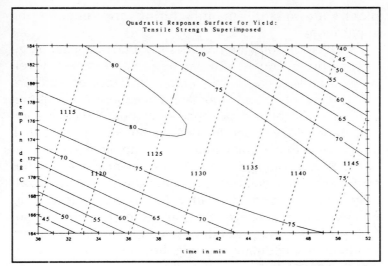

Fig. 24.3.3 An example of response surface modeling, produced by RS/Discover software.

tisticians through complete statistical analyses. It provides automated reporting of results, particularly useful for looking at long term trends and analyzing large amounts of production data. RS/Explore software leads users through the analytical process using graphical displays and a series of menus (see Figure 24.3.4).

RS/Decision™ Software

RS/Decision software is a tool kit for creating expert systems. It enables manufacturers to store and distribute critical manufacturing knowledge for consistent and immediate decision making — especially when experts are not available. The software is fully integrated with industry standard RS/1 data analysis software for developing integrated,

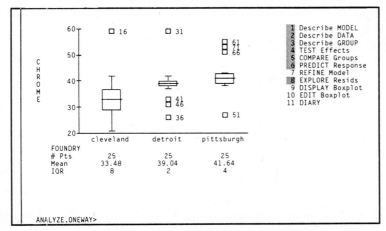

Fig. 24.3.4 An analysis of historical data for three foundries, with RS/Explore® software. This graph is called a summary boxplot, a one-way analysis of variance.

plant-wide quality manufacturing and engineering applications.

RS/Decision software provides manufacturing and engineering professionals with practical expert system development tools for decision support. RS/Decision software assists the capture and automation of expertise — without artificial intelligence experience — by both the novice and experienced developer. In addition, the RS/Decision shell integrates with existing databases and applications.

The RS/Decision package provides menu-driven utilities for building, debugging, maintaining and accessing knowledge bases. It is ideally suited for applications such as equipment maintenance and troubleshooting, quality control, process control, product selection and formulation, production scheduling, training, hotline support and on-line documentation.

For more information about the RS Series of products, contact BBN Software Products Corporation at 10 Fawcett Street, Cambridge, MA 02138.

25. DIGITAL EQUIPMENT CORPORATION AND COMPUTER INTEGRATED MANUFACTURING

25.1 DIGITAL COMPANY PROFILE

Digital Equipment Corporation (Digital) is the world's leading manufacturer of networked computer systems and associated peripheral equipment. Digital is the leader in systems integration with its networks, communications, services, and software products. A Fortune 50 company, Digital employs over 125,000 people worldwide.

Digital's products integrate the enterprise, from the individual and work group level to the whole organization. Digital's computer systems are used worldwide in such manufacturing applications as statistical process control, quality management, data analysis, cell control, shop floor management, process control, and MRP.

Digital manages its own worldwide business using networked VAX systems. The largest nonmilitary network in the world, Digital's Easynet network links over 34,000 systems with more than 80,000 users in 33 countries. The Easynet network, based on DECnet/OSI, integrates all levels of Digital employees in all departments, allowing them to exchange information instantaneously across the entire enterprise.

25.2 DIGITAL'S COMMITMENT TO COMPUTER INTEGRATED MANUFACTURING (CIM)

A growing number of leading edge manufacturers are adopting enterprise-wide, networked computing strategies. These manufacturers realize that to remain competitive in domestic and global markets, they need computer integrated manufacturing (CIM) as a major component of their business strategy.

Those companies that plan for CIM are making a major commitment: CIM is an evolutionary process — a long-term journey. It is not a specific product that can be bought "off the shelf"; it is not available from any single vendor. Rather, CIM is a combination of elements, all working together to provide an integrated solution. In today's multivendor manufacturing environment, open systems, standards and vendor cooperation are essential. These requirements fit hand in glove with Digital's corporate product strategy: to provide distributed computing resources when and where they are needed throughout the organization, integrating people, technology and information across the entire enterprise.

Digital is the leading supplier of open systems computing products and services for manufacturing, with a 30% share of the manufacturing computing market and nearly 50% of the manufacturing network market. Compatible architectures that span a wide breadth of price/performance, strong multi-vendor communication capabilities, and a history of protecting previous investments have enabled Digital to earn 15 to 20% of its revenue per year from sales to the manufacturing market.

Quality Information Systems

The need for open systems to permit integration and data sharing is no more evident than in the area of quality. Typically, existing homegrown quality solutions evolved over time, without a well thought-out plan. Consequently, they consist of a variety of sensors, gages and controls on the manufacturing lines, and many different computing systems ranging from personal computers on the factory floor to large computers at corporate levels. The wide variety of devices and computers involved makes integration of these systems very difficult. And, though plant floor information systems for quality are now affordable, highly functional and readily available, there remains a strong need to integrate existing hardware and software to prevent loss of investment and ensure the continuation of normal business activity.

Quality information systems combine hardware, software and networks for use by operators, inspectors, quality engineers and managers. Currently, quality information systems are spread across multiple pieces of hardware and application software. The overriding need today is to link these separate systems or niches into an integrated quality information system. Key niches in the quality software market include statistical process control, quality data analysis (sophisticated statistical graphing, analyzing and reporting capabilities, as in Figure 25.2.1) and quality management (linking elements of the quality information system through scheduled or ad hoc reporting capabilities).

Digital's approach to the quality information system market is to recruit a select portfolio of quality application vendors, and work with them to develop integrated solutions on Digital platforms. This is done by promoting integration:

- Among the quality software market niches.

- Between quality solutions and related areas of CIM (e.g. lab, maintenance).

- Among shop floor sources of quality data.

Fig. 25.2.1   RS Series software, from BBN Software Products, operating on a VAX computer system.

Digital provides tools for comprehensive data-sharing across hardware platforms and between applications. Quality information systems based on Digital's products can be migrated from low-end single-point solutions to more complete networked, distributed solutions.

In summary, Digital's product strategy is based on networking, integration tools, and computer families that allow you to link all parts of the enterprise more readily than any other vendor. This provides the ease of integration, investment protection and very competitive overall cost of ownership critical to the Digital manufacturing customer. Digital combines this with cooperative partnerships and alliances with key third party solution suppliers — including application software, service providers and industrial control vendors such as BBN, Allen-Bradley, and DataMyte.

25.3   UNDERSTANDING DIGITAL'S ADVANTAGES

Digital is the leading computer supplier to manufacturing. Digital is intent on continuing to be the system of choice for helping manufacturers become more competitive.

Digital products are used to create integrated solutions on which manufacturers can build CIM solutions today and in the future. The products provide the foundation for connecting the factory floor with the business information systems.

Key Digital advantages include:

- Close cooperation with third parties to develop applications software (wide array of manufacturing applications).
- Compatibility among Digital's families of computers (software can run on all computers within a family).
- Ability to grow systems cost-effectively (incremental addition of processors, memory, disks and computers to a network or cluster as needed).
- Computer systems offering a wide choice of price/performance (wide breadth within VAX and DECsystem families).
- Strong multi-vendor communications (extensive networking capabilities).
- Strong multi-vendor integration (data sharing and application integration across vendors).

Third Party Developers

Third party companies, such as BBN and DataMyte, are very important to Digital's CIM strategy. Digital relies on third parties to develop applications software, working with a large number of third parties since no one vendor can supply all the necessary components of a CIM solution. Digital's computer systems, networking and integration capabilities, coupled with third party application software, provides the enterprise-wide integrated solutions manufacturing customers are seeking.

Digital has made a commitment to provide the necessary services to support CIM solutions. Digital has also made a major investment in establishing architectures or frameworks to deliver the widest range of compatible computer hardware, software, integration and networking products. The following sections further examine these key components to Digital's CIM solutions, and provide a better understanding of Digital's advantages.

25.4   DIGITAL COMPUTER SYSTEMS

Digital offers two families of computers. The DECsystem family is a broad range of UNIX-based RISC systems. It consists of general purpose computers (DECsystem 3100, 5400 and 5800) and workstations (DECstation 2100 and 3100).

The VAX family, the computer used most frequently in manufacturing, is a VMS-based CISC family. The VAX family is used by the DataMyte OVERVIEW™ system.

All models within the VAX family are compatible. Regardless of how a customer's VAX system may grow or be upgraded, application software can be run without modification. There is no need for major conversions or retraining of users. This compatibility ensures investment protection for the customer, in both the VAX system and the software.

The VAX family includes workstations, low-end computers, mid-range systems, high-end systems and "coupled" systems (VAXcluster systems and Local Area VAXcluster systems). Figure 25.4.1 indicates some of the key differences between the low-end and mid-range VAX models. Figure 25.4.2 shows the relative performance of the various

See Chapter 18 for the DataMyte OVERVIEW system.

Fig. 25.4.1 Key differences of low-end and mid-range VAX models.

MicroVAX Family	Relative CPU Performance	Max. Memory Capacity	Max. Local Disk Capacity	Max. I/O Through-put	Footprint	Ethernet Adapters
MicroVAX 2000	0.9	14 MB	318 MB	3.3 MB/s	0.08 m² (0.98 ft²)	1 optional
MicroVAX II	0.9	16 MB	2.8 GB	3.3 MB/s	0.17 m² (1.9 ft²)	1 optional
MicroVAX 3300	2.4	52 MB	1.5 GB	3.3 MB/s	0.14 m² (2.4 ft²)	2 optional
MicroVAX 3400	2.4	52 MB	2.4 GB	3.3 MB/s	0.23 m² (2.4 ft²)	2 optional
MicroVAX 3800	3.8	64 MB	2.4 GB	3.3 MB/s	0.23 m² (2.4 ft²)	2 optional
MicroVAX 3900	3.8	64 MB	9.7 GB	3.3 MB/s	0.44 m² (4.7 ft²)	2 optional
VAXstation Family						
VAXstation 3100	2.7	32 MB	1.3 MB	1.2 MB/s	0.19 m² (2.0 ft²)	1
VAXstation 3200	2.7	16 MB	318 MB	3.3 MB/s	0.16 m² (1.8 ft²)	1
VAXstation 3500	2.7	64 MB	650 MB	3.3 MB/s	0.09 m² (2.7 ft²)	1
VAXstation 3520	Up to 5.4	64 MB	1.3 GB	30 MB/s	0.09 m² (2.7 ft²)	1
VAXstation 3540	Up to 10.8	64 MB	1.3 GB	30 MB/s	0.09 m² (2.7 ft²)	1
VAX 6000 Series						
VAX 6000 Model 210	2.8	256 MB	58.2 GB	60 MB/s	0.6 m² (6.4 ft²)	up to 4
VAX 6000 Model 310	3.8	256 MB	58.2 GB	60 MB/s	0.6 m² (6.4 ft²)	up to 4
VAX 6000 Model 410	7	256 MB	58.2 GB	60 MB/s	0.6 m² (6.4 ft²)	up to 4
VAX 6000 Model 420	Up to 13	256 MB	58.2 GB	60 MB/s	0.6 m² (6.4 ft²)	up to 4
VAX 6000 Model 430	Up to 19	256 MB	58.2 GB	60 MB/s	0.6 m² (6.4 ft²)	up to 4
VAX 6000 Model 440	Up to 25	256 MB	58.2 GB	60 MB/s	0.6 m² (6.4 ft²)	up to 4
VAX 6000 Model 450	Up to 31	192 MB	38.8 GB	40 MB/s	0.6 m² (6.4 ft²)	up to 4
VAX 6000 Model 460	Up to 36	192 MB	38.8 GB	40 MB/s	0.6 m² (6.4 ft²)	up to 4

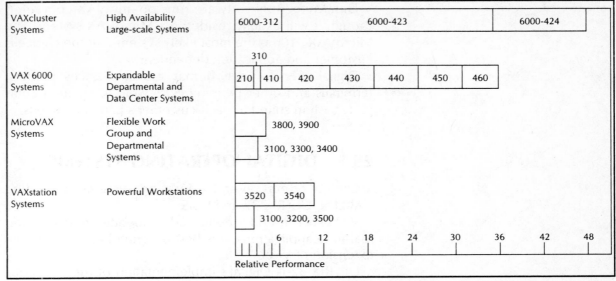

VAXcluster Systems	High Availability Large-scale Systems	6000-312	6000-423					6000-424	

		310						
VAX 6000 Systems	Expandable Departmental and Data Center Systems	210	410	420	430	440	450	460

MicroVAX Systems	Flexible Work Group and Departmental Systems	3800, 3900
		3100, 3300, 3400

VAXstation Systems	Powerful Workstations	3520	3540
		3100, 3200, 3500	

Relative Performance: 6 12 18 24 30 36 42 48

Fig. 25.4.2   Relative performance of VAX models.

systems. Note that one Ethernet adaptor is required when using a VAX computer in the DataMyte OVERVIEW™ system since part of the communications to data collectors involves an Ethernet network.

A unique Digital strength is the ability to cost-effectively go beyond single processor computing. Digital can do this through many forms of incremental growth, such as:

- Added processors within one computer — creating a multiple processor system, greatly increasing its power.
- Networked VAX computers — enabling the application and data to be distributed across multiple computers.
- VAXcluster systems — tightly linking several computers together, such that the system becomes a powerful machine.

VAXcluster systems are the alternative to traditional mainframes. The linked independent computers cooperate as though they were a single system. Additional computers can be added as needed in a cost-effective manner. Local Area VAXcluster systems are a similar concept for sharing resources such as printers or disk drives, and sharing data or loads.

See Chapter 18 for the DataMyte OVERVIEW system.

DataMyte/BBN software runs on any VAX computer, though it will usually reside on a MicroVAX system. The MicroVAX 3100 is the most likely system, but the choice is customer and application dependent.

Digital offers a variety of mass storage devices, printers, terminals and other peripherals as well, to create a complete system suited to each customer's particular needs.

25.5 DIGITAL OPERATING SYSTEMS

Digital offers several operating systems, including VAXELN, ULTRIX-32 and VMS.

VAXELN is used for dedicated, embedded or distributed real-time applications. It is best described as the real-time extension of VMS.

ULTRIX-32 is Digital's implementation of the UNIX operating system. It supports standard UNIX features, as well as additional capabilities, and conforms to all relevant standards. ULTRIX-32 is Digital's offering for markets requiring an open standard operating system.

VMS has been the most functional, fully integrated operating system in the industry for over 12 years. It is a general purpose, disk-based time sharing operating system. VMS runs on the entire VAX family of products. This enables application software — such as OVERVIEW software — to run on any VAX system.

25.6 DIGITAL AND NETWORKING

One of Digital's greatest strengths is its networking capabilities. Just a few of those capabilities, relevant to manufacturing in general and to the DataMyte OVERVIEW™ system in particular, are discussed on the following pages.

DECnet/OSI Architecture

See Chapter 18 for the DataMyte OVERVIEW system.

Digital's primary network architecture is DECnet/OSI. DECnet/OSI networks include the hardware and software required to enable computers to communicate with each other. A 1989 report by Advanced Manufacturing Research found·that within manufacturing plants, the leading computer and networking vendor is Digital. Further, the report found that DECnet has a predominant position on the shop floor. It represented close to 50% of 1988 installations.

Digital offers the network components as well as a full range of products and services for network management and control. These products allow configuration flexibility and the ability to "debug" and tune" the network for optimal performance.

Ethernet Networks

In manufacturing plants, typically the 802.3/Ethernet Local Area Network and DECnet/OSI protocols are employed. Along with DECnet/OSI, 802.3/Ethernet has proven to be a highly reliable, cost effective, and flexible networking approach in manufacturing environments. 802.3/Ethernet is employed in the DataMyte OVERVIEW system.

See Chapter 18 for the DataMyte OVERVIEW system.

The specification for the Ethernet network was developed in the 1970's by Xerox Corporation and published in 1980 as a public standard by Xerox, Digital and Intel Corporations. It is a standard network that has been used with most computers, including personal computers.

802.3/Ethernet transmission speed is relatively fast, up to 10 Megabits per second. This is 2½ to 10 times the speed of most competitive networks. A study done by Digital shows that 802.3/Ethernet is not affected by the electromagnetic interference (EMI) typically found in manufacturing facilities. Performance measurements were taken at automotive, aerospace, heavy machinery, chemical, pulp & paper, metals and printing facilities.

There are a variety of 802.3/Ethernet transmission mediums, or cables, that can be used. These include broadband CATV coaxial, standard baseband coaxial, twisted pair, and fiber optic. Coaxial cable is the most common medium used in manufacturing facilities.

25.7 DIGITAL'S INTEGRATION CAPABILITIES

Digital has made a commitment to integrate multi-vendor environments. Network Application Support (NAS) is a set of application services that assist in the communication, data sharing and integration of applications distributed across multiple systems.

NAS supports microcomputers from Digital, IBM, Compaq, Olivetti; the Apple Macintosh; Digital VMS and UNIX-based computers; terminals from Digital and IBM; and connection to the other OSI vendors and IBM mainframes.

Apple Macintosh Interconnect

As part of the NAS effort, Digital and Apple Computer are working together to provide a jointly developed and endorsed environment for common communication. This will be based on the AppleTalk and DECnet/OSI networking foundations. The development effort between the two companies is designed to provide developers with standard technologies for Macintosh and VAX integration.

Apple and Digital have outlined the common communication foundation and the core network services under development. Also outlined are the tools and capabilities for building distributed applications. Some of the planned capabilities are integrated AppleTalk and DECnet/OSI networks, file services, print services, database access and network management. Currently many third party packages are available to integrate Macintosh computers and VAX systems.

The following are trademarks of Digital Equipment Corporation: DEC, DECsystem, MicroVAX, ULTRIX, VAX, VAXELN and VMS.

APPENDIX

TABLE A-1   CONSTANTS FOR CALCULATING CONTROL LIMITS

The formulas for calculating control limits are shown below. They are based on the normal distribution curve.

Number of Observations in subgroup	Constants for $\bar{x}$ Chart		Divisors for Estimate of Standard Deviation	Range Chart Lower Control Limit	Range Chart Upper Control Limit	Constant for s Chart	Divisors for Estimate of Standard Deviation	Sigma Chart Lower Control Limit	Sigma Chart Upper Control Limit
n	A	A_2	d_2	D_3	D_4	A_3	c_4	B_3	B_4
2	2.12	1.880	1.128	—	3.267	2.659	0.7979	—	3.267
3	1.73	1.023	1.693	—	2.574	1.954	0.8862	—	2.568
4	1.50	0.729	2.059	—	2.282	1.628	0.9213	—	2.266
5	1.34	0.577	2.326	—	2.114	1.427	0.9400	—	2.089
6	1.22	0.483	2.534	—	2.004	1.287	0.9515	0.030	1.970
7	1.13	0.419	2.704	0.076	1.924	1.182	0.9594	0.118	1.882
8	1.06	0.373	2.847	0.136	1.864	1.099	0.9650	0.185	1.815
9	1.00	0.337	2.970	0.184	1.816	1.032	0.9693	0.239	1.761
10	0.95	0.308	3.078	0.223	1.777	0.975	0.9727	0.284	1.716
11	0.90	0.285	3.173	0.256	1.744	0.927	0.9754	0.321	1.679
12	0.87	0.266	3.258	0.283	1.717	0.886	0.9776	0.354	1.646
13	0.83	0.249	3.336	0.307	1.693	0.850	0.9794	0.382	1.618
14	0.80	0.235	3.407	0.328	1.672	0.817	0.9810	0.406	1.594
15	0.77	0.223	3.472	0.347	1.653	0.789	0.9823	0.428	1.572
16	0.75	0.212	3.532	0.363	1.637	0.763	0.9835	0.448	1.552
17	0.73	0.203	3.588	0.378	1.622	0.739	0.9845	0.466	1.534
18	0.71	0.194	3.640	0.391	1.608	0.718	0.9854	0.482	1.518
19	0.69	0.187	3.689	0.403	1.597	0.698	0.9862	0.497	1.503
20	0.67	0.180	3.735	0.415	1.585	0.680	0.9869	0.510	1.490
21	0.65	0.173	3.778	0.425	1.575	0.663	0.9876	0.523	1.477
22	0.64	0.167	3.819	0.434	1.566	0.647	0.9882	0.534	1.466
23	0.63	0.162	3.858	0.443	1.557	0.633	0.9887	0.545	1.455
24	0.61	0.157	3.895	0.451	1.548	0.619	0.9892	0.555	1.445
25	0.60	0.153	3.931	0.459	1.541	0.606	0.9896	0.565	1.435

FORMULAS

$\bar{x}$ Control Limits based on $\bar{R} = \bar{\bar{x}} \pm A_2\bar{R}$

$\bar{x}$ Control Limits based on $\bar{s} = \bar{\bar{x}} \pm A_3\bar{s}$

$\bar{x}$ Control Limits based on the sigma of the data $= \bar{x}' \pm A\sigma'$

$\bar{x}$ Control Limits based on the sigma of the x-bars $= \bar{\bar{x}} \pm 3\sigma_{\bar{x}}$

$\bar{x}$ Control Limits based on moving range $= \bar{\bar{x}} \pm 2.66m\bar{R}$

Range Upper Control Limit $= D_4\bar{R}$

Range Lower Control Limit $= D_3\bar{R}$

Sigma Upper Control Limit $= B_4\bar{\sigma}$

Sigma Lower Control Limit $= B_3\bar{\sigma}$

Estimated Sigma Hat based on $\bar{R} = \hat{\sigma} = \bar{R}/d_2$

Estimated Sigma Hat based on $\bar{s} = \hat{\sigma} = \bar{s}/c_4$

*Adapted from ASTM publicaton STP-15D, **Manual on the Presentation of Data and Control Chart Analysis**, 1976; pp 134-136. Copyright ASTM, 1916 Race Street, Philadelphia, Pennsylvania 19103. Reprinted, with permission.

TABLE A-2 E_2 TABLE

Number of observations in subgroup	E_2
2	2.660
3	1.772
4	1.457
5	1.290
6	1.184
7	1.109
8	1.054
9	1.010
10	0.975
11	0.946
12	0.921
13	0.899
14	0.881
15	0.864

TABLE A-3   AREAS UNDER THE NORMAL CURVE

$\dfrac{X_i - \bar{X}'}{\sigma'}$	0.09	0.08	0.07	0.06	0.05	0.04	0.03	0.02	0.01	0.00
−3.5	0.00017	0.00017	0.00018	0.00019	0.00019	0.00020	0.00021	0.00022	0.00022	0.00023
−3.4	0.00024	0.00025	0.00026	0.00027	0.00028	0.00029	0.00030	0.00031	0.00033	0.00034
−3.3	0.00035	0.00036	0.00038	0.00039	0.00040	0.00042	0.00043	0.00045	0.00047	0.00048
−3.2	0.00050	0.00052	0.00054	0.00056	0.00058	0.00060	0.00062	0.00064	0.00066	0.00069
−3.1	0.00071	0.00074	0.00076	0.00079	0.00082	0.00085	0.00087	0.00090	0.00094	0.00097
−3.0	0.00100	0.00104	0.00107	0.00111	0.00114	0.00118	0.00122	0.00126	0.00131	0.00135
−2.9	0.0014	0.0014	0.0015	0.0015	0.0016	0.0016	0.0017	0.0017	0.0018	0.0019
−2.8	0.0019	0.0020	0.0021	0.0021	0.0022	0.0023	0.0023	0.0024	0.0025	0.0026
−2.7	0.0026	0.0027	0.0028	0.0029	0.0030	0.0031	0.0032	0.0033	0.0034	0.0035
−2.6	0.0036	0.0037	0.0038	0.0039	0.0040	0.0041	0.0043	0.0044	0.0045	0.0047
−2.5	0.0048	0.0049	0.0051	0.0052	0.0054	0.0055	0.0057	0.0059	0.0060	0.0062
−2.4	0.0064	0.0066	0.0068	0.0069	0.0071	0.0073	0.0075	0.0078	0.0080	0.0082
−2.3	0.0084	0.0087	0.0089	0.0091	0.0094	0.0096	0.0099	0.0102	0.0104	0.0107
−2.2	0.0110	0.0113	0.0116	0.0119	0.0122	0.0125	0.0129	0.0132	0.0136	0.0139
−2.1	0.0143	0.0146	0.0150	0.0154	0.0158	0.0162	0.0166	0.0170	0.0174	0.0179
−2.0	0.0183	0.0188	0.0192	0.0197	0.0202	0.0207	0.0212	0.0217	0.0222	0.0228
−1.9	0.0233	0.0239	0.0244	0.0250	0.0256	0.0262	0.0268	0.0274	0.0281	0.0287
−1.8	0.0294	0.0301	0.0307	0.0314	0.0322	0.0329	0.0336	0.0344	0.0351	0.0359
−1.7	0.0367	0.0375	0.0384	0.0392	0.0401	0.0409	0.0418	0.0427	0.0436	0.0446
−1.6	0.0455	0.0465	0.0475	0.0485	0.0495	0.0505	0.0516	0.0526	0.0537	0.0548
−1.5	0.0559	0.0571	0.0582	0.0594	0.0606	0.0618	0.0630	0.0643	0.0655	0.0668
−1.4	0.0681	0.0694	0.0708	0.0721	0.0735	0.0749	0.0764	0.0778	0.0793	0.0808
−1.3	0.0823	0.0838	0.0853	0.0869	0.0885	0.0901	0.0918	0.0934	0.0951	0.0968
−1.2	0.0985	0.1003	0.1020	0.1038	0.1057	0.1075	0.1093	0.1112	0.1131	0.1151
−1.1	0.1170	0.1190	0.1210	0.1230	0.1251	0.1271	0.1292	0.1314	0.1335	0.1357
−1.0	0.1379	0.1401	0.1423	0.1446	0.1469	0.1492	0.1515	0.1539	0.1562	0.1587
−0.9	0.1611	0.1635	0.1660	0.1685	0.1711	0.1736	0.1762	0.1788	0.1814	0.1841
−0.8	0.1867	0.1894	0.1922	0.1949	0.1977	0.2005	0.2033	0.2061	0.2090	0.2119
−0.7	0.2148	0.2177	0.2207	0.2236	0.2266	0.2297	0.2327	0.2358	0.2389	0.2420
−0.6	0.2451	0.2483	0.2514	0.2546	0.2578	0.2611	0.2643	0.2676	0.2709	0.2743
−0.5	0.2776	0.2810	0.2843	0.2877	0.2912	0.2946	0.2981	0.3015	0.3050	0.3085
−0.4	0.3121	0.3156	0.3192	0.3228	0.3264	0.3300	0.3336	0.3372	0.3409	0.3446
−0.3	0.3483	0.3520	0.3557	0.3594	0.3632	0.3669	0.3707	0.3745	0.3783	0.3821
−0.2	0.3859	0.3897	0.3936	0.3974	0.4013	0.4052	0.4090	0.4129	0.4168	0.4207
−0.1	0.4247	0.4286	0.4325	0.4364	0.4404	0.4443	0.4483	0.4522	0.4562	0.4602
−0.0	0.4641	0.4681	0.4721	0.4761	0.4801	0.4840	0.4880	0.4920	0.4960	0.5000

Reproduced with permission from Eugene L. Grant and Richard S. Leavenworth, "Statistical Quality Control," 5th ed., 1980, McGraw-Hill Book Company.

HOW TO USE THIS TABLE: Subtract the mean ($\bar{x}$) from the specification (x_i) and divide by sigma (σ'). Negative numbers indicate areas to the left of the mean and positive to the right. Use the value in the table that corresponds to units and tenths (first column), and hundredths (top row) digits of the number. For example, .38 corresponds to 0.6480. Subtract from 1.0 to get the percent area from the specification out to infinity on a normal curve. For .38, it is 35%.

TABLE A-3, CONTINUED

$\dfrac{X_i - \bar{X}'}{\sigma'}$	0.00	0.01	0.02	0.03	0.04	0.05	0.06	0.07	0.08	0.09
+0.0	0.5000	0.5040	0.5080	0.5120	0.5160	0.5199	0.5239	0.5279	0.5319	0.5359
+0.1	0.5398	0.5438	0.5478	0.5517	0.5557	0.5596	0.5636	0.5675	0.5714	0.5753
+0.2	0.5793	0.5832	0.5871	0.5910	0.5948	0.5987	0.6026	0.6064	0.6103	0.6141
+0.3	0.6179	0.6217	0.6255	0.6293	0.6331	0.6368	0.6406	0.6443	0.6480	0.6517
+0.4	0.6554	0.6591	0.6628	0.6664	0.6700	0.6736	0.6772	0.6808	0.6844	0.6879
+0.5	0.6915	0.6950	0.6985	0.7019	0.7054	0.7088	0.7123	0.7157	0.7190	0.7224
+0.6	0.7257	0.7291	0.7324	0.7357	0.7389	0.7422	0.7454	0.7486	0.7517	0.7549
+0.7	0.7580	0.7611	0.7642	0.7673	0.7704	0.7734	0.7764	0.7794	0.7823	0.7852
+0.8	0.7881	0.7910	0.7939	0.7967	0.7995	0.8023	0.8051	0.8079	0.8106	0.8133
+0.9	0.8159	0.8186	0.8212	0.8238	0.8264	0.8289	0.8315	0.8340	0.8365	0.8389
+1.0	0.8413	0.8438	0.8461	0.8485	0.8508	0.8531	0.8554	0.8577	0.8599	0.8621
+1.1	0.8643	0.8665	0.8686	0.8708	0.8729	0.8749	0.8770	0.8790	0.8810	0.8830
+1.2	0.8849	0.8869	0.8888	0.8907	0.8925	0.8944	0.8962	0.8980	0.8997	0.9015
+1.3	0.9032	0.9049	0.9066	0.9082	0.9099	0.9115	0.9131	0.9147	0.9162	0.9177
+1.4	0.9192	0.9207	0.9222	0.9236	0.9251	0.9265	0.9279	0.9292	0.9306	0.9319
+1.5	0.9332	0.9345	0.9357	0.9370	0.9382	0.9394	0.9406	0.9418	0.9429	0.9441
+1.6	0.9452	0.9463	0.9474	0.9484	0.9495	0.9505	0.9515	0.9525	0.9535	0.9545
+1.7	0.9554	0.9564	0.9573	0.9582	0.9591	0.9599	0.9608	0.9616	0.9625	0.9633
+1.8	0.9641	0.9649	0.9656	0.9664	0.9671	0.9678	0.9686	0.9693	0.9699	0.9706
+1.9	0.9713	0.9719	0.9726	0.9732	0.9738	0.9744	0.9750	0.9756	0.9761	0.9767
+2.0	0.9773	0.9778	0.9783	0.9788	0.9793	0.9798	0.9803	0.9808	0.9812	0.9817
+2.1	0.9821	0.9826	0.9830	0.9834	0.9838	0.9842	0.9846	0.9850	0.9854	0.9857
+2.2	0.9861	0.9864	0.9868	0.9871	0.9875	0.9878	0.9881	0.9884	0.9887	0.9890
+2.3	0.9893	0.9896	0.9898	0.9901	0.9904	0.9906	0.9909	0.9911	0.9913	0.9916
+2.4	0.9918	0.9920	0.9922	0.9925	0.9927	0.9929	0.9931	0.9932	0.9934	0.9936
+2.5	0.9938	0.9940	0.9941	0.9943	0.9945	0.9946	0.9948	0.9949	0.9951	0.9952
+2.6	0.9953	0.9955	0.9956	0.9957	0.9959	0.9960	0.9961	0.9962	0.9963	0.9964
+2.7	0.9965	0.9966	0.9967	0.9968	0.9969	0.9970	0.9971	0.9972	0.9973	0.9974
+2.8	0.9974	0.9975	0.9976	0.9977	0.9977	0.9978	0.9979	0.9979	0.9980	0.9981
+2.9	0.9981	0.9982	0.9983	0.9983	0.9984	0.9984	0.9985	0.9985	0.9986	0.9986
+3.0	0.99865	0.99869	0.99874	0.99878	0.99882	0.99886	0.99889	0.99893	0.99896	0.99900
+3.1	0.99903	0.99906	0.99910	0.99913	0.99915	0.99918	0.99921	0.99924	0.99926	0.99929
+3.2	0.99931	0.99934	0.99936	0.99938	0.99940	0.99942	0.99944	0.99946	0.99948	0.99950
+3.3	0.99952	0.99953	0.99955	0.99957	0.99958	0.99960	0.99961	0.99962	0.99964	0.99965
+3.4	0.99966	0.99967	0.99969	0.99970	0.99971	0.99972	0.99973	0.99974	0.99975	0.99976
+3.5	0.99977	0.99978	0.99978	0.99979	0.99980	0.99981	0.99981	0.99982	0.99983	0.99983

TABLE A-4 STUDENTS' t DISTRIBUTION

ν \ α	\multicolumn{7}{c}{Two-tail Critical Values}						
ν	0.50	0.25	0.10	0.05	0.025	0.01	0.005
1	1.00000	2.4142	6.3138	12.706	25.452	63.657	127.32
2	0.81650	1.6036	2.9200	4.3027	6.2053	9.9248	14.089
3	0.76489	1.4226	2.3534	3.1825	4.1765	5.8409	7.4533
4	0.74070	1.3444	2.1318	2.7764	3.4954	4.6041	5.5976
5	0.72669	1.3009	2.0150	2.5706	3.1634	4.0321	4.7733
6	0.71756	1.2733	1.9432	2.4469	2.9687	3.7074	4.3168
7	0.71114	1.2543	1.8946	2.3646	2.8412	3.4995	4.0293
8	0.70639	1.2403	1.8595	2.3060	2.7515	3.3554	3.8325
9	0.70272	1.2297	1.8331	2.2622	2.6850	3.2498	3.6897
10	0.69981	1.2213	1.8125	2.2281	2.6338	3.1693	3.5814
11	0.69745	1.2145	1.7959	2.2010	2.5931	3.1058	3.4966
12	0.69548	1.2089	1.7823	2.1788	2.5600	3.0545	3.4284
13	0.69384	1.2041	1.7709	2.1604	2.5326	3.0123	3.3725
14	0.69242	1.2001	1.7613	2.1448	2.5096	2.9768	3.3257
15	0.69120	1.1967	1.7530	2.1315	2.4899	2.9467	3.2860
16	0.69013	1.1937	1.7459	2.1199	2.4729	2.9208	3.2520
17	0.68919	1.1910	1.7396	2.1098	2.4581	2.8982	3.2225
18	0.68837	1.1887	1.7341	2.1009	2.4450	2.8784	3.1966
19	0.68763	1.1866	1.7291	2.0930	2.4334	2.8609	3.1737
20	0.68696	1.1848	1.7247	2.0860	2.4231	2.8453	3.1534
21	0.68635	1.1831	1.7207	2.0796	2.4138	2.8314	3.1352
22	0.68580	1.1816	1.7171	2.0739	2.4055	2.8188	3.1188
23	0.68531	1.1802	1.7139	2.0687	2.3979	2.8073	3.1040
24	0.68485	1.1789	1.7109	2.0639	2.3910	2.7969	3.0905
25	0.68443	1.1777	1.7081	2.0595	2.3846	2.7874	3.0782
26	0.68405	1.1766	1.7056	2.0555	2.3788	2.7787	3.0669
27	0.68370	1.1757	1.7033	2.0518	2.3734	2.7707	3.0565
28	0.68335	1.1748	1.7011	2.0484	2.3685	2.7633	3.0469
29	0.68304	1.1739	1.6991	2.0452	2.3638	2.7564	3.0380
30	0.68276	1.1731	1.6973	2.0423	2.3596	2.7500	3.0298
40	0.68066	1.1673	1.6839	2.0211	2.3289	2.7045	2.9712
60	0.67862	1.1616	1.6707	2.0003	2.2991	2.6603	2.9146
120	0.67656	1.1559	1.6577	1.9799	2.2699	2.6174	2.8599
∞	0.67449	1.1503	1.6449	1.9600	2.2414	2.5758	2.8070
ν	0.25	0.125	0.05	0.025	0.0125	0.005	0.0025
α	\multicolumn{7}{c}{One-tail Critical Values}						

TABLE A-5 X^2 DISTRIBUTION

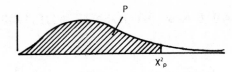

VALUES to X^2_P CORRESPONDING TO P

DF	$X^2.005$	$X^2.01$	$X^2.025$	$X^2.05$	$X^2.10$	$X^2.90$	$X^2.95$	$X^2.975$	$X^2.99$	$X^2.995$
1	0.000039	0.00016	0.00098	0.0039	0.0158	2.71	3.84	5.02	6.63	7.88
2	0.0100	0.0201	0.0506	0.1026	0.2107	4.61	5.99	7.38	9.21	10.60
3	0.0717	0.115	0.216	0.352	0.584	6.25	7.81	9.35	11.34	12.84
4	0.207	0.297	0.484	0.711	1.064	7.78	9.49	11.14	13.28	14.86
5	0.412	0.554	0.831	1.15	1.61	9.24	11.07	12.83	15.09	16.75
6	0.676	0.872	1.24	1.64	2.20	10.64	12.59	14.45	16.81	18.55
7	0.989	1.24	1.69	2.17	2.83	12.02	14.07	16.01	18.48	20.28
8	1.34	1.65	2.18	2.73	3.49	13.36	15.51	17.53	20.09	21.96
9	1.73	2.09	2.70	3.33	4.17	14.68	16.92	19.02	21.67	23.59
10	2.16	2.56	3.25	3.94	4.87	15.99	18.31	20.48	23.21	25.19
11	2.60	3.05	3.82	4.57	5.58	17.28	19.68	21.92	24.73	26.76
12	3.07	3.57	4.40	5.23	6.30	18.55	21.03	23.34	26.22	28.30
13	3.57	4.11	5.01	5.89	7.04	19.81	22.36	24.74	27.69	29.82
14	4.07	4.66	5.63	6.57	7.79	21.06	23.68	26.12	29.14	31.32
15	4.60	5.23	6.26	7.26	8.55	22.31	25.00	27.49	30.58	32.80
16	5.14	5.81	6.91	7.96	9.31	23.54	26.30	28.85	32.00	34.27
18	6.26	7.01	8.23	9.39	10.86	25.99	28.87	31.53	34.81	37.16
20	7.43	8.26	9.59	10.85	12.44	28.41	31.41	34.17	37.57	40.00
24	9.89	10.86	12.40	13.85	15.66	33.20	36.42	39.36	42.98	45.56
30	13.79	14.95	16.79	18.49	20.60	40.26	43.77	46.98	50.89	53.67
40	20.71	22.16	24.43	26.51	29.05	51.81	55.76	59.34	63.69	66.77
60	35.53	37.48	40.48	43.19	46.46	74.40	79.08	83.30	88.38	91.95
120	83.85	86.92	91.58	95.70	100.62	140.23	146.57	152.21	158.95	163.64

TABLE A-6   METRIC SYSTEM

METRIC SYSTEM
LENGTH (METERS)

unit	abbr.	
myriameter	mym	10,000
kilometer	km	1,000
hectometer	hm	100
decameter	dkm	10
meter	m	1
decimeter	dm	0.1
centimeter	cm	0.01
millimeter	mm	0.001

AREA (SQUARE METERS)

square kilometer	sq km *or* km^2	1,000,000
hectare	ha	10,000
are	a	100
centare	ca	1
square centimeter	sq cm *or* cm^2	0.0001

VOLUME (CUBIC METERS)

decastere	dks	10
stere	s	1
decistere	ds	0.10
cubic centimeter	cu cm *or* cm^3 *also* cc	0.000001

CAPACITY (LITERS)

kiloliter	kl	1,000
hectoliter	hl	100
decaliter	dkl	10
liter	l	1
deciliter	dl	0.10
centiliter	cl	0.01
milliliter	ml	0.001

MASS AND WEIGHT (GRAMS)

metric ton	MT *or* t	1,000,000
quintal	q	100,000
kilogram	kg	1,000
hectogram	hg	100
decagram	dkg	10
gram	g *or* gm	1
decigram	dg	0.10
centigram	cg	0.01
milligram	mg	0.001

TABLE A-7   SI SYSTEM

Characteristic	Unit of measure	Symbol	Formula
	Fundamental units		
Length	meter	m	
Mass	kilogram	kg	
Time	second	s	
Electric current	ampere	A	
Temperature	degree Kelvin	K	
Luminous intensity	candela	cd	
	Supplementary units		
Plane angle	radian	rad	
Solid angle	steradian	sr	
	Derived units		
Area	square meter	m^2	
Volume	cubic meter	m^3	
Frequency	hertz	Hz	(s^{-1})
Density	kilogram per cubic meter	kg/m^3	
Velocity	meter per second	m/s	
Angular velocity	radian per second	rad/s	
Acceleration	meter per second squared	m/s^2	
Angular acceleration	radian per second squared	rad/s^2	
Force	newton	N	$(kg\text{-}m/s^2)$
Pressure	newton per sq meter	N/m^2	
Kinematic viscosity	sq meter per second	m^2/s	
Dynamic viscosity	newton-second per sq meter	$N\text{-}s/m^2$	
Work, energy, quantity of heat	joule	J	(N-m)
Power	watt	W	(J/s)
Electric charge	coulomb	C	(A-s)
Voltage, potential difference, electromotive force	volt	V	(W/A)
Electric field strength	volt per meter	V/m	
Electric resistance	ohm		(V/A)
Electric capacitance	farad	F	(A-s/V)
Magnetic flux	weber	Wb	(V-s)
Inductance	henry	H	(V-s/A)
Magnetic flux density	tesla	T	(Wb/m^2)
Magnetic field strength	ampere per meter	A/m	
Magnetomotive force	ampere	A	
Luminous flux	lumen	lm	(cd-sr)
Luminance	candela per sq meter	cd/m^2	
Illumination	lux	lx	(lm/m^2)

TABLE A-8   CONVERSION FACTORS

Unless otherwise specified, the units *oz* and *lb* in the following tables are *units of force.*

<table>
<tr><td rowspan="3">LENGTH</td><td>m
km
cm
mm
μm
nm
Å</td><td>metre
kilometre
centimetre
millimetre
micrometre
naometre
angstrom</td><td colspan="2">$1\text{ m} = 10^{-3}\text{ km}$
10^2 cm
10^3 mm
$10^6\ \mu\text{m}$
10^9 nm
10^{10} Å</td></tr>
</table>

m metre km kilometre cm centimetre mm millimetre μm micrometre nm naometre Å angstrom	$1\text{ m} = 10^{-3}\text{ km}$ 10^2 cm 10^3 mm $10^6\ \mu\text{m}$ 10^9 nm 10^{10} Å	$1\text{ m} = 39.370\text{ in}$ 3.2808 ft 1.0936 yd

LENGTH

$1\text{ in} = 25.4\text{ mm (exactly)}$
$1\text{ ft} = 12\text{ in} = 0.3048\text{ m}$
$1\text{ yd} = 3\text{ ft} = 0.9144\text{ m}$
$1\text{ mi (statute mile)} = 1609.344\text{ m} = 5280\text{ ft}$
$1\text{ nautical mile} = 1852\text{ m}$
$1\text{ mil} = 10^{-3}\text{ in} = 0.0254\text{ mm}$

AREA

m^2 square metre
km^2 square kilometre
ha hectare
a are
cm^2 square centimetre
mm^2 square millimetre

$1\text{ m}^2 = 10^4\text{ cm}^2$ 10^6 mm^2	$1\text{ m}^2 = 1550.0\text{ in}^2$ 10.764 ft^2
	$1\text{ cm}^2 = 0.155\text{ in}^2$

$1\text{ in}^2 = 645.16\text{ mm}^2 = 6.4516\text{ cm}^2$
$6.9444 \times 10^{-3}\text{ ft}^2$
$1\text{ ft}^2 = 144\text{ in}^2 = 0.0929\text{ m}^2$

$1\text{ km}^2 = 100\text{ ha}$ 10^4 a 10^6 m^2 0.3861 mi^2	$1\text{ ha} = 10^4\text{ m}^2 = 100\text{ a}$ 2.471 acres
	$1\text{ a} = 100\text{ m}^2$ 0.01 ha
$1\text{ mi}^2 = 640\text{ acres}$ 2.59 km^2	$1\text{ acre} = 0.40469\text{ ha}$ 4046.9 m^2

VOLUME

m^3 cubic metre
dm^3 cubic decimetre
ℓ litre $(=dm^3)$
$m\ell$ millilitre $(=cm^3)$
cm^3 cubic centimetre
mm^3 cubic millimetre

$1\text{ m}^3 = 10^3\text{ dm}^3$ $10^3\ \ell$ 10^6 cm^3 $10^6\text{ m}\ell$ 10^9 mm^3	$1\text{ m}^3 = 6.1024 \times 10^4\text{ in}^3$ 35.315 ft^3
	$1\text{ cm}^3 = 6.1024 \times 10^{-2}\text{ in}^3$

$1\text{ in}^3 = 16.387\text{ cm}^3$
$1\text{ ft}^3 = 1728\text{ in}^3 = 2.8317 \times 10^{-2}\text{ m}^3$

$1\text{ m}\ell = 0.0338\text{ fl oz}$	$1\text{ fl oz} = 29.5737\text{ m}\ell$
$1\ \ell = 33.8\text{ fl oz}$ 1.0567 qt 0.26417 gal	1.8047 in^3
	$1\text{ qt} = 0.94636\ \ell$
	$1\text{ gal} = 3.7854\ \ell$

LINEAR VELOCITY	m/s — metre per second mm/s — millimetre per second km/h — kilometre per hour	1 m/s = 10^3 mm/s 3.6 km/h	1 m/s = 29.370 in/s 3.2808 ft/s
		1 in/s = 2.54×10^{-2} m/s = 25.4 mm/s	
		1 mph = 1.6093 km/h	1 km/h = 0.6214 mph
LINEAR ACCELER-ATION	m/s^2 — metre per square second	1 m/s^2 = 39.370 in/s^2 3.2808 ft/s^2	
		1 in/s^2 = 2.54×10^{-2} m/s^2	
PLANE ANGLE	rad — radian r — revolution o — angular degree ' — angular minute '' — angular second	1 rad = 57.296^o = $(360/2\pi)^o$ = 0.15915 r	
		1° = 60' 3600'' 1.7453×10^{-2} rad 2.7778×10^{-3} r	
		1 r = 360° 6.2832 rad = (2π) rad	
ANGULAR VELOCITY	rad/s — radian per second r/s — (rps) revolution per second r/min — (rpm) revolution per minute o/s — angular degree per second	1 rad/s = 0.15915 rps 9.5493 rpm	
		1 rpm = 10^{-3} krpm 1.6667×10^{-2} rps 6 o/s 0.10472 rad/s	1 krpm = 104.72 rad/s
		1 rps = 60 rpm 360 o/s 6.2832 rad/s	
ANGULAR ACCELERATION	rad/s^2 — radian per square second r/s^2 — revolution per square second rpm/s — revolution per minute per second o/s^2 — angular degree per square second	1 rad/s^2 = 0.15915 r/s^2 9.5493 rpm/s	
		1 rpm/s = 10^{-3} krpm/s 1.6667×10^{-2} r/s^2 6 o/s^2 0.10472 rad/s^2	
		1 r/s^2 = 60 rpm/s 360 o/s^2 6.2832 rad/s^2	

MASS	kg kilogram
	g gram
	t tonne
	     (metric ton)
		1 kg = 10^3 g = 10^{-3} t
		    35.274 oz (mass)
		    2.2046 lb (mass)
		    6.8522×10^{-2} slug	

		1 oz (mass) = 28.3495 g	
		1 lb (mass) = 16 oz (mass) 0.45359 kg	
		1 slug = 14.5939 kg	

		1 t = 10^3 kg
		   2204.6 lb
		     (mass)
		  1.1023 short ton
			1 short
			  ton = 2000 lb (mass)
			    907.185 kg
			    0.907185 t

FORCE	N newton
	kp kilopond
	kgf (= kp) kilogram-force
	p pond
	gf (=p) gram-force
		1 N = 0.10197 kp
		     0.22481 lb
		     3.5969 oz
		     7.2330 poundals	

		1 kp = 9.80665 N
		     2.2046 lb
		     35.274 oz	

		1 oz = 0.27801 N
		     2.83495×10^{-2} kp
		     28.3495 p	

		1 lb = 16 oz
		     4.4482 N
		     0.45359 kp	

		1 poundal = 0.138255 N	

PRESSURE, STRESS	Pa pascal
	kPa kilopascal
	MPa megapascal
	kp/cm^2 kilopond per
	      square centimetre
	at ($=kp/cm^2$) technical
	      atmosphere
	kp/mm^2 kilopond per
	      square millimetre
		1 Pa = 1.0197×10^{-5} at
		    1.45034×10^{-4}
		    lb/in^2
			1 at = 1 kp/cm^2
			    98.0665 kPa
			    14.223 lb/in^2

		1 kPa = 1.0197×10^{-2} at
		    0.145034
		    lb/in^2
			1 kp/mm^2 = 100 at
			    9.80665 MPa
			    1422.3 lb/in^2

		1 MPa = 10.197 at 145.034 lb/in^2	
		1 lb/in^2 = 6.895 kPa = 0.07031 at	

TORQUE	Nm newton metre	1 Nm = 0.10197 kpm 0.73756 lb-ft 8.85075 lb-in 141.612 oz-in
	kpm kilopond metre	1 kpm = 9.80665 Nm 1.3887 x 10^3 oz-in
		1 oz-in = 7.0615 x 10^{-3} Nm 7.2008 x 10^{-4} kpm
		1 lb-ft = 192 oz-in = 12 lb-in 1.3558 Nm 0.13825 kpm
MOMENT OF INERTIA	kg m^2 kilogram-square metre kg cm^2 kilogram-square centimetre kg mm^2 kilogram-square millimetre g cm^2 gram-square centimetre	1 kg m^2 = 10^4 kg cm^2 = 10^6 kg mm^2 = 10^7 g cm^2 8.85075 lb-in-s^2 141.612 oz-in-s^2
		1 kg cm^2 = 0.01416 oz-in-s^2
		1 oz-in-s^2 = 6.25 x 10^{-2} lb-in-s^2 7.06155 x 10^{-3} kg m^2 70.6155 kg cm^2
ENERGY, WORK, AMOUNT OF HEAT	J joule Ws wattsecond kWh kilowatthour kpm kilopond metre kcal kilocalorie (nutrition calorie) cal calorie	1 J = 1 Ws 2.7778 x 10^{-7} kWh 0.10197 kpm 2.38846 x 10^{-4} kcal 9.4781 x 10^{-4} Btu
		1 kcal = 10^3 cal 4186.8 J 1.1630 x 10^{-3} kWh 3.9683 Btu
		1 kWh = 3.6 x 10^6 J 859.845 kcal 3.4121 x 10^3 Btu
		1 Btu = 1055.06 J 2.9307 x 10^{-4} kWh 0.251997 kcal

TABLE A-9 TORQUE CONVERSION FACTORS

A \ B	Nm	kpm (kg*-m)	g*-cm	oz-in	lb-in	lb-ft
Nm	1	0.101972	1.01972×10^4	141.612	8.85075	0.737562
kpm (kg*-m)	9.80665	1	10^5	1.38874×10^3	86.7962	7.23301
g*-cm	9.80665×10^{-5}	10^{-5}	1	1.38874×10^{-2}	8.67962×10^{-4}	7.23301×10^{-5}
oz-in	7.06155×10^{-3}	7.20077×10^{-4}	72.0077	1	6.25×10^{-2}	5.20833×10^{-3}
lb-in	0.112985	1.15212×10^{-2}	1.15212×10^3	16	1	8.33333×10^{-2}
lb-ft	1.35582	0.138255	1.38255×10^4	192	12	1

*units of force

To convert from A to B, multiply by entry in table.

Example: 1 oz-in = 7.06155×10^{-3} Nm

TABLE A-10   RANDOM NUMBERS

```
07  28  68  61  81  38  11  98  34  74  64  03  48  09  18  10  15  25  98  80
29  24  86  11  41  21  16  12  96  17  56  61  49  32  48  35  43  29  34  12
76  05  58  54  35  55  35  59  07  19  00  92  65  95  34  88  26  32  61  36
95  01  20  28  66  31  15  92  14  33  39  98  55  85  71  35  82  04  51  64
73  89  25  53  83  33  75  79  98  20  09  06  76  92  43  42  55  86  41  67
41  58  46  41  68  72  73  78  34  65  87  08  10  93  46  00  32  48  29  68
53  46  33  57  86  99  47  87  14  55  98  93  72  15  77  23  13  26  37  20
39  46  65  77  16  92  33  65  57  49  18  41  87  68  05  23  73  33  55  49
40  98  58  06  54  13  55  31  86  06  34  94  43  59  08  54  86  44  59  84
06  45  65  80  97  46  95  38  82  01  88  12  28  75  93  39  33  60  00  48
84  72  36  35  94  11  36  23  17  09  95  90  26  46  90  70  81  40  77  38
61  14  68  60  77  44  75  28  56  67  36  58  03  82  16  76  39  12  73  70
07  47  15  19  64  62  17  97  36  08  22  55  58  81  17  77  83  65  75  05
70  43  84  46  41  98  44  54  23  72  39  79  53  16  88  04  66  00  66  43
57  10  02  26  17  12  56  48  43  97  65  06  21  97  65  97  95  77  93  01
95  01  58  34  51  77  89  80  79  72  60  94  43  05  89  83  88  15  09  58
53  00  18  66  58  39  02  95  62  79  35  52  01  06  50  18  98  88  87  81
51  86  20  34  89  54  54  61  15  00  96  89  11  34  05  18  26  77  17  23
38  63  42  41  87  99  37  18  91  08  55  42  27  51  69  48  94  14  70  96
47  77  39  28  14  56  98  96  73  22  31  67  20  90  85  04  01  87  42  17
26  20  46  66  36  28  98  66  97  56  78  29  19  53  46  08  20  30  55  61
58  58  28  68  36  45  83  66  12  05  17  37  74  90  81  86  99  04  17  90
80  83  75  20  32  63  09  41  69  12  43  82  63  40  08  89  71  89  68  44
40  90  05  68  85  00  90  91  49  16  23  00  26  56  52  66  71  22  63  40
77  38  50  26  29  57  56  31  37  52  88  88  37  72  14  52  73  79  23  79
51  62  77  67  70  21  17  88  22  26  66  77  78  55  87  14  39  07  31  67
66  81  52  18  87  47  01  60  71  73  90  72  90  39  37  64  44  26  82  07
67  72  78  24  07  12  61  67  78  85  92  68  95  24  69  57  74  13  28  64
14  29  00  91  50  43  64  63  85  17  54  46  92  58  58  52  97  54  84  09
30  89  99  07  56  26  49  27  83  67  52  35  36  93  63  60  15  71  16  34
26  42  43  27  81  79  67  35  84  28  64  59  79  16  11  54  85  34  01  49
98  05  34  47  71  14  87  98  70  21  53  51  01  46  60  71  19  33  62  43
02  82  10  42  11  62  87  83  16  96  34  46  04  25  33  69  55  37  82  29
99  88  34  85  46  77  12  00  89  17  04  48  85  62  32  77  08  24  88  65
83  59  57  38  84  22  08  75  21  10  58  75  87  70  19  07  94  83  09  37
76  27  52  23  67  14  39  88  57  00  72  71  21  68  81  49  24  94  19  37
03  80  24  56  17  64  66  90  80  09  62  03  65  61  66  39  83  87  41  95
40  86  98  74  63  72  14  00  08  38  25  25  37  93  89  96  74  66  36  06
38  02  78  20  39  15  04  67  68  27  46  22  43  79  26  45  45  17  66  13
19  51  85  12  56  95  63  15  44  74  88  26  02  10  68  09  84  86  26  81
```

TABLE A-11 ASCII CHARACTER CODES

Decimal	Hexadecimal	ASCII Code	Decimal	Hexadecimal	ASCII Code	Decimal	Hexadecimal	ASCII Code
032	20	(space)	064	40	@	096	60	
033	21	!	065	41	A	097	61	a
034	22	"	066	42	B	098	62	b
035	23	#	067	43	C	099	63	c
036	24	$	068	44	D	100	64	d
037	25	%	069	45	E	101	65	e
038	26	&	070	46	F	102	66	f
039	27	'	071	47	G	103	67	g
040	28	(	072	48	H	104	68	h
041	29	)	073	49	I	105	69	i
042	2A	*	074	4A	J	106	6A	j
043	2B	+	075	4B	K	107	6B	k
044	2C	'	076	4C	L	108	6C	l
045	2D	-	077	4D	M	109	6D	m
046	2E	.	078	4E	N	110	6E	n
047	2F	/	079	4F	O	111	6F	o
048	30	0	080	4G	P	112	70	p
049	31	1	081	50	Q	113	71	q
050	32	2	082	51	R	114	72	r
051	33	3	083	53	S	115	73	s
052	34	4	084	54	T	116	74	t
053	35	5	085	55	U	117	75	u
054	36	6	086	56	V	118	76	v
055	37	7	087	57	W	119	77	w
056	38	8	088	58	X	120	78	x
057	39	9	089	59	Y	121	79	y
058	3A	:	090	5A	Z	122	7A	z
059	3B	;	091	5B	[	123	7B	{
060	3C	<	092	5C	\	124	7C		
061	3D	=	093	5D	]	125	7D	}
062	3E	>	094	5E	^	126	7E	~
063	3F	?	095	5F	—			

INDEX